# XIV Mexican Workshop on Particles and Fields 2013



Journal of Physics: Conference Series Volume 651

Oaxaca, Mexico
24 - 29 October 2013

ISBN: 978-1-5108-1477-6
ISSN: 1742-6588

**Printed from e-media with permission by:**

Curran Associates, Inc.
57 Morehouse Lane
Red Hook, NY 12571

**Some format issues inherent in the e-media version may also appear in this print version.**



Printed by Curran Associates, Inc. (2015)

For permission requests, please contact the Institute of Physics
at the address below.

Institute of Physics
Dirac House, Temple Back
Bristol BS1 6BE UK

Phone: 44 1 17 929 7481
Fax: 44 1 17 920 0979

techtracking@iop.org

**Additional copies of this publication are available from:**

Curran Associates, Inc.
57 Morehouse Lane
Red Hook, NY 12571 USA
Phone: 845-758-0400
Fax: 845-758-2634
Email: curran@proceedings.com
Web: www.proceedings.com

## *Table of contents*

# Volume 651

**XIV Mexican Workshop on Particles and Fields**
**24–29 November 2013, Oaxaca, Mexico**



## Papers

Journal of Physics: Conference Series **651** (2015) 011001 doi:10.1088/1742-6596/651/1/011001

# Preface

**David Delepine**

Physics Department, Guanajuato University, Loma del Bosque 103, col. Loma del Campestre, 37150, Leon, Mexico

E-mail: delepine@fisica.ugto.mx

**Mauro Napsuciale**

Physics Department, Guanajuato University, Loma del Bosque 103, col. Loma del Campestre, 37150, Leon, Mexico

E-mail: mauro@fisica.ugto.mx

**Humberto Salazar Ibarguen**

Facultad de Ciencias Físico Matematicas, Benemérita Universidad Autónoma de Puebla, 18 sur y Av. San Claudio, C.P. 72570 Puebla, Pue., Mexico

E-mail: hsalazar@fcfm.buap.mx

The Mexican Workshop on Particles and Fields (MWPF) is a biennial meeting organized by the Division of Particles and Fields of the Mexican Physical Society designed to gather specialists in different areas of high energy physics to discuss the latest developments in the field. The fourteenth edition of this meeting was held from November 25 to 29, 2013, at the colonial city of Oaxaca de Juárez, Oaxaca, Mexico.

The XIV Mexican Workshop on Particles and Fields consisted of invited lectures, discussion and poster sessions. Experimental and theoretical developments were presented by distinguished physicists, addressing the most recent results in the field. The invited review talks included topics on collider physics, neutrino physics, physics beyond the Standard Model, flavor and hadronic physics, astroparticle physics, dark matter physics and effective theories, among others. The highlight topic of the conference was the presentation of the most resent results from the most popular high energy experiments in the world. The discovery of a particle consistent with the long-sought Higgs boson, considered one of the most important discoveries of the 21st century, was fully addressed by José Benítez and Kirill Prokofiev from CERN. The overview of the results of ALICE on the first run of the LHC was extensivly covered by Antonio Ortiz, from Lund University, and Daniel Tapia, from Université Paris-Sud. The prospects and status of the new Belle II experiment were presented by Yoshi Sakai from KEK. The plans and projects of Tevatron on the new era of accelerators were explained by Gene Fisk from FERMILAB. Eric Vázquez from SNOLAB presented a wonderful explanation about the Dark Matter detection and the most resent results about the searches for it. The largest high energy cosmic rays detector, the Pierre Auger, was presented by Luis Villaseñor from University of Michoacán.

On Friday 29th of November, we had an excursion to the archeological site of Mitla and to Santa María del Tule. Mitla is a pre-Hispanic site, located 25 miles to the south of the city of Oaxaca, founded around 100 years a.c. by the native Zapotecos. Mitla had its golden age between 950-1500 a.c., after the dawn of the city of Monte Albán. It was the main city and center of the Zapotecos society in the Oaxaca central valley. Santa Mará del Tule is a nice town located 7 miles to the south of Oaxaca, famous for the so called "Árbol del Tule". This is an ancient tree belonging to the family commonly known as *Ahuehuete* or *Sabino* and whose scientific name is *Taxodium mucronatum*. It is more than 2000 years old and has a trunk with a circumference bigger than 45 meters.

The workshop and these Proceedings would have not been possible without the efforts of many institutions: Consejo Nacional de Ciencia y Tecnología (CONACyT) through the Red Nacional de Física de Altas Energías and individual research projects, Universidad de Guanajuato, Universidad Nacional Autónoma de México, Centro de Investigación y de Estudios Avanzados del IPN (CINVESTAV), Universidad Michoacana de San Nicolás de Hidalgo and Benemérita Universidad Autónoma de Puebla. We thank all these institutions for their support. Especially, we we wish to thank Universidad de Guanajuato for the financial support via PIFI for the publication of these proceedings.

We thank all the speakers for making this a timely and informative workshop. Thanks are due to all who encourage the discussions and stimulate the flow of information during the question and the discussion sessions.

**Organizing committee**

- Humberto Salazar (FCFM-BUAP)
- Mauro Napsuciale (DCI-UG)
- David Delepine (DCI-UG)
- Eduard de la Cruz (CINVESTAV)
- Saul Ramos (IF-UNAM)
- Arturo Fernández (FCFM-BUAP)
- Lorenzo Díaz (FCFM-BUAP)
- Omar Miranda (CINVESTAV)
- Wolfang Bietenholz (ICN-UNAM)
- Heriberto Castilla (CINVESTAV)
- Rodrigo Pelayo (UPIITA-IPN)

**List of participants**

1 ALESSANDRO, Bruno
2 ARANDA SÁNCHEZ, Jorge Isidro
3 ARROYO UREÑA, Marco Antonio
4 ARTEAGA VELAZQUEZ, Juan Carlos
5 BARRANCO MONARCA, Juan
6 BASHIR, Adnan
7 BAUTISTA, Irais
8 BEDOLLA, Marco Antonio
9 BELLO, Héctor
10 BENITEZ GUZMÁN, Lucero Guadalupe
11 BENITEZ, Jose

Journal of Physics: Conference Series **651** (2015) 011001 doi:10.1088/1742-6596/651/1/011001

12 BERMDEZ RAMOS, Rocio
13 BIETENHOLZ, Wolfgang
14 BIJKER, Roelof
15 CABALLERO-MORA, Karen Salomé
16 CARPINTEYRO, Severiano
17 CASTILLA, Alejandro
18 CASTILLO, Mario
19 COBOS-MARTINEZ, Javier
20 CONDE, Ruben
21 CORTES GONZALEZ, Arely
22 CORTES MALDONADO, Ismael
23 COURTOY, Aurore
24 CRUZ, Rigoberto
25 DE ANDA, Francisco J.
26 DE LA CRUZ BURELO, Eduard
27 DIAZ-CRUZ, Lorenzo
28 DIAZ, Enrique
29 DORAME, LUIS
30 EPIFANIO, Ponce
31 ESPINOSA GÓMEZ, David
32 FERNANDEZ TELLEZ, Arturo
33 FERRO, Rodolfo
34 FONT, Anamaria
35 GAITÁN, R.
36 GAONA, Alejandro
37 GARCIA GONZALEZ, Jose Andre
38 GARCIA, Jose Didino
39 GARCÉS, E. A.
40 GARCÍA GUDIÑO, David
41 GARCÍA TECOCOATZI, Hugo
42 GERBER, Urs
43 GONZÁLEZ CANALES , Flix Francisco
44 GONZÁLEZ MACÍAS, Vannia
45 GONZÁLEZ-SOLÍS, Sergi
46 JUAN MANUEL, Grados Luyando
47 GUEVARA, Adolfo
48 GUTIERREZ GUERRERO, Xiomara
49 HENTSCHINSKI, Martin
50 HERNÁNDEZ CHAHÍN, Karim Gibrán
51 HERNÁNDEZ-TOMÉ, Gerardo
52 HOFMANN, Christoph Peter

53 HONORATO, Carlos
54 LARIOS, Francisco
55 LOPEZ , Gabriel
56 LOPEZ, Lao Tse
57 MALDONADO, Félix
58 MARTINEZ HERNANDEZ, Mario Ivan
59 MARTÍNEZ, Humberto
60 MARTÍNEZ, Oscar
61 MAURY CUNA, Humberto
62 MAVROMATOS , Nick
63 MEJIA DIAZ, Hector
64 MIRANDA, O. G.
65 MITSOU, Vasiliki
66 MONTES DE OCA Y., J. H.
67 NAPOLES, Gibraham
68 NAPSUCIALE, Mauro
69 ORDUZ-DUCUARA, Javier-Andres
70 ORTIZ VELASQUEZ, Antonio
71 PARADA, Alexander
72 PARRA, Alejandra
73 PEDRAZA MORALES, Maria Isabel
74 PEINADO, Eduardo
75 PELAYO, Rodrigo
76 PEREZ LORENZANA, Abdel
77 PEREZ, Miguel Angel
78 PONCE, Epifanio
79 PROKOFIEV, Kirill
80 QUINTERO POVEDA, Néstor
81 RAMOS SANCHEZ, Saul
82 RAMÍREZ ZAVALETA, Fernando
83 RAYA MONTAÑO, Khépani
84 REJÓN-BARRERA, Fernando G.
85 RIVERA AGUDELO, Diana Carolina
86 RODRÍGUEZ CAHUANTZI, Mario
87 RODRIGUEZ LIMA, Josue
88 ROIG, Pablo
89 ROMERO, Juan Manuel
90 ROSADO, Alfonso
91 RUBEN, Conde
92 SALAZAR ARIAS, José Germán
93 SALAZAR, Humberto

Journal of Physics: Conference Series **651** (2015) 011001 doi:10.1088/1742-6596/651/1/011001

94 SARKIS MOBARAK, Youssef
95 SHTEJER DIAZ, Katherin
96 SIES RAMOS, LAURA
97 SOSA SANCHEZ, Citlalli Teresa
98 TAPIA TAKAKI, Daniel
99 TOLEDANO, Ivan
100 TOLEDO, Genaro
101 TOSTADO ROBLEDO, Sergio
102 TUTUTI HERNÁNDEZ, Eduardo Salvador
103 VAQUERA, Carlos
104 VARELA, Enrique
105 VAZQUEZ-JAUREGUI, Eric
106 VASILIKI, Mitsou
107 VILLASEÑOR, Luis
108 WEBER, Axel
109 ZANOTTI, James

Journal of Physics: Conference Series **651** (2015) 011002 doi:10.1088/1742-6596/651/1/011002

# Peer review statement

All papers published in this volume of *Journal of Physics: Conference Series* have been peer reviewed through processes administered by the proceedings Editors. Reviews were conducted by expert referees to the professional and scientific standards expected of a proceedings journal published by IOP Publishing.

Journal of Physics: Conference Series **651** (2015) 012001 doi:10.1088/1742-6596/651/1/012001

# The KASCADE-Grande observatory and the composition of very high-energy cosmic rays

**J.C. Arteaga-Velázquez[1,*], W.D. Apel[2], K. Bekk[2], M. Bertaina[3], J. Blümer[2,4], H. Bozdog[2], I.M. Brancus[5], E. Cantoni[3,6,a], A. Chiavassa[3], F. Cossavella[4,b], K. Daumiller[2], V. de Souza[7], F. Di Pierro[3], P. Doll[2], R. Engel[2], J. Engler[2], B. Fuchs[4], D. Fuhrmann[8,c], H.J. Gils[2], R. Glasstetter[8], C. Grupen[9], A. Haungs[2], D. Heck[2], J.R. Hörandel[10], D. Huber[4], T. Huege[2], K.-H. Kampert[8], D. Kang[4], H.O. Klages[2], K. Link[4], P. Łuczak[11], M. Ludwig[4], H.J. Mathes[2], H.J. Mayer[2], M. Melissas[4], J. Milke[2], B. Mitrica[5], C. Morello[6], J. Oehlschläger[2], S. Ostapchenko[2,d], N. Palmieri[4], M. Petcu[5], T. Pierog[2], H. Rebel[2], M. Roth[2], H. Schieler[2], F.G. Schröder[2], O. Sima[12], G. Toma[5], G.C. Trinchero[6], H. Ulrich[2], A. Weindl[2], J. Wochele[2], J. Zabierowski[11]**

[1]Universidad Michoacana, Instituto de Física y Matemáticas, Morelia, Mexico
[2]Institut für Kernphysik, KIT - Karlsruher Institut für Technologie, Germany
[3]Dipartimento di Fisica, Università degli Studi di Torino, Italy
[4]Institut für Experimentelle Kernphysik, KIT - Karlsruher Institut für Technologie, Germany
[5]Horia Hulubei National Institute of Physics and Nuclear Engineering, Bucharest, Romania
[6]Osservatorio Astrofisico di Torino, INAF Torino, Italy
[7]Universidade São Paulo, Instituto de Física de São Carlos, Brasil
[8]Fachbereich Physik, Universität Wuppertal, Germany
[9]Department of Physics, Siegen University, Germany
[10]Deptartment of Astrophysics, Radboud University Nijmegen, The Netherlands
[11]National Centre for Nuclear Research, Department of Cosmic Ray Physics, Lódź, Poland
[12]Department of Physics, University of Bucharest, Bucharest, Romania
[a]now at: Istituto Nazionale di Ricerca Metrologia, INRIM, Torino
[b]now at: DLR Oberpfaffenhofen, Germany
[c]now at: University of Duisburg-Essen, Duisburg, Germany
[d]now at: University of Trondheim, Norway
[*] Corresponding author

E-mail: `J.C.Arteaga: arteaga@ifm.umich.mx`

**Abstract.** KASCADE-Grande is an air-shower observatory devoted to the detection of cosmic rays with energies in the range of $10^{16}$ to $10^{18}$ eV. This energy region is of particular interest for the cosmic ray astrophysics, since it is the place where some models predict the existence of a transition from galactic to extragalactic origin of cosmic rays and the presence of a break in the flux of its heavy component. The detection of these features requires detailed and simultaneous measurements of the energy and composition of cosmic rays with sufficient statistics. These kinds of studies are possible for the first time in KASCADE-Grande due to the accurate measurements of several air-shower observables, i.e., the number of charged particles, electrons and muons in the shower, using the different detector systems of the observatory. In this contribution, a detailed look into the composition of $10^{16} - 10^{18}$ eV cosmic rays with KASCADE-Grande is presented.

*Keywords*: Cosmic rays, KASCADE-Grande, Air showers, Energy spectrum, Composition

## 1. Introduction

A long standing problem in astroparticle physics has been the identification of the sources and the acceleration mechanism of galactic cosmic rays. Possible scenarios range from supernova remnants (SNR) [1], superbubbles [2] and magnetars [3] to cannonballs from supernovae [4] and blast waves from an early gamma-ray burst in our galaxy [5], being the SNR model the accepted paradigm. Modern $\gamma$-ray observations seems to support the SNR and the superbubble scenarios [6], however, the question whether these are the accelerators of the very high-energy cosmic rays ($E > 10^2$ TeV) is still open. Undoubtedly, the start of neutrino astronomy with ICECUBE [7], the arrival of new gamma-ray detectors, such as HAWC [8] and CTA [9], together with detailed measurements of galactic cosmic rays will pave the way to settle down the question as to the origin and acceleration of these energetic particles.

Nowadays, there are different models to explain how cosmic rays can reach high-energies at different sources. However, there exist some common characteristics shared among them: they involve astrophysical plasma motions and magnetic fields and predict the existence of individual knee-like structures or breaks in the corresponding spectra of cosmic nuclei, which are associated with the efficiency of the accelerator. Such structures lead to the formation of a knee at around $(3-5)\cdot 10^{15}$ eV in the all-particle energy spectrum, which is a prominent feature discovered in the cosmic ray flux caused by a steepening of the spectrum [10]. According to the theory of cosmic rays, the positions of the knees are distributed inside the energy interval from $E = 10^{15}$ to $10^{18}$ eV and, depending of the model, they scale with the electric charge, Z, of the ion ($E_Z \propto Z \times B \times R$, with $B$: the magnitude of the magnetic field and $R$: the size of the confinement region) [1, 5, 11], or the atomic mass, $A$, of the nucleus being accelerated [4]. Further differences among the models can also give rise to distinct predictions on the relative abundances of each element in the primary flux and the existence of fine structures on the overall cosmic ray energy spectrum between $10^{15}$ and $10^{18}$ eV.

The spectrum observed at Earth is not only shaped by the acceleration mechanisms working at the sources but also by the propagation process of cosmic rays inside the galaxy, which is also a matter that is not completely understood (see [12, 13] and references therein). An important ingredient of the propagation models of cosmic rays is the interstellar magnetic field, which has a deep influence on the paths of galactic cosmic rays and their spectrum. In general, the particular effects of the intergalactic magnetic fields on the shape of the all-particle spectrum depend on the model under consideration. Models can be found in which the knee-like feature in the total energy spectrum of cosmic rays is explained as a natural consequence of the loss of efficiency of the galactic magnetic field at confining energetic charged particles in the Milky Way. In this context, the leakage of cosmic rays from the galaxy would imply also the presence of kinks in the spectra of the different kinds of cosmic ray nuclei. The kinks would be characterized by a rigidity dependent position.

Some clues to elucidate among the aforementioned scenarios are to be found in the study of the properties of cosmic rays, i.e., the analysis of their energies, arrival directions and composition. First measurements performed by the KASCADE experiment[1] in the energy range $E = 10^{15} - 10^{17}$ eV revealed in fact the existence of individual knees in the spectra of the light and intermediate mass groups of cosmic rays ($Z \leq 14$) [14], which was confirmed by the EAS-TOP [15], Tibet [16] and GAMMA [17] observations. The knees are formed because the fluxes of the corresponding primaries begin to decrease with the energy. Interestingly, KASCADE discovered that the observed depletions, particularly, in the energy spectra of the

[1] The KASCADE data taken from the period 1996-2003 is already available at the website of the KCDC (KASCADE Cosmic ray Data Centre) at https://kcdc.ikp.kit.edu. A data base with more than 158 million cosmic ray air-shower events is at disposal for the general public.

light mass groups ($Z < 6$) were responsible for the origin of the *knee* feature of the all-particle cosmic ray flux [14]. From the results of the experiment a pattern emerged, in which the positions of the individual knees are shifted to higher energies as the atomic/mass number of the primaries increases. However, these data were not enough to rule out a $Z$- or $A$- dependence of the individual knees due to the uncertainties in the measured locations of the breaks [2]. A convincing argument in favor of one or another type of dependence has to come from the measurement of the knee of the heavy mass group (for which Fe is the representative element) in cosmic rays. That structure was missing in the KASCADE data, presumably due to the lack of statistics at the energies of interest. Assuming $E_H \approx 3 \cdot 10^{15}\,\mathrm{eV}$ [14] for the position of the knee of the hydrogen mass group and supposing that the atomic or mass dependence holds for the location of the individual knees, the break for the iron nuclei is expected at energies in the interval $E_{Fe} = (Z \cdot E_H, A \cdot E_H) \approx (7.8 \cdot 10^{16}\,\mathrm{eV}, 1.7 \cdot 10^{17}\,\mathrm{eV})$, outside the sensitivity range of the KASCADE experiment. To explore this energy regime in search for the so called *iron-knee*, KASCADE was upgraded to KASCADE-Grande [18] using the components of the EAS-TOP observatory. During the period from 2003 until the date of its shut-down, in 2012, measurements of cosmic ray events were performed with the KASCADE-Grande experiment in the interval $10^{16} - 10^{18}\,\mathrm{eV}$. So far several dedicated and careful analyses have been applied to the data concerning the study of the chemical composition and the all-particle energy spectrum of very high-energy cosmic rays. The effort has result in very interesting results [19, 20, 21, 22]. In this contribution, we will take a fast look into the KASCADE-Grande instrument and will sum up the details of some of the aforementioned analyses and their corresponding results.

## 2. Detection systems of KASCADE-Grande

Above $\sim 1\,\mathrm{PeV}$, cosmic rays are studied by means of indirect techniques as a consequence of the low flux at the high-energy regime. The techniques are based on the detection of the extensive air showers (EAS) that cosmic rays induce in the Earth's atmosphere. In order to derive significant information about composition and primary energy the simultaneous observation of several EAS parameters [23, 24] is imperative, e.g., the total number of electrons/muons in the shower, the density of particles in the shower front, the hadron content, the muon production height and the depth of the shower maximum. The measurements are done combining different kinds of particle detectors such as plastic scintillator detectors, tracking detectors, fluorescence and Cherenkov telescopes, water Cherenkov detectors and radio antennas among others [23, 24]. The devices are arranged in arrays and are deployed over large surfaces on the ground in order to achieve big collecting areas. However, measurements are not only bounded to the ground. Fluorescence and Cherenkov telescopes, can be put in orbit around the Earth in order to study extensive air showers from space [25].

In case of KASCADE-Grande, we speak of a ground-based EAS array, which consisted of different detection systems. This instrument was designed to measure and separate the muon and electromagnetic components of the EAS [18]. The KASCADE-Grande detector (located at 49.1° N, 8.4° E, 110 m a.s.l.) was the successor of the KASCADE experiment [26] and incorporated the original electromagnetic detectors and muon devices of KASCADE to a bigger system of detectors, called Grande, composed of a $700 \times 700\,\mathrm{m}^2$ array with $37 \times 10\,\mathrm{m}^2$ scintillator stations regularly spaced by an average distance of 137 m [18]. A smaller array, named Piccolo was also added and was used to coordinate the KASCADE and Grande triggers. The main characteristics of the KASCADE-Grande detectors are displayed in table 1 and the layout of the experiment is presented in figure 1.

The Grande array was used to sample the density of charged particles of the shower front at ground level and to measure the particle arrival times of the EAS. The core position, the

[2] Part of these uncertainties are associated with the hadronic interaction models.

Table 1: Some components of KASCADE-Grande: total sensitive areas and threshold kinematic energies for vertically incident particles are presented [18].

| Detector | Particle | Area($m^2$) | Threshold |
|---|---|---|---|
| Grande array (plastic scintillators) | Charged | 370 | 3 MeV |
| Piccolo array (plastic scintillators) | Charged | 80 | 3 MeV |
| KASCADE array (liquid scintillators) | $e/\gamma$ | 490 | 5 MeV |
| KASCADE array (shielded plast. scint.) | $\mu$ | 622 | 230 MeV |
| MTD (muon tracking detectors) | $\mu$ | $4 \times 128$ | 800 MeV |

number of charged particles ($N_{ch}$) and the arrival direction of the shower were extracted from the Grande data through an iterative fit and a careful modeling of the EAS front. To reconstruct the arrival direction of the shower from the arrival times a $\chi^2$ fit was applied assuming a curved shower front as suggested by CORSIKA/QGSJET II simulations. On the other hand, to obtain the core position and the shower size from the density of charged particles of the EAS at ground, a modified NKG lateral distribution function was fitted to the data by means of a maximum-likelihood procedure [18].

An important component of this experiment was the KASCADE muon array, composed by $192 \times 3.2\,\mathrm{m}^2$ shielded scintillator detectors, which are sensitive to muons with threshold energy above 230 MeV for vertical incidence. With the Grande information and the measurements of the shielded array of the lateral distribution of muons in the shower front, the muon size ($N_\mu$) was reconstructed event-by-event at KASCADE-Grande. The procedure involved the maximum log-likelihood technique along with a Lagutin-Raikin distribution function [18]. Once the muon content was known, the number of electrons, $N_e$, in the EAS was estimated subtracting $N_\mu$ from $N_{ch}$.

Buried in a tunnel under several layers of soil, iron and concrete, the muon tracking detector (MTD) was installed close to the center of the KASCADE array. The MTD was composed of streamer tubes grouped in modules. It was used to reconstruct the individual tracks of penetrating particles of a section of the EAS and to measure the muon pseudorapidities [27]. By applying a triangulation procedure on the reconstructed tracks the mean muon production height of each event was also derived [28].

As it can be seen, different EAS parameters are at disposal for the research of high-energy cosmic rays at the KASCADE-Grande data base. However, due to the lack of space, only those analyses based on the muon and charged particle numbers will be considered for the present work.

## 3. Accuracy of EAS reconstruction

Systematic uncertainties for the core, $N_{ch}$ and arrival direction of the EAS are studied directly [18] by comparing the results of the Grande and KASCADE reconstructions, which work independently. Comparisons were performed for a subset of data with cores located inside a common area for both detectors and shower sizes in the interval $\log_{10} N_{ch} = 5.8-7.2$. Accuracies of the EAS core positions and arrival directions are found to be of the order of 5 m and 0.7°, respectively. Meanwhile, for the total number of charged particles a resolution $\leq 15\%$ is achieved. Those values are in full agreement with expectations from Monte Carlo simulations.

To study the accuracy of the reconstruction of the muon size, Monte Carlo simulations have to be used. The reason is that shielded detectors were only available for KASCADE. In this case, the results show that, for $N_{ch} > 10^{4.7}$, the $N_\mu$ systematic uncertainties are $\leq 25\%$, decreasing for high-energies. They exhibit a dependence with the core position and the arrival direction [19]. Since $N_\mu$ is reconstructed with less accuracy than the charged particle number a muon correction

Journal of Physics: Conference Series **651** (2015) 012001 doi:10.1088/1742-6596/651/1/012001

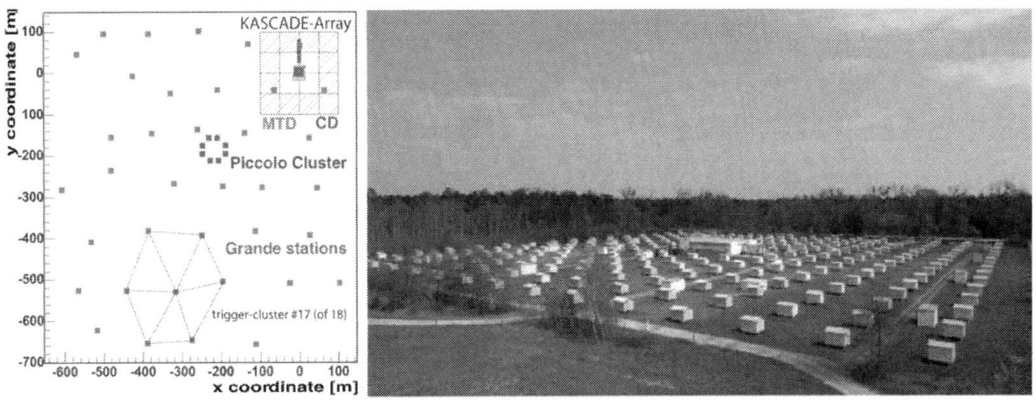


Figure 1: *Left panel*: The KASCADE-Grande experiment. Small squares represent the Grande stations. The KASCADE array is seen at the upper right hand of the figure. Close to the center of KASCADE, the MTD is shown. KASCADE detectors are arranged in 16 clusters (big squares). The outer 12 clusters contain the muon detectors. One example of a Grande-trigger is shown. It consists of 6 activated stations in hexagonal configuration plus a central one. *Right panel*: Photograph of the KASCADE experiment showing the array of 252 detector stations.

function is applied to the data. This is possible because the behavior of muon uncertainties are well understood. The function is built from Monte Carlo simulations (using the high-energy hadronic interaction model QGSJET II-02 [29]) for a mixed composition assumption (H, He, C, Si, and Fe in equal abundances) and is parameterized as a function of the muon size, the EAS arrival direction and the distance of the EAS core to the KASCADE center.

## 4. Analyses and results

The available shower parameters to study the properties of cosmic rays with the KASCADE-Grande experiment are the muon and electron contents, the number of charged particles, the muon production height, the muon pseudorapidities and the muon/charged particle density at the shower front and at a given distance from the core. These parameters can be used in different ways to get insight into the energy and composition of cosmic rays (see, e.g., [19]-[22] and [30]-[33]) and, even more, to test the available high-energy hadronic interaction models [33]-[36]. In the next subsections, the main concern will be on the all-particle energy spectrum and the energy spectra of mass groups of cosmic ray particles in the sensitive range of KASCADE-Grande ($E = 10^{16} - 10^{18}$ eV). In particular, the analysis techniques and the results based on the study of the EAS observables $N_{ch}$ and $N_{\mu}$ will be summarized [19, 20, 21, 22, 30]. These analyses were applied over a subset of selected data that passed several quality cuts.

Since unfortunately these cosmic-ray analyses must rely on MC simulations to interpret the EAS events several sets of simulated data had to be built. They included both the development of the EAS and its interaction with the detectors. These MC data were analyzed in the same way as the experimental ones. The production and development of the showers were simulated with CORSIKA [37] and the low and high-energy hadronic interaction models FLUKA [38] and QGSJET II [29]. Events were sampled from a differential spectrum described by a power law distribution ($\Phi \propto E^{\gamma}$) with spectral index $\gamma = -2$, which was later re-weighted to simulate a $\gamma = -3$ spectrum. The EAS were generated for the zenith angle interval $\theta = 0° - 40°$ using several types of primaries (H, He, C, Si and Fe). Finally, smaller sets of simulations were also created using EPOS 1.99 [39], SIBYLL 2.1 [40] and QGSJET II-04 [41] to check the sensitivity of the results to the high-energy hadronic interaction model.

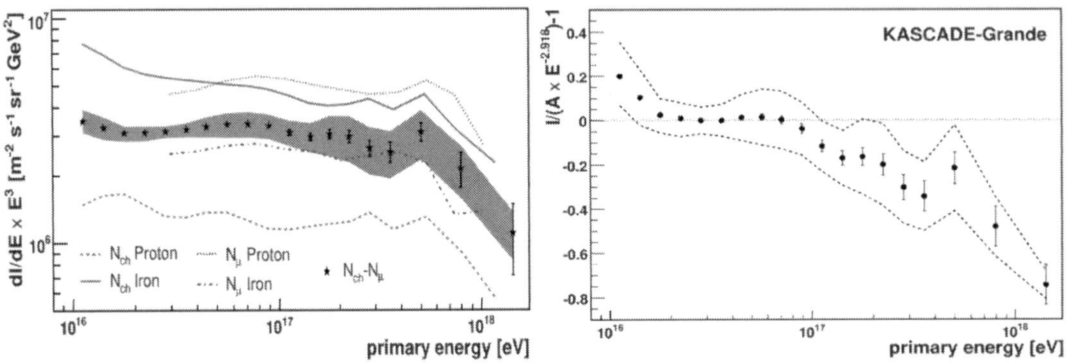


Figure 2: *Left panel*: Reconstructed all-particle energy spectrum, multiplied by $E^3$, as obtained from the KASCADE-Grande $N_{ch}$ and $N_\mu$ data and the different methods described in the present paper. The coloured band represents the preliminary systematic uncertainty for the spectrum reconstructed through the $N_{ch} - N_\mu$ method. *Right panel*: Residual plot after fitting the reconstructed all-particle energy spectrum derived from the $N_{ch} - N_\mu$ method with a power law function in the interval $E = (2-8) \cdot 10^{16}$ eV. The systematic error band is also shown (dotted lines) [19].

### *4.1. The all-particle energy spectrum*

*4.1.1. $N_{ch} - N_\mu$ technique* [19]. The reconstruction method in this case makes use of the correlation between $N_{ch}$ and $N_\mu$ to reduce the uncertainties introduced in the result by the composition of the primaries. For the analysis, data with $\theta < 40°$ is divided in different zenith angle intervals. Only EAS events with cores located in a central area of KASCADE-Grande were selected. The assignment of energy is done event-by-event according to a MC derived formula,

$$\log_{10}(E) = [b_H + (b_{Fe} - b_H) \cdot k] + [a_H + (a_{Fe} - a_H) \cdot k] \cdot \log_{10}(N_{ch}), \quad (1)$$

which depends on the magnitudes of $N_{ch}$ and the parameter $k = k(N_{ch}, N_\mu)$. The latter is defined as:

$$k = \frac{\log_{10}(N_{ch}/N_\mu) - \log_{10}(N_{ch}/N_\mu)_H}{\log_{10}(N_{ch}/N_\mu)_{Fe} - \log_{10}(N_{ch}/N_\mu)_H}. \quad (2)$$

These functions are calibrated for each zenith angle interval according to the predictions of the MC program CORSIKA and the FLUKA/QGJSET II hadronic interaction models for pure proton and iron nuclei (an $E^{-3}$ energy spectrum is assumed). From this procedure, the energy spectrum for each zenith angle interval is derived. Before adding up all the spectra to obtain the final result, each flux is corrected for the effect due to the bin-to-bin migration of events. That is done building a response matrix from MC simulations and applying an unfolding procedure. The reconstructed all-particle energy spectrum is shown at the left panel of figure 2. It is based on data measured by KASCADE-Grande during the period from 2003 to 2009 with total exposure $\approx 2 \cdot 10^{13}$m$^2$ s sr after applying selection cuts. On the other hand, a residual plot of the spectrum is presented on the right panel of figure 2. It reveals the presence of two significant structures in the all-particle energy spectrum, specifically, a concavity around $10^{16}$ eV and a weak knee-like feature at $8.3 \cdot 10^{16}$ eV, where the spectrum changes from a power law behavior with spectral index $\gamma = -2.92 \pm 0.02$ to another one with $\alpha = -3.39 \pm 0.07$. Both structures are significant with respect to the systematic and statistical uncertainties. A statistical analysis reveals, for example, that the spectrum above $\log_{10}(E/\text{eV}) = 10^{16.2}$ can not be described by a single power law at a significance level of $2.1\sigma$. Subsequent analyses with EPOS 1.99, SIBYLL 2.1 and QGSJET II-04, following the foregoing strategy, have revealed that the structures on

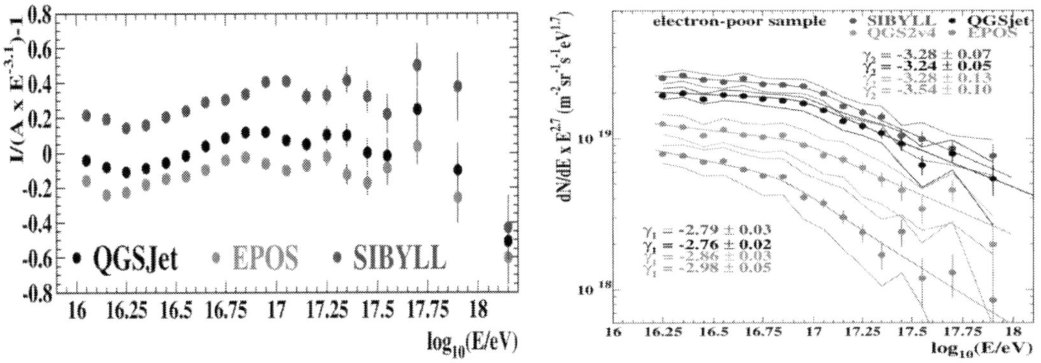


Figure 3: *Left panel*: Residual plot of the experimental all-particle energy spectrum reconstructed with the $N_{ch} - N_{\mu}$ method and using different hadronic interaction models for the energy assignment to the data [42, 43]. *Right panel*: Experimental spectrum for the heavy mass group of cosmic rays derived from the $N_{ch} - N_{\mu}$ technique using four hadronic interaction models. The error bars represent the statistical uncertainties and the bands, the systematic ones. Spectral indexes obtained from a power-law fit to the energy spectra before and after the corresponding knees are indicated on the corners of the plot (left and right, respectively) [43].

the all-particle energy spectrum are not an artifact of the hadronic interaction model employed for the energy calibration [42, 43]. In all cases, the features remain stable. Just a shift in the energy scale is observed among the spectra as seen on the left panel of fig. 3. On the other hand, recent results of the TUNKA [44], GAMMA [45] and ICETOP [46] experiments have also shown the presence of the concavity in their spectra around the energy scale reported by KASCADE-Grande. In addition, the TUNKA [44] and ICETOP [46] collaborations have reported the existence of the small break in the spectrum close to $10^{17}$ eV. A change from a light mass group ($Z = 1 - 2$) to the intermediate one ($Z = 6 - 12$) in the all-particle energy spectrum could explain the concave behavior of the spectrum [47]. However, a transition between two different kinds of galactic sources could also have the same effect [48]. On the other hand, the origin of the observed break is also intriguing. That feature is related with the structure known as the *iron-knee*, as it will be seen later on in the section devoted to the composition studies.

*4.1.2. $N_{ch}$ and $N_{\mu}$ techniques* [19]. The $N_{ch}$ and $N_{\mu}$ observables can be employed individually for the task of reconstructing the energy spectrum. However, the procedure has the disadvantage that it is more sensitive to the primary composition, which is reflected in the form of a wide systematic error band in the energy spectrum. Such analyses are important for several reasons, for instance, in order to study the influence of systematic uncertainties, to cross-check the reconstruction methods and to investigate the validity of the used hadronic interaction model. The technique goes as follows: Applying the constant intensity cut method on EAS with $\theta < 40°$, the attenuation curves for the $N_{ch}$ and $N_{\mu}$ parameters in the atmosphere are obtained. These curves are used to calculate the equivalent muon/charged particle number of the showers at a reference angle, $\theta_{ref} = 20°(22°)$ - chosen as the mean of the angular distributions of the $N_{ch}$ ($N_{\mu}$) parameters. Then, the corresponding energy of the event is estimated according to the equivalent number of muons or charged particles in the shower. For this purpose, calibration formulae of the form $E = \alpha_{\mu}[N_{\mu}(\theta_{ref})]^{\delta_{\mu}}$ and $E = \alpha_{ch}[N_{ch}(\theta_{ref})]^{\delta_{ch}}$ are employed, respectively. The power-law behavior of the expressions is not arbitrary, but it is suggested by MC simulations. The parameters $\alpha$ and $\delta$ are constants and are obtained from fits to MC data around $\theta_{ref}$.

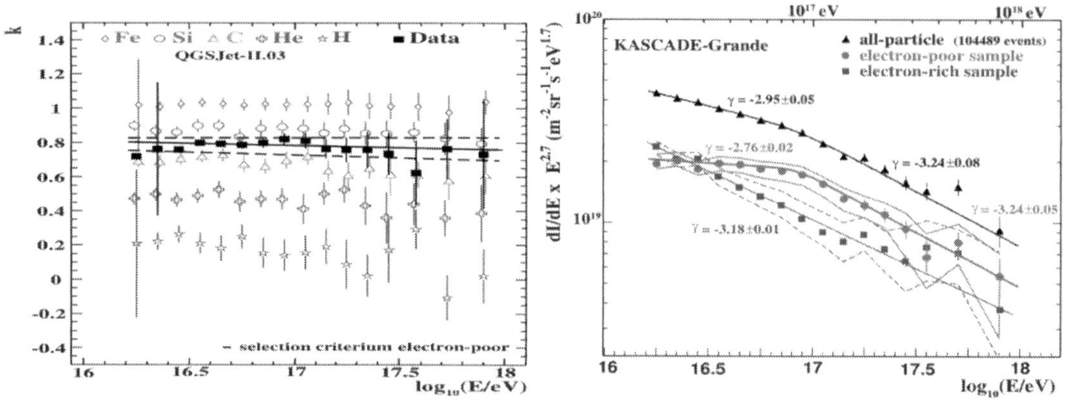


Figure 4: *Left panel*: Mean $k$ parameter vs reconstructed energy for different primaries according to MC simulations (CORSIKA/QGSJET II). The thick line represents the frontier line used for classification of experimental data into different mass groups and the thin lines, the corresponding uncertainties on the definitions. Statistical and reconstruction uncertainties of $k$ are shown as error bars. *Right panel*: All-particle energy spectrum and energy spectra (without unfolding) of the heavy (electron poor) and the light and medium (electron rich) mass groups derived by applying the $k$ technique to the KASCADE-Grande data. Error bands are obtained by considering the changes in the spectra introduced by varying the frontier lines inside their respective interval of uncertainties [19].

Since the composition is *a priori* unknown, fits are performed for two extreme cases: pure protons and iron nuclei. To end the procedure, data derived from each observable under a given composition assumption are combined to construct the associated energy spectrum. The reconstructed spectra for the data from the period 2003−2009 are shown in the left panel of figure 2 (migration effects are taken into account). From the widths of the bands generated by the H/Fe solutions, it is clear that the $N_{ch}$ variable is more sensitive to the primary composition than the $N_{\mu}$ case. For the $N_{ch}$ case, the region of the expected solution covers a bigger area than that obtained from the $N_{\mu}$ analysis technique. However, it is seen some room for a common solution, where in fact the $N_{ch} - N_{\mu}$ spectrum is found. That speaks in favor of the internal consistency of the results when interpreted in the framework of the QGSJET II hadronic interaction model.

### *4.2. The energy spectra of mass groups*

*4.2.1. k parameter technique* [20]. One way to take a look into the composition of cosmic rays from the KASCADE-Grande data is by using the $k$ parameter, already described in the last section. This quantity is sensitive to the composition of the primary particles (H, He, C, Si, Fe) and can be employed to separate the data into a light/medium and a heavy mass groups, at least. To achieve the separation, a plot with the mean value of $k$ for different primaries as a function of the reconstructed energy is created from MC simulations for every zenith angle interval in which data is divided. Here, energy is estimated through formula (1). An example, of such a graph can be seen in the left panel of figure 4. It happens that, in general, the mean $k$ value grows with the mass of the particle. That is not surprising, since the $k$ parameter is defined in such a way that it returns the biggest values for iron nuclei and the lowest, for protons. With the above plots, data is divided into mass groups by defining a frontier line, $\bar{k}_{Si-C}$, which is obtained from a fit to the average between the mean $k$ values of Si and C. If $k > \bar{k}_{Si-C}$, events are classified as part of the heavy component (also called *electron poor sample*). In other case, events are included in the light and medium group (*electron rich sample*). The energy spectra for

Journal of Physics: Conference Series **651** (2015) 012001 doi:10.1088/1742-6596/651/1/012001

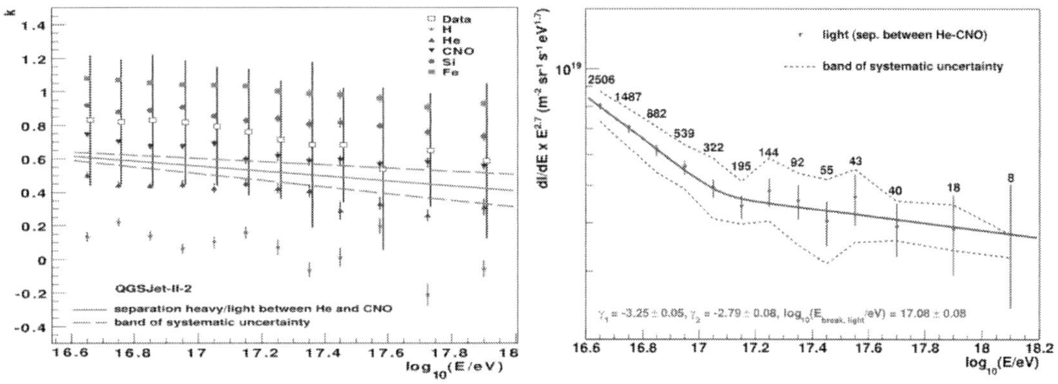


Figure 5: *Left panel*: Mean $k$ parameter vs reconstructed energy for different primaries according to MC simulations (CORSIKA/QGSJET II) for the analysis of [21]. The thick line represents the frontier line used for classification of experimental data into different mass groups and the thin lines, the corresponding uncertainties on the definitions. Statistical and reconstruction uncertainties of $k$ are shown as error bars. *Right panel*: Experimental energy spectrum for light primaries according to the analysis of [21] based on the $k$ technique. Error bands are obtained by considering the changes in the spectra introduced by varying the frontier lines inside their respective interval of uncertainties. Statistical uncertainties appear as error bars.

the light/medium and heavy mass primaries corresponding to the data of the 2003 – 2009 period are presented in figure 4 (right panel) along with the all-particle energy spectrum obtained by adding the individual spectra. In this plot, the knee around $10^{17}$ eV in the all-particle energy spectrum is seen again. But now, it is more clear what is its origin. The answer must be track down to the existence of a knee in the energy spectrum of the heavy component, whose relative abundance around $10^{17}$ eV is bigger than that of the other mass groups. A fit performed to the spectrum of the heavy component using a broken power-law function with a smooth knee [49],

$$\frac{\mathrm{d}\Phi(E)}{\mathrm{d}E} = \Phi_0 \; E^{\gamma_1} \left[1 + \left(\frac{E}{E_{\mathrm{knee}}}\right)^{\varepsilon}\right]^{(\gamma_2-\gamma_1)/\varepsilon} \quad , \tag{3}$$

indicates a change $\Delta\gamma = \gamma_2 - \gamma_1 = -0.48$ of the spectral index and locates the knee at $\log_{10}(E_{knee}/\mathrm{eV}) = 16.92 \pm 0.04$. The significance that this spectrum is not described by a single power-law is of $3.5\sigma$. Applying the same fit to the all-particle energy spectrum results in $\Delta\gamma = -0.29$ and $\log_{10}(E_{knee}/\mathrm{eV}) = 16.92 \pm 0.10$. It is worth to mention that the presence of the knee-feature at the spectrum of the heavy component is independent of the high-energy hadronic interaction model (although relative abundances are strongly affected). That was checked out by reproducing the analysis within the framework of EPOS 1.99, SIBYLL 2.1 and QGSJET II-04 (see right panel of figure 4) [42, 43]. In these analyses, the significance levels for the existence of the heavy knee in the data goes from 2.4, for QGSJET II-04, up to 7.4, in case of SIBYLL 2.1.

Turning now the attention to the light/medium spectrum, the data seems to reveal that this component is present even up to $10^{18}$ eV. That is an important issue. Some astrophysical models predict the existence of an extragalactic component dominated by protons in the cosmic ray flux, which would take over the galactic contribution in the energy regime of $10^{17} - 10^{19}$ eV [50]. Therefore, somewhere in this energy range, part of it accessible to KASCADE-Grande, clues of a possible *galactic-extragalactic transition* may be buried. One way to look for them is by examining the light/medium spectrum intently. For this purpose, a larger data sample was selected from a bigger data acquisition period 2003 – 2012 and for an extended fiducial area as described in [21]. The steps from the $k$-technique were also applied here. But additionally, to

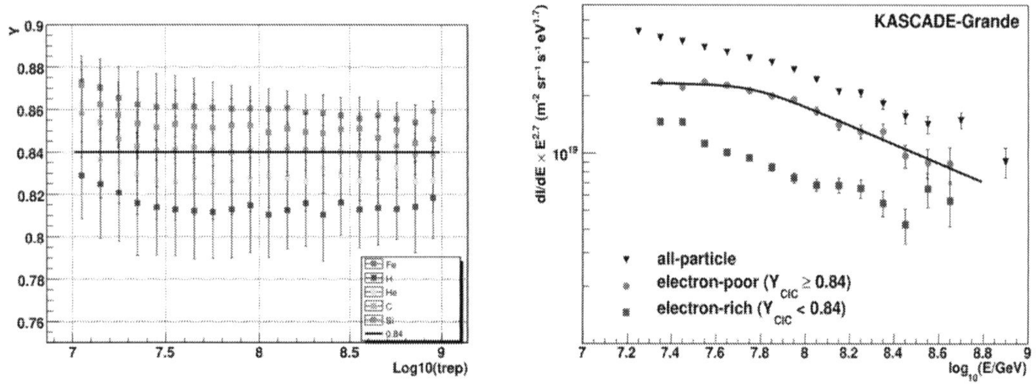


Figure 6: *Left panel*: Estimations of the $Y$ ratio for five primaries (from bottom to top: H, He, C, Si and Fe) shown as a function of the reconstructed energy using MC simulations (CORSIKA/QGSJET II).The thick line represents the $Y$ cut used for the $Y$-based analysis. Error bars represents the RMS of the $Y$ distributions. *Right panel*: Light and heavy energy spectra derived from the KASCADE-Grande data using the $Y$ ratio as a mass estimator. The all-particle energy spectrum derived by adding the individual mass group fluxes is also shown [51].

separate the light from the medium contribution a new cut was introduced by defining another frontier line $\bar{k}_{C-He}$, which is obtained by averaging the mean k values of C and He at a given energy (see figure figure 5, left side). The resulting energy spectra for the light component is shown in figure 5 (right side). An *ankle-like feature* in the light component of cosmic rays can be clearly observed. A double power-law fit with formula (3) indicates that the hardening of the spectrum of the light component is characterized by a change of spectral index from $\gamma = -3.25 \pm 0.05$ to $\gamma = -2.79 \pm 0.08$ and occurs at $\log_{10}(E_{ankle}/\text{eV}) = 17.08 \pm 0.08$, slightly above the position of the heavy-knee. Analyses conducted in [21, 43] point out that the changes in the spectrum of heavy and light primaries around $10^{17}$ eV are not connected by a bias introduced by the classification procedure or reconstruction techniques. The appearance of the ankle-like feature is striking and may be indicative of an early transition from galactic to extragalactic cosmic rays.

*4.2.2. Y-cut technique* [20, 51, 52]. This technique makes use of the ratio between the equivalent muon and charged numbers at a zenith angle of reference ($\theta_{ref} = 22°$),

$$Y = \log_{10} N_\mu(\theta_{ref}) / \log_{10} N_{ch}(\theta_{ref}), \tag{4}$$

which is also a sensitive quantity to the composition of cosmic ray particles. In equation (4) the equivalent number of particles of an EAS is estimated with the CIC method, as explained in the subsection 4.1.2. However, the energy of the event is calculated using the expression (1). As it was the case for the $k$ parameter, plots for the $Y$ ratio as a function of the reconstructed energy are produced for several elements in order to be used in the classification of events (c.f figure 6). The difference with the $k$ parameter is that for the plots of the $Y$ ratio, data from all zenith angles are considered to build up the graph. It is worth to mention that for this analysis a smaller data sample was selected ($\theta < 30°$) in order to have events with lower systematic uncertainties (errors tend to increase with the value of the zenith angle). As seen in figure 6, for a fixed energy, the $Y$ ratio grows with the mass of the primary nuclei. The effect can be understood as a result of the fact that physics of hadronic interactions favors the production of more muons and less electrons for heavy nuclei. From the same figure, it is also noted that the mean $Y$ distributions are almost energy independent above the threshold ($\log_{10}(E/\mathrm{GeV}) \geq 7.4$). Taking advantage of the observed behavior of $Y$ a cut on this parameter is applied at $Y_{cut} = 0.84$ to divide data into two different sets: the *electron rich* ($Y < 0.84$) and *electron poor* ($Y \geq 0.84$) groups. Using figure 6, it is clear that the first set covers the mean $Y$ values of light mass elements (here H and He), while the second set contains the expected region for heavy mass nuclei (Si and Fe). In this way, the *electron rich* and *electron poor* samples become representative of the light and heavy mass groups (the fraction of misclassified events inside each group is less than 20% in the region of maximum efficiency).

The results for the energy spectra obtained from the 2003 – 2009 data sample are presented in the right panel of figure 6. Plots were not corrected for migration effects. From this figure, the knee-like structure of the heavy component of cosmic rays is immediately recognized. Applying again a fit with a broken power-law spectrum to the spectrum of the heavy mass group a change in the spectral index $\Delta\gamma = -0.47$ at $\log_{10}(E_{knee}/\mathrm{GeV}) = 7.77 \pm 0.06$ is obtained. The same analysis for the all-particle energy spectrum gives the following values: $\Delta\gamma = -0.29$ at $\log_{10}(E_{knee}/\mathrm{GeV}) = 7.92 \pm 0.10$. In this analysis, the light component of cosmic rays is also present. Relative abundances, of course, change with the position of $Y_{cut}$, however, the presence of the knee-like future in the heavy mass group is independent of the cut, i.e. it is inherent to the cosmic ray data [52].

*4.2.3. $N_\mu/N_{ch}$ distributions* [30]. In order to avoid the influence of increasing systematic uncertainties with $\theta$ in the results, the present analysis is restricted to vertical showers ($\theta < 24°$). From experimental data a bidimensional histogram of $N_\mu/N_{ch}$ vs $\log_{10}(N_{ch})$ is constructed. Similarly, two-dimensional histograms for simulated data organized into three mass groups (light, intermediate and heavy) are created. The light MC data set is composed solely of protons, the intermediate one, of He (50%) and C (50%), while the heavy one, of Si (50%) and Fe (50%). Then, inside a single bin $\log_{10} N_{ch,j}$, each experimental $N_\mu/N_{ch}$ distribution, $n_j^{exp}$, is fitted with a function, $n_j^{MC}$, formed by the linear combination of those simulated distributions derived for the three mass groups (see, for example, left panel of figure 7). Here

$$n_{ji}^{MC} = \sum_{\ell} \alpha_j^{\ell} n_{ji}^{MC,\ell}, \tag{5}$$

where $i$ runs over all $N_\mu/N_{ch}$-bins and $\ell$, over each mass component. The fitting parameters are the $\alpha_j^\ell$ coefficients, which represent the relative abundances of the different mass groups inside each $\log_{10} N_{ch,j}$ channel. Hence, the parameters must be constraint to the interval $0 \leq \alpha_j^\ell \leq 1$ and satisfy $\sum_\ell \alpha_j^\ell = 1$. The coefficients are extracted by minimizing the $\chi^2$ functions:

$$\chi_j^2 = \sum_i \left( \frac{n_{ji}^{exp} - n_{ji}^{MC}}{\sigma_i} \right)^2, \tag{6}$$

with $\sigma_i$ defined as the statistical error on the corresponding $n_{ji}^{MC}$ entry. With the relative abundances, the number of events per mass group for each charged particle number interval is found. Finally, $N_{ch}$ is used to estimate the primary energy by using a power-law expression ($E = \zeta N_{ch}^{\epsilon}$) derived from MC simulations accordingly to each mass group.

The energy spectra obtained from this method for data from the period 2003 – 2009 (exposure $\approx 8 \cdot 10^{11} \mathrm{m}^2\,\mathrm{s\,sr}$) are presented in the right panel of figure 8. Again, in this plot, the outstanding knee-like feature in the spectrum of the heavy mass group is visible. The change in the spectral index and the position of the knee-like feature are $\Delta\gamma = -0.62$ and $\log_{10}(E_{knee}/\mathrm{eV}) = 16.79 \pm 0.04$, respectively. This analysis also confirms the non-vanishing

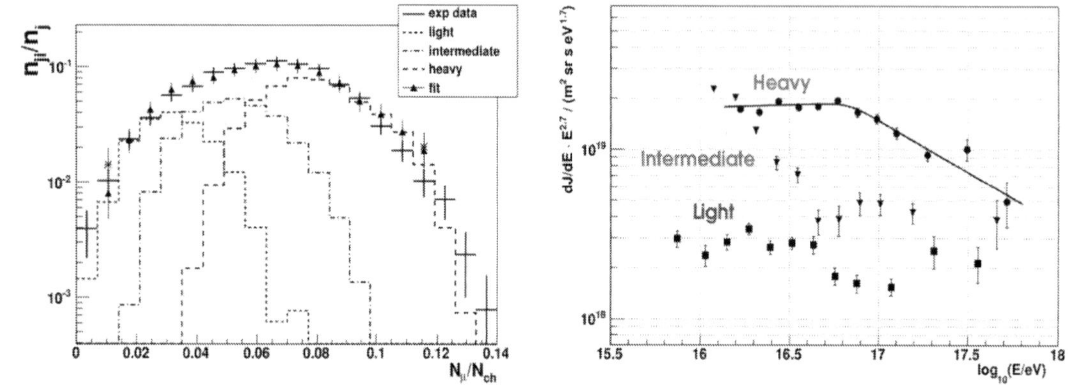


Figure 7: *Left panel*: An example of an experimental $N_\mu/N_{ch}$ distribution (solid line) for the interval $7.15 \leq \log_{10} N_{ch} < 7.24$ (data has been normalized dividing by the total number of events in the histogram). The result of the fit (triangles) with the simulated distributions (based on CORSIKA/QGSJET II-03 [29]) for the light (thin-dashed line), intermediate (dot-dashed line) and heavy (thick-dashed line) mass groups is shown. *Right panel*: Energy spectra for the light, intermediate and heavy mass groups derived from experimental data using the technique described in section 4.2.3 [30]. Data was interpreted using QGSJET II-03.

contribution of the intermediate and light components and the dominance of the heavy mass group at high-energies. However, due to the limited statistics no significant conclusions can be achieved regarding the shape and the details of the corresponding spectra. From the present analysis an additional result is derived: the fits of the experimental $N_\mu/N_{ch}$ distributions point out that the intermediate mass group is also needed in addition to the light and heavy ones in order to describe the observed data.

*4.2.4. Unfolding with Gold's algorithm* [22]. The basis for this analysis is the two-dimensional histogram $\log_{10} N_{ch}$ vs $\log_{10} N_\mu$. The purpose is to unfold from that distribution the cosmic ray fluxes, $dJ/d\log_{10} E$, for five different mass groups (represented by H, He, C, Si and Fe). The problem is formulated in terms of the following coupled equations:

$$n_{ji}^{exp} = \sum_{\ell} \int \frac{dJ_\ell}{d\log_{10} E} P_\ell(\log_{10} N_{ch,j}, \log_{10} N_{\mu,i} | \log_{10} E) \, d\log_{10} E \, \cos\theta dA \, d\Omega \, dt, \qquad (7)$$

where $P_\ell$ represents the conditional probability that an EAS with a given combination of charged particle and muon numbers, $(\log_{10} N_{ch,j}, \log_{10} N_{\mu,i})$, will be observed given that the primary cosmic ray belongs to the mass group $\ell$ and has an energy $E$, $n_{ji}^{exp}$ is the number of entries in the bin $(\log_{10} N_{ch,j}, \log_{10} N_{\mu,i})$ of the experimental distribution, $t$ is the observation time, $\Omega$ is the solid angle of observation, $\theta$ is the arrival zenith angle of the EAS and $A$ is the fiducial area. $P_\ell$ must be known *apriori* for the unfolding procedure. These probabilities comprise all relevant information regarding the properties of the observables of the shower, e.g. EAS fluctuations, and the efficiency as well as the reconstruction accuracy of the instrument/experiment. They are calculated using MC simulations (based on FLUKA 2002.4 and QGSJET II-02). To unfold the bidimensional distributions, Gold's unfolding algorithm [53] is applied. Solutions are checked out using also the Bayes' algorithm [54] and a method based on the principle of reduced cross-entropy [55]. The corresponding bidimensional distribution used in this analysis (with $\theta < 18°$) and the respective unfolded results are shown in figure 8 (for an effective time of 1318 days and

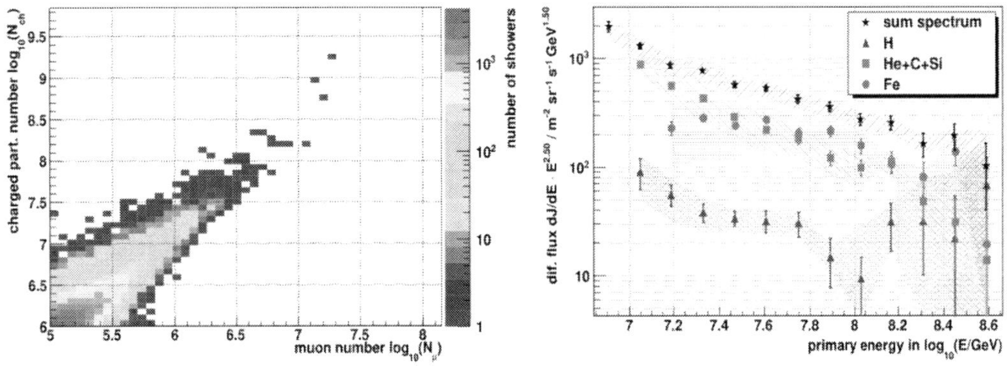


Figure 8: *Left panel*: The experimental $\log_{10} N_{ch}$ vs $\log_{10} N_{\mu}$ distribution used for the unfolding procedure (see section 4.2.4). *Right panel*: Resulting energy spectra for the light, intermediate and heavy mass groups obtained by unfolding of the KASCADE-Grande data with Gold's algorithm. The sum of all individual/elemental spectra, i.e. the all-particle energy spectrum, is also shown [22]. Results are based on FLUKA 2002.4 and QGSJET II-02.

an exposure of $\approx 5 \cdot 10^{12}\mathrm{m}^2\,\mathrm{s\,sr}$). Final spectra have been grouped into three different mass groups: light (H), intermediate (He+C+Si) and heavy (Fe).

A knee-like feature in the spectrum of the Fe component is visible in figure 8. Remarkably, the kink of the heavy component obtained from this analysis is more pronounced[3]. A double power-law fit applied to the iron spectrum - see equation (3) - reveals a change of spectral index equivalent to $\Delta\gamma = -1.08$ at the position $\log_{10}(E_{knee}/\mathrm{eV}) = 16.9 \pm 0.1$. Moreover, the analysis shows that the iron primaries dominate in the high-energy regime (under the framework of QGSJET II). On the other hand, the all-particle energy spectrum does not exhibit the small knee around $10^{17}$ eV. That could be the result of both the lower statistics of the subsample used for this analysis and the uncertainties that affect the unfolding of the five different mass groups, which are propagated to the all-particle spectrum.

With the position of the iron-knee at around $63 - 100$ PeV and that for the proton-knee between 3 PeV and 5 PeV (according to the newest analysis of the KASCADE data, see first reference of [22] and figure 9) a rigidity-dependent picture for the position of the knees of the elemental primary cosmic rays seems to be preferred.

## 5. Conclusions

In this contribution, the composition of cosmic rays in the energy interval $10^{16} - 10^{18}$ eV has been explored with the KASCADE-Grande experiment. Composition results from several analyses based on the study of the $N_{ch}$ and $N_{\mu}$ observables of EAS were presented. The main conclusions of this research are that the spectrum of the heavy component of cosmic rays exhibits a knee-like feature around $10^{17}$ eV, which is responsible for a subtle break in the all-particle energy spectrum around the same energy value. The spectrum of the light component of cosmic rays presents a hardening around $10^{17}$ eV. Two mass groups, light and heavy (represented by H and Fe+Si nuclei, respectively), are not enough to describe the $N_{\mu}/N_{ch}$ distributions,

[3] Here, it is important to mention one important difference with the composition analyses already described and it is that the separation of the mass groups by using the unfolding technique is based on the properties of each of the five elemental groups. In case of the other techniques, mass groups with a different degree of mixing of the elemental groups are employed. This difference could explain why the iron-knee is more pronounced in the unfolding analysis.

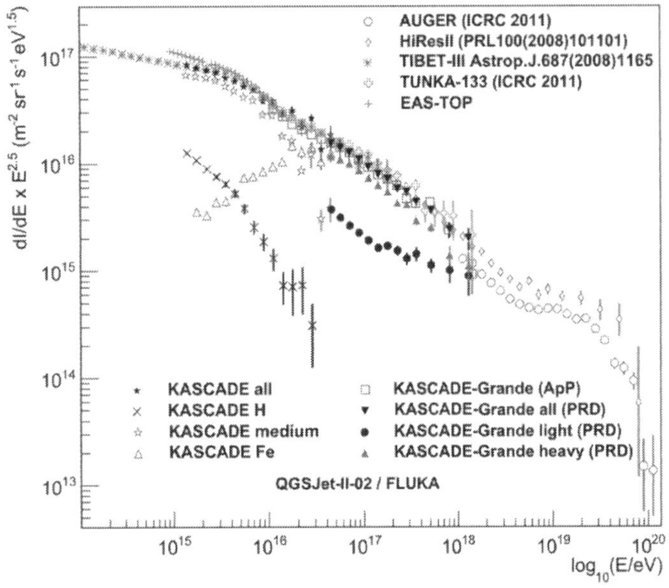


Figure 9: The all-particle energy spectrum [19, 21] and spectra of individual mass groups obtained with KASCADE [22] and KASCADE-Grande data [21]. Results are compared with the energy spectra measured with other experiments (figure taken from [56]).

an intermediate mass group is needed (at least for a QGSJET II-03 based analysis). From the different components, the heavy one is dominant inside the framework of the QGSJET II hadronic interaction model. Finally, relative abundances at very high-energies depend on the high-energy hadronic interaction model. Nevertheless, the main features observed in the all-particle and the heavy and light mass group spectra remain stable.

## Acknowledgments

The authors would like to thank the members of the engineering and technical staff of the KASCADE-Grande Collaboration, who contributed to the success of the experiment. The KASCADE-Grande experiment is supported by the BMBF and the Helmholtz Alliance for Astroparticle Physics - HAP of Germany, the MIUR and INAF of Italy, the Polish Ministry of Science and Higher Education, the Romanian Authority for Scientific Research UEFISCDI (PNII-IDEI grants 271/2011 and 17/2011), and the German-Mexican bilateral collaboration grant (DAAD-Proalmex 2009-2012). J.C.A.V. acknowledges the partial support of CONACyT and the Coordinación de la Investigación Científica de la Universidad Michoacana.

## References

[1] V.S. Ptuskin et al., Astron. Astrophys. 268 (1993) 726.
[2] W. R. Binns et al., Proc. of the 33rd ICRC, Rio de Janeiro, Brazil (2013), ID0646; A. F. Barghouty and D. A. Schnee, ApJ 749 (2012) 178; A. M. Bykov and G. D. Fleishman, MNRAS 255 (1992) 269-275; W. R. Binns et al., New Astronomy Reviews 52 (2008) 427-430.
[3] P. Blasi, R.L. Epstein, and A.V. Olinto, Ap J 533 (2000) L123.
[4] A. De Rújula, Proc. of the 20th International Symposium on Lattice Field Theory, Nucl. Phys. B, Proc. Suppl. 151 (2005) 23-32.
[5] S.D. Wick, C.D. Dermer and A. Atoyan, Astropart.Phys.21 (2004) 125-148.
[6] L. Tibaldo et al (Fermi-LAT Coll.), Proc. of the 33rd ICRC, Rio de Janeiro, Brazil (2013), ID1299, astro-ph/1311.2896.
[7] M. G. Aartsen et al. (ICECUBE Coll.), Science 342, 1242856 (2013), astro-ph/1311.5238.
[8] M. Mostafa et al. (HAWC Coll.), Proc. of the 33rd ICRC, Rio de Janeiro, Brazil (2013), ID933, astro-ph/1310.7237.
[9] CTA Consortium, arXiv:1008.3703.

[10] G.V. Kulikov and G.B. Khristiansen, J. Exper. Theor. Phys. 35 (1958) 635; G.V. Kulikov, G.B. Khristiansen, Sov. Phys. JETP 35 (1959) 441.
[11] B. Peters, Il Nuo. Cim. 22 (1961) 800.
[12] J. R. Hoerandel, Astropart.Phys. 21 (2004) 241-265.
[13] V.S. Ptuskin, AIP Conf. Proc. 745, 14 (2005).
[14] W.D. Apel et al. (KASCADE Coll.), Astrop. Phys. 24 (2005) 1.
[15] M. Aglietta et al. (EAS-TOP Coll.), Astrop. Phys. 21 (2004) 583.
[16] M.Amenomori et al. (Tibet Coll.), J. Phys. Soc. Jpn. 78A (2009) 206.
[17] A.P. Garyaka et al.(GAMMA Coll.), Astropart.Phys.28 (2007) 169.
[18] W.D. Apel et al. (KASCADE-Grande Coll.), NIMA 620 (2010) 202.
[19] W. Apel et al. (KASCADE-Grande Coll.), Astropart. Phys. 36, 183 (2012).
[20] W.D. Apel et al. (KASCADE-Grande Coll.), PRL 107 (2011) 171104.
[21] W.D. Apel et al. (KASCADE-Grande Coll.), PRD 87(2013)081101. S. Schoo et al. (KASCADE-Grande Coll.), Proc. of the 33rd ICRC, Rio de Janeiro, Brazil (2013), ID0527, astro-ph/1308.1858v1.
[22] D. Fuhrmann, PhD thesis, University of Wuppertal (2012); D. Fuhrmann et al.(KASCADE-Grande Coll.), Proc. of the 33rd ICRC, Rio de Janeiro, Brazil (2013), ID0531, astro-ph/1308.2098v1.
[23] A. Haungs, H. Rebel and M. Roth, Rep. Prog. Phys. 66 (2003) 1145-1206.
[24] M. Nagano and A.A. Watson, Rev. Mod. Phys. 72 (2000) 689.
[25] J.H. Adams et al. (JEM-EUSO Coll.), Astrop. Phys. 44 (2013) 76-90.
[26] T. Antoni et al. (KASCADE Coll.), NIMA 513 (2003) 490.
[27] P. Doll et al. (KASCADE Coll.), NIMA 488 (2002) 517.
[28] W.D. Apel et al. (KASCADE-Grande Coll.), Astropart. Phys.34 (2011) 476485.
[29] S.Ostapchenko, Nucl.Phys.B (Proc. Suppl.) 151 (2006) 143; S. Ostapchenko, Phys.Rev.D 74, 014026 (2006).
[30] E. Cantoni et al. (KASCADE Coll.), Proc. of the 32nd ICRC, Beijing, China, vol.11 (2012) 259.
[31] M.R. Finger, PhD thesis, KIT Campus North, Karlsruhe, Germany (2011).
[32] G. Toma et al. (KASCADE Coll.), Proc. of the 33rd ICRC, Rio de Janeiro, Brazil (2013), ID517.
[33] V. Souza et al. (KASCADE Coll.), Proc. of the 32nd ICRC, Beijing, China, vol.1 (2012) 295.
[34] P. Doll et al. (KASCADE Coll.), Proc. of the 31st ICRC, Lodz, Poland (2009) ID429; W.D. Apel et al., Astrop. Phys. 34, 476 (2011); P. Luczak et al. (KASCADE Coll.), Proc. of the 33rd ICRC, Rio de Janeiro, Brazil (2013), ID528.
[35] D. Kang et al. (KASCADE-Grande Coll.), Proc. of the 33rd ICRC, Rio de Janeiro, Brazil (2013), ID521, astro-ph/1309.4295.
[36] J.C. Arteaga-Velázquez et al. (KASCADE Coll.), Proc. of the 32nd ICRC, Beijing, China, vol.1 (2012) 275; J.C. Arteaga-Velázquez et al. (KASCADE Coll.), Proc. of the 33rd ICRC, Rio de Janeiro, Brazil (2013), ID772, astro-ph/1308.3202.
[37] D. Heck et al., Report FZKA 6019, Forschungszentrum Karlsruhe (1998).
[38] A. Fassò et al., Report CERN-2005-10, INFN/TC-05/11, SLAC-R-773 (2005).
[39] T. Pierog et al., Report FZKA 7516, Forschungszentrum Karlsruhe 133 (2009).
[40] E.J. Ahn, et al., Phys. Rev.D 80 (2009) 094003.
[41] S.S. Ostapchenko, Phys. Rev. D 83 (2011) 014018.
[42] W.D. Apel et al. (KASCADE-Grande Coll.), J. Adv. Space Res. (2013) in press dx.doi.org/10.1016/j.asr.2013.05.008.
[43] M. Bertaina et al. (KASCADE-Grande Coll.), Proc. of the 33rd ICRC, Rio de Janeiro, Brazil (2013), ID0196.
[44] L.G. Sveshnikova (Tunka Coll.), Proc. of the 33rd ICRC, Rio de Janeiro, Brazil (2013), ID0304. S. Epimakhov et al. (Tunka Coll.), Proc. of the 33rd ICRC, Rio de Janeiro, Brazil (2013), ID0326.
[45] A. Garyaka et al. (Gamma Coll.), Journal of Physics: Conference Series, 409, (2013) 012081.
[46] M. G. Aartsen et al. (IceCube Coll.), Phys. Rev. D 88, (2013) 042004.
[47] C. De Donato and G.A. Medina-Tanco, Astrop. Phys. 32 (2009) 253.
[48] A.M. Hillas, J. Phys. G: Nulc. Part. Phys. 31 (2005) 95.
[49] J.R. Hörandel, Astropart. Phys. 19 (2003) 193-220.
[50] D. R. Bergman and J. W. Belz, J. Phys. G 34, R359 (2007).
[51] J.C. Arteaga-Velázquez et al. (KASCADE-Grande Coll.), Proc. of the 32nd ICRC, Beijing, China, vol.1 (2012) 271.
[52] A. Chiavassa et al. (KASCADE-Grande Coll.), Proc. of the 33rd ICRC, Rio de Janeiro, Brazil (2013), ID092.
[53] R. Gold, Report ANL-6984, Argonne (1964).
[54] G.D'Agostini, NIMA 362 (1995) 487.
[55] M. Schmelling, NIMA 340 (1994) 400.
[56] A. Haungs et al. (KASCADE-Grande Coll.), Proc. of the 33rd ICRC, Rio de Janeiro, Brazil (2013), ID300, arxiv: 1308.1485.

Journal of Physics: Conference Series **651** (2015) 012002 doi:10.1088/1742-6596/651/1/012002

# A study of $\pi-$ and $\rho-$ mesons with a non-perturbative approach

**Rocio Bermudez**

Departamento de Investigación en Fsica, Universidad de Sonora, Blvd. Transversal S/N,83000, Hermosillo, Mexico

E-mail: rabermudezr@gmail.com

**J.J. Cobos-Martnez**

Departamento de Física, Universidad de Sonora, Blvd. Transversal S/N,83000, Hermosillo, Mexico

E-mail: j.j.cobos.martinez@gmail.com

**M.E. Tejeda-Yeomans**

Departamento de Física, Universidad de Sonora, Blvd. Transversal S/N,83000, Hermosillo, Mexico

E-mail: elena.tejeda@correo.fisica.uson.mx

**Abstract.** In this poster, we give a summary of work in progress regarding the study of meson's PDAs. We explore the consequences of a momentum-independent interaction as an ad-hoc tool to rebuild the pion distribution amplitude (PDA). The PDA is obtained through its moments using the Schwinger-Dyson formalism within QCD in the rainbow-ladder approximation.

## 1. Introduction

We can say that and mesons are the simplest bound-states to study in QCD. This can be achieved by describing light-quark confinement and dynamical chiral symmetry breaking (DCSB), and admits a symmetry-preserving truncation scheme. QCDs Dyson-Schwinger equations (DSEs) provide a picture of these mesons in hadron physics. We assume that u/dmesons are produced by a vector-vector current-current interaction that is mediated by a momentum-independent boson propagator, i.e., by the symmetry preserving regularisation of a contact interaction.

## 2. Pion Distribution Amplitude

Consider S the dressed-quark propagator, which is obtained from the gap equation:

$$S^{-1}(p) = i\gamma \cdot p + m + \int \frac{d^4q}{(2\pi)^4} g^2 D_{\mu\nu}(p-q)\frac{\lambda^a}{2}\gamma_\mu S(q)\frac{\lambda^a}{2}\Gamma_\nu(q,p) \quad (1)$$

Journal of Physics: Conference Series **651** (2015) 012002 doi:10.1088/1742-6596/651/1/012002

where $m$ is the Lagrangian current-quark mass, $D_{\mu\nu}$ is the vector-boson propagator, and $\Gamma_\nu$ is the quark vector-boson vertex which contains the non-perturbative information.

In order to build a stock of material that can be used to identify unambiguous signals in experiment for the pointwise behavior of interaction between light-quarks, their mass-functions and other quantities, we use a description through a contact interaction. We thus define

$$g^2 D_{\mu\nu}(p-q) = \frac{1}{m_G^2}\delta_{\mu\nu} \tag{2}$$

where $m_G$ is a gluon mass-scale, and preceed by embedding this interaction in a rainbow-ladder truncation of the DSEs, which implies $\Gamma_\nu(k,p) = \gamma_\nu$. This leads to the quark propagator equation:

$$S^{-1}(p) = i\gamma\cdot p + M \tag{3}$$

where $M$ is momentum independent and is defined as:

$$M = m + \frac{4}{3}\frac{1}{m_G^2}\int\frac{d^4q}{(2\pi)^4}\gamma_\mu S(q)\gamma_\mu \tag{4}$$

The regularisation procedure used is

$$\frac{1}{s+M^2} = \int_0^\infty d\tau e^{-\tau(s+M^2)} \to \int_{\tau_{uv}^2}^{\tau_{ir}^2} d\tau e^{-\tau(s+M^2)} \tag{5}$$

where $\tau_{uv,ir}$ are, respectively, infrared and ultraviolet regulators. A nonzero value of $\tau_{ir} = 1/\Lambda_{ir}$ implements confinement by ensuring the absence of quark production thresholds. $\tau_{ir} = 1/\Lambda_{ir}$ plays a dynamical role and sets the scale of all dimensioned quantities.

The homogeneous Bethe-Salpeter Equation (BSE) for the pseudoscalar meson for this interaction is

$$\Gamma_\pi(P) = -\frac{4}{3}\frac{1}{m_G^2}\int\frac{d^4q}{(2\pi)^4}\gamma_\mu\chi_\pi(q_+,q_-)\gamma_\mu \tag{6}$$

where $q_+ = q + P$ and $q = q$ are the gluons relative momentum. With a symmetry-preserving regularisation of the interaction, the Bethe-Salpeter amplitude cannot depend on relative momentum. We use

$$\Gamma_\pi(P) = \gamma_5[iE_\pi(P) + \frac{1}{M}\gamma\cdot PF_\pi(P)] \tag{7}$$

We want to build the pion distribution amplitude (PDA) from a moments calculation of it. The PDA is

$$f_\pi\phi_\pi(x) = Z_2N_c\int\frac{d^4q}{(2\pi)^4}\delta(n\cdot q - x\, n\cdot P)\ tr[\gamma_5\gamma\cdot nS(q_+)\Gamma_\pi(q;P)S(q_-)] \tag{8}$$

The PDAs moments are

$$\langle x^m\rangle = \int_0^1 dx\ x^m\phi_\pi(x) \tag{9}$$

So, we need to solve

$$f_\pi\langle x^m\rangle = Z_2N_c\int\frac{d^4q}{(2\pi)^4}\frac{(n\cdot q)^m}{(n\cdot P)^{m+1}}\ tr[\gamma_5\gamma\cdot nS(q_+)\Gamma_\pi(q;P)S(q_-)] \tag{10}$$

where $f_\pi$ is the pions leptonic decay constant define as the zero-moment of the PDA.

Journal of Physics: Conference Series **651** (2015) 012002 doi:10.1088/1742-6596/651/1/012002

### *2.1. Results*

Weve obtained the following expression for the PDAs moments:

$$f_\pi \langle x^m \rangle = \frac{1}{4\pi^2} Z_2 N_c \frac{1}{M} \int_0^1 d\alpha (-\alpha)^m \Big[ E_\pi M^2 \, \mathcal{C}_1(\mathcal{M}^2) + \frac{1}{2} F_\pi \left[ (\mathcal{M}^2 - 4M^2)\mathcal{C}_1(\mathcal{M}^2) + \mathcal{C}(\mathcal{M}^2) \right] \Big] \tag{11}$$

where $\mathcal{M}^2 = \alpha(1-\alpha)P^2 + M^2$ and the constants

$$\begin{aligned} \mathcal{C}_1(\mathcal{M}^2; \tau_{uv}^2, \tau_{ir}^2) &= \Gamma(0; \mathcal{M}^2\tau_{uv}^2) - \Gamma(0; \mathcal{M}^2\tau_{ir}^2) \\ \mathcal{C}(\mathcal{M}^2; \tau_{uv}^2, \tau_{ir}^2) &= \mathcal{M}^2[\Gamma(-1; \mathcal{M}^2\tau_{uv}^2) - \Gamma(-1; \mathcal{M}^2\tau_{ir}^2)] \end{aligned} \tag{12}$$

The $f_\pi$, define as $m = 0$, is

$$f_\pi = \frac{1}{4\pi^2} Z_2 N_c \frac{1}{M} \int_0^1 d\alpha \Big[ E_\pi M^2 \, \mathcal{C}_1(\mathcal{M}^2) + \frac{1}{2} F_\pi \left[ (\mathcal{M}^2 - 4M^2)\mathcal{C}_1(\mathcal{M}^2) + \mathcal{C}(\mathcal{M}^2) \right] \Big] \tag{13}$$

This means $\langle x^0 \rangle = 1$, i.e. PDAs moments describe a point-like particle.

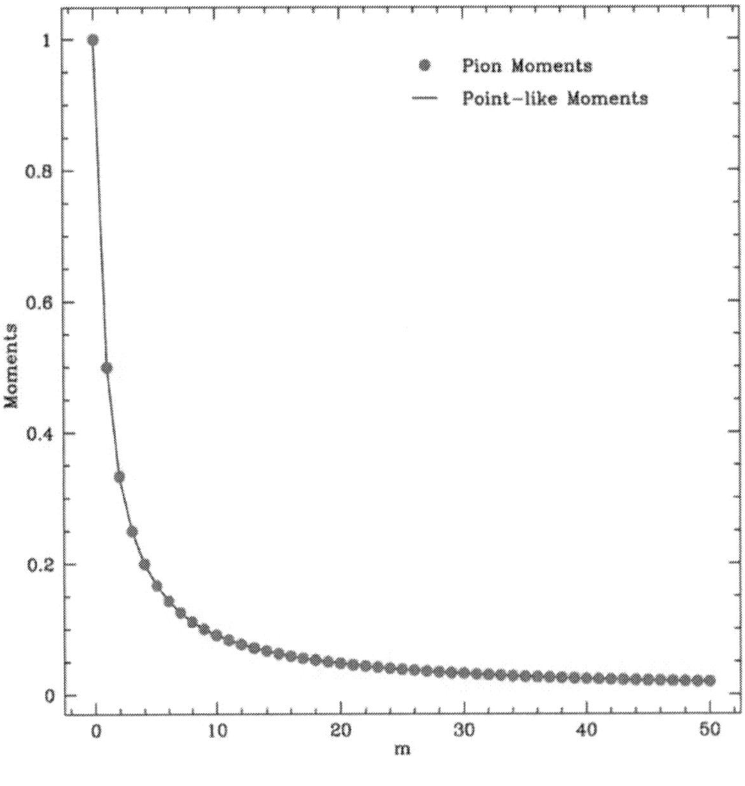


**Figure 1.** Results for the Pions moments and the point-like behavior of the PDA.

## 3. Pion Distribution Amplitude

We want to build a distribution amplitude for the $\rho$-meson by getting a procedure analogous to the PDAs moments calculation. In this case, we need to change the Bethe-Salpeter amplitude

Journal of Physics: Conference Series **651** (2015) 012002 doi:10.1088/1742-6596/651/1/012002

in order to have a description of a vector rather than a pseudoscalar meson. The vector BSA takes the form

$$\Gamma^\rho_\mu(P) = \gamma^T_\mu E_\rho(P) \tag{14}$$

with

$$P^\mu \gamma^T_\mu = 0 \qquad \gamma^L_\mu + \gamma^T_\mu = \gamma_\mu \tag{15}$$

The first attempt for a $\rho$ distribution amplitude ($\rho$DA) from a moment calculation is

$$f_\rho \phi_\rho(x) = \frac{1}{3} Z_2 N_c \int \frac{d^4q}{(2\pi)^4} \delta(n \cdot q - x\, n \cdot P) tr[\gamma_\mu S(q_+) \Gamma^\rho_\mu(q;P) S(q_-)] \tag{16}$$

The moments are given by

$$\begin{aligned}
f_\rho \langle x_L^m \rangle &= \frac{1}{3} Z_2 N_c \int \frac{d^4q}{(2\pi)^4} \frac{(n\cdot q)^m}{(n\cdot P)^{m+1}}\; tr[\gamma_\mu S(q_+) \Gamma^\rho_\mu(q;P) S(q_-)] &(17)\\
f_\rho \langle x_T^m \rangle &= \frac{1}{3} Z_2 N_c \int \frac{d^4q}{(2\pi)^4} \frac{(n\cdot q)^m}{(n\cdot P)^{m+1}}\; tr[\sigma_{\mu\nu} n_\nu S(q_+) \Gamma^\rho_\mu(q;P) S(q_-)] &(18)\\
f_\rho \langle x'^m \rangle &= \frac{-i}{3} Z_2 N_c \int \frac{d^4q}{(2\pi)^4} \frac{(n\cdot q)^m}{(n\cdot P)^{m+1}}\; tr[n\cdot\gamma \gamma_\mu S(q_+) \Gamma^\rho_\mu(q;P) S(q_-)] &(19)\\
& &(20)
\end{aligned}$$

where $f_\rho$ is the $\rho$'s leptonic decay constant define as the zero-moment of the $\rho$DA. The test moments $\langle x_L^m \rangle$ and $\langle x_T^m \rangle$ are a longitudinal-like and a transverse-like, respectively, forms inspired from [4]. On the other hand, test moment $\langle x'^m \rangle$ it's a fit to the point-like form that we expect.

The results of these test expressions are presented in the following graph

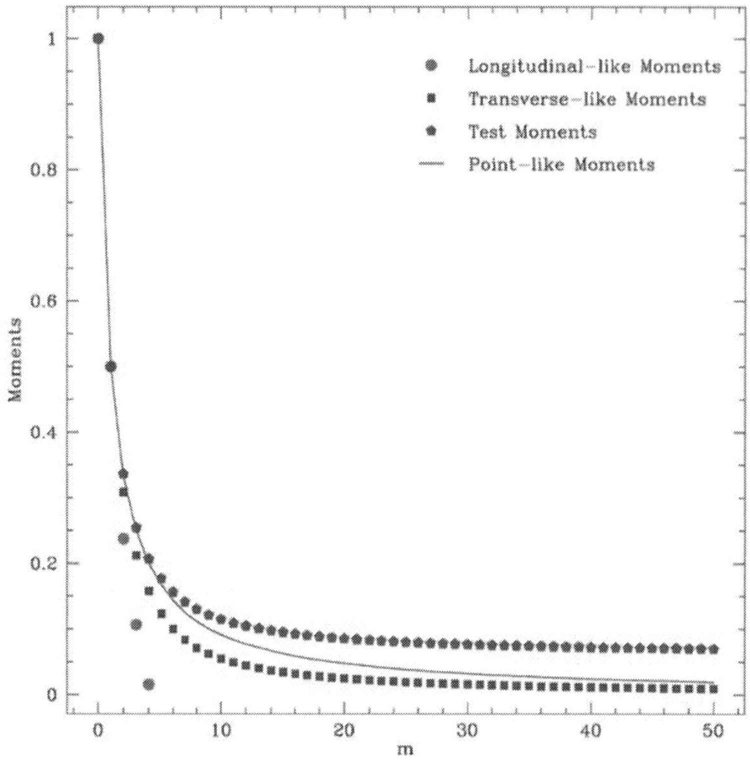


**Figure 2.** Results for the $\rho$'s moments and the point-like behavior of the $\rho$DA.

## 4. Summary

Weve obtained expressions for the DAs moments using the contact interaction. This represent an easy way to describe the behavior of the $\pi$- and $\rho$- mesons as point-like particles. Its important to note that weve only tested expressions that still need to be explored and fixed for a better description. It is also necessary to find a way to reconstruct $\rho$DA using the moments for the complete picture.

## References

[1] Pion form factor from a contact interaction. *Phys. Rev.* C 81 065202
[2] pi- and rho-mesons, and their diquark partners, from a contact interaction. *Phys. Rev.* C 83 065206
[3] Pion and kaon valence-quark parton distribution functions. *Phys. Rev.* C 83 062201
[4] The rho- meson light-cone distribution amplitudes of leading twist revisited. *Phys. Rev.* D 54 2182
[5] Bethe-Salpeter study of vector meson masses and decay constants. *Phys. Rev.* C 60 055214

# Scalar fields in a non-commutative space

**Wolfgang Bietenholz**[a], **Frank Hofheinz**[b], **Héctor Mejía-Díaz**[a] **and Marco Panero**[c]

[a] Instituto de Ciencias Nucleares, Universidad Nacional Autónoma de México A.P. 70-543, C.P. 04510 México, Distrito Federal, Mexico
[b] Institute of Radiopharmaceutical Cancer Research Helmholtz-Zentrum Dresden-Rossendorf, Germany
[c] Instituto de Física Teórica UAM/CSIC, Universidad Autónoma de Madrid Ciudad Universitaria de Cantoblanco, 28049 Madrid, Spain

E-mail: wolbi@nucleares.unam.mx

**Abstract.** We discuss the $\lambda\phi^4$ model in 2- and 3-dimensional non-commutative spaces. The mapping onto a Hermitian matrix model enables its non-perturbative investigation by Monte Carlo simulations. The numerical results reveal a phase where stripe patterns dominate. In $d = 3$ we show that in this phase the dispersion relation is deformed in the IR regime, in agreement with the property of UV/IR mixing. This "striped phase" also occurs in $d = 2$. For both dimensions we provide evidence that it persists in the simultaneous limit to the continuum and to infinite volume ("Double Scaling Limit"). This implies the spontaneous breaking of translation symmetry.

## 1. Quantum physics in a non-commutative space

Since in standard quantum mechanics the operators of space and momentum coordinates do not commute, it seems like an obvious idea to "quantise further" by also introducing non-zero commutators among space coordinates (or among momentum coordinates) in different directions. Hence it is not surprising that this idea dates back to the 1940s. Its pre-history involves famous names like Heisenberg, Pauli, Peierls and Oppenheimer, and in 1947 the first papers on this subject were published [1].

However, the consequences of non-commutative (NC) geometry in quantum field theory are extremely involved. Here we consider only the simplest case of two NC spatial coordinates, with a constant non-commutativity "tensor" $\Theta$,

$$[\hat{x}_i, \hat{x}_j] = \mathrm{i}\Theta_{ij} = \mathrm{i}\theta\epsilon_{ij} \ , \tag{1.1}$$

where $\hat{x}_i$ are Hermitian operators, $i, j \in \{1, 2\}$, and $\theta$ is the non-commutativity parameter.

In the 1980s deep mathematical work was carried out about the formal formulation of field theories on such spaces (for a review, see Ref. [2]). Applications were discussed in solid state physics, in particular related to the quantum Hall effect, see *e.g.* Ref. [3]. Of course, in this context one actually deals with the usual geometry, but a magnetic background field $B$ can be interpreted as $\theta \propto 1/B$, which leads to NC canonical coordinates.

In the period from 1996 to 1998 a boom of interest in NC field theories set in, which led to about 3000 papers on this subject up to now [4]. This boom was triggered by the observation

Journal of Physics: Conference Series **651** (2015) 012003 doi:10.1088/1742-6596/651/1/012003

that low energy string theory can be related to NC field theory [5], following the spirit of the re-interpretation of the magnetic background field.

Here we are going to address NC field theory as such; no strings attached. A qualitative difference from usual field theory is its *non-locality;* fields interact at distinct points over a characteristic distance $\sim \sqrt{\|\Theta\|}$. This entails frightening conceptual problems. On the other hand, from a very optimistic point of view, this is just what it takes for a proposal to formulate quantum gravity.

In this regard, we mention a simple *Gedankenexperiment:* assume some event to be measured with extremely tiny space-time uncertainties $\Delta x_1$, $\Delta x_2$, $\Delta x_3$, $\Delta t$, on the order of the Planck length $l_{\rm Planck} \simeq 1.6 \cdot 10^{-35}$ m. This requires a huge energy density, so gravitation should be taken into consideration. In the extreme case this could yield an event horizon, which is larger than the Heisenberg uncertainties, so the event is invisible. According to Ref. [6], avoiding that scenario requires

$$\begin{aligned} \Delta x_1 \Delta x_2 + \Delta x_1 \Delta x_3 + \Delta x_2 \Delta x_3 &\geq l_{\rm Planck}^2 \ , \\ (\Delta x_1 + \Delta x_2 + \Delta x_3)\Delta t &\geq l_{\rm Planck}^2 \ . \end{aligned} \tag{1.2}$$

Such space-time uncertainties are characteristic for an NC space. Is this a natural framework for the conciliation of quantum theory and gravity? We should add, however, that much of the literature on this subject keeps the time commutative. That deviates from the above consideration, but it may save unitarity and reflection positivity, and it alleviates the problems related to causality.

The historic motivation, however, was different. People hoped that washing out the space-time points in this way[1] could remove (or at least weaken) the notorious UV divergences in quantum field theory, and avoid (or simplify) renormalisation. This turned out to be wrong: renormalisation is not getting easier, but much harder due to non-commutativity. First of all, in planar diagrams of a perturbative expansion, the UV divergences simply persist [7]. Second, in the non-planar diagrams they tend to "mix" with IR divergences. This type of singularities does not occur in the commutative world, and it is very difficult to deal with. For a simple intuitive picture, we start again from the Heisenberg uncertainty $\Delta x_i \sim 1/\Delta p_i$, and we combine it with

$$\Delta x_i \sim \Theta_{ij}/\Delta x_j \sim \Theta_{ij}\Delta p_j \qquad (i \neq j) \ . \tag{1.3}$$

Therefore an attempt to squeeze $\Delta p_i \to 0$ makes $\Delta p_j$ diverge. Due to such mixed singularities, the renormalisation of multi-loop diagrams is mysterious. Hence it is highly motivated to adopt a fully *non-perturbative approach.*

In many models, the lattice regularisation enables the non-perturbative treatment of quantum field theory. So let us introduce a lattice structure also on the NC plane of eq. (1.1) (this concept is reviewed in Ref. [8]). This is achieved — at least in a fuzzy form — if we impose the operator identity

$$\exp\Big( {\rm i}\frac{2\pi}{a}\hat x_i \Big) = \hat{1\!\!1} \ . \tag{1.4}$$

For lattice spacing $a$ we expect periodicity of the (commutative) momentum components over the Brillouin zone,

$$\exp\Big( {\rm i}\sum_i k_i \hat x_i \Big) = \exp\Big( {\rm i}\sum_i (k_i + \frac{2\pi}{a})\hat x_i \Big) \ , \quad i = 1,2 \ . \tag{1.5}$$

[1] Some people denote it as "pointless geometry".

Journal of Physics: Conference Series **651** (2015) 012003 doi:10.1088/1742-6596/651/1/012003

Multiplication with the inverse factor $\exp(-\mathrm{i}\sum_j k_j \hat{x}_j)$ from the right, and applying the Baker-Campbell-Hausdorff formula, leads to

$$\hat{1\!\!1} = \hat{1\!\!1} \exp\left(\frac{\mathrm{i}\pi}{a}\theta(k_2 - k_1)\right) \quad \Rightarrow \quad \frac{\theta}{2a}k_i \in \mathbb{Z} \,. \tag{1.6}$$

Therefore, in contrast to the commutative space, the NC lattice is automatically *periodic*.

If we now assume periodicity over the lattice volume $N \times N$, we have discrete momenta $k^{(n)} = \frac{2\pi}{aN}n$, $n \in \mathbb{Z}^2$, and we arrive at the relation

$$\theta = \frac{1}{\pi}Na^2 \,. \tag{1.7}$$

In order to keep $\theta$ finite, in particular for $\theta = \text{const.}$, we have to take the limits to the continuum, $a \to 0$, and to infinite volume, $Na \to \infty$, *simultaneously*. This is the *Double Scaling Limit*,

$$\left.\begin{array}{ccc} a & \to & 0 \\ N & \to & \infty \end{array}\right\} \quad \text{such that} \quad Na^2 = \text{const.} \,, \tag{1.8}$$

which leads to a NC plane of infinite extent. Clearly, this requirement is again related to the property of UV/IR mixing. Taking these limits differently, one would usually end up with $\theta = 0$ or $\theta = \infty$, which are both (different) cases of commutative field theory. For a consistent study of NC field theory, we have to follow the instruction (1.8).

## 2. The non-commutative $\lambda\phi^4$ model

### 2.1. Formulation

NC field theory can be formulated such that the fields are functions of the standard (commutative) coordinates $x_\mu$, if all field multiplications are performed by *star products* (or *Moyal products*). A prototype reads

$$\phi(x) \star \psi(x) := \phi(x) \exp\left(\frac{\mathrm{i}}{2}\overleftarrow{\partial}_\mu \Theta_{\mu\nu} \overrightarrow{\partial}_\nu\right)\psi(x) \tag{2.1}$$

(for instance $[x_\mu, x_\nu]_\star = \mathrm{i}\Theta_{\mu\nu}$). This can be justified with a plane wave decomposition.

For bilinear terms under a space-time integral (without boundary terms) the star product is equivalent to a simple product, because of the anti-symmetry of $\Theta$. Hence the action of the $\lambda\phi^4$ model in NC Euclidean space can be written as

$$S[\phi] = \int d^d x \left[\frac{1}{2}\partial_\mu\phi\,\partial_\mu\phi + \frac{m^2}{2}\phi^2 + \frac{\lambda}{4}\phi \star \phi \star \phi \star \phi\right] \,. \tag{2.2}$$

This shows that the parameter $\lambda$ does not only determine the strength of the self-interaction, but also the extent of NC effects.

The perturbative expansion of this model has been discussed extensively in the literature. Regarding the 1-loop diagrams, there is a planar contribution, which takes the standard form [7], as we mentioned before. On the other hand, the non-planar diagrams pick up a phase factor due to the non-commutativity. For the moment, let us introduce a momentum cutoff $\Lambda$. Then the 1-loop integrals and their leading divergences in $d = 4$ take the form [9]

$$\text{planar}: \quad \int d^d k \, \frac{1}{k^2 + m^2} \propto \Lambda^2 \,, \quad \text{non-planar}: \quad \int d^d k \, \frac{\exp(\mathrm{i}k_\mu \Theta_{\mu\nu} p_\nu)}{k^2 + m^2} \propto \frac{1}{1/\Lambda^2 + p_\mu \Theta_{\mu\nu} p_\nu} \,.$$

We see that a finite $\Theta$ does indeed allow us to take the limit $\Lambda \to \infty$ in the non-planar part. However, then a singularity emerges for external momentum $p \to 0$, which illustrates the UV/IR mixing. Moreover, even at finite $p$, the limit $\Theta \to 0$ is not smooth; therefore the meaning of a truncated expansion in small $\|\Theta\|$ is questionable. Finally we also confirm that the opposite limit $\Theta \to \infty$ is commutative, but different from $\Theta = 0$.

Now we consider $d = 3$, so that the scalar field $\phi(\vec{x}, t)$ lives on a NC plane $(x_1, x_2)$, plus a commutative Euclidean time $t$. We assume a lattice structure, for the NC plane as described in Section 1, and for the time in a regular form. The action on a $N^2 \times T$ lattice can be mapped (with identical algebras) onto a matrix model with twisted boundary conditions [10],

$$S[\bar{\phi}] = \mathrm{Tr} \sum_{t=1}^{T} \Big[ \frac{1}{2} \sum_{i=1}^{2} \Big( \Gamma_i \bar{\phi}(t) \Gamma_i^\dagger - \bar{\phi}(t) \Big)^2 + \frac{1}{2} \Big( \bar{\phi}(t+1) - \bar{\phi}(t) \Big)^2 + \frac{m^2}{2} \bar{\phi}^2(t) + \frac{\lambda}{4} \bar{\phi}^4(t) \Big] , \quad (2.3)$$

where $\bar{\phi}(t)$ are Hermitian $N \times N$ matrices, located at $t = 1, \dots, T$ (we are using lattice units). The kinetic term has the standard lattice form in time direction, but in the NC plane it is constructed by means of so-called twist eaters $\Gamma_i$. They arrange for a shift by one lattice unit, if they obey the 't Hooft-Weyl algebra

$$\Gamma_i \Gamma_j = Z_{ji} \Gamma_j \Gamma_i \ . \quad (2.4)$$

Here the twisted boundary conditions enter, and we choose the corresponding phase factor as

$$Z_{21} = Z_{12}^* = \exp\Big( i\pi (N+1)/N \Big) \ , \quad (2.5)$$

where the size $N$ is assumed to be odd. Then we insert a unitary solution for $\Gamma_1$ and $\Gamma_2$, which is known as clock- and shift-matrix (they are written down explicitly *e.g.* in Refs. [10, 11]). The crucial property, however, is relation (2.4).

### *2.2. Phase diagram*

Some years ago, Gubser and Sondhi performed a 1-loop calculation in the Hartree-Fock approximation [12] and conjectured the following properties of the phase diagram of the NC $\lambda\phi^4$ model in $d = 3$ and 4:

- At small $\theta$, there is an Ising-type phase transition between a disordered and a uniform phase at some critical value $m_c^2 < 0$ (a strongly negative parameter $m^2$ can be interpreted as low temperature).
- At large $\theta$ and some $m_c^2 < 0$, there is another phase transition, but now between a disordered and a *striped phase*.

Further considerations were added with Renormalisation Group techniques [13], and with the Cornwall-Jackiw-Tomboulis effective action approach [14]. They are consistent with the qualitative picture by Gubser and Sondhi.

A *quantitative* study was based on Monte Carlo simulations, which probed the phase diagram in the $(m^2, \lambda)$ plane, for the 3d matrix formulation described in Subsection 2.1, at $N = T = 15 \dots 45$ [11]. Thus the picture by Gubser and Sondhi is converted into a uniform/disordered transition at small $\lambda$, and a striped/disordered transition at large $\lambda$. This is in fact observed, as the phase diagram in Figure 1 shows. Figure 2 adds the features of typical configurations in the four sectors (after mapping back the matrices to a lattice scalar field).

The phases, and their transitions, were identified with the momentum dependent order parameter

$$M(k) = \frac{1}{NT} \max_{\frac{N}{2\pi}|\vec{p}|=k} \Big| \sum_t \tilde{\phi}(\vec{p}, t) \Big| \ . \quad (2.6)$$

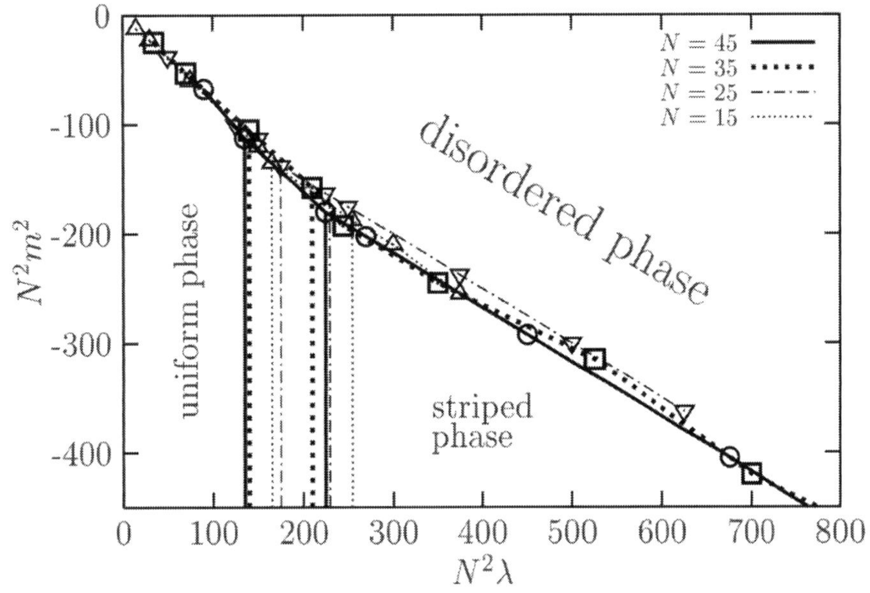


**Figure 1.** The phase diagram of the 3d $\lambda\phi^4$ model on an NC plane but with a commutative Euclidean time coordinate. The ordered regime — at strongly negative $N^2m^2$ — is divided into phases of uniform and of striped order.

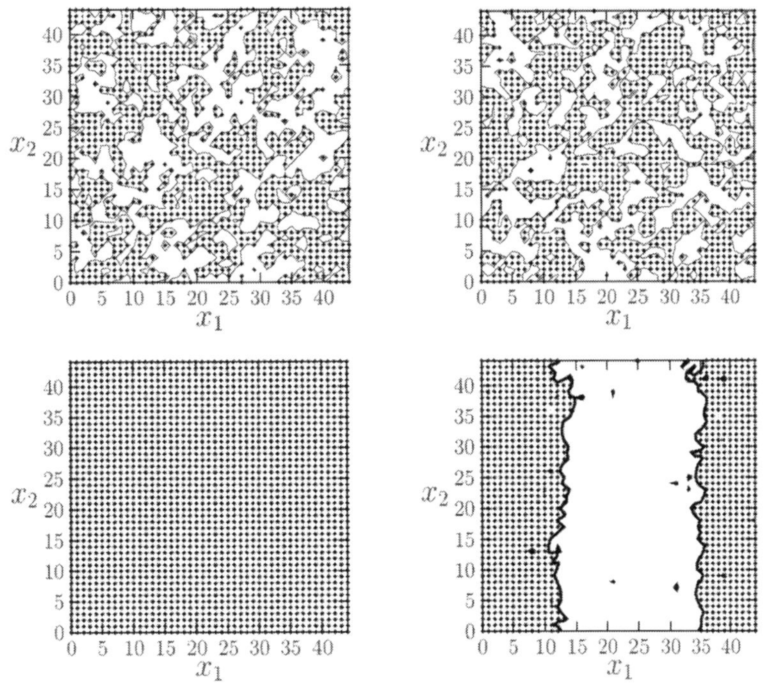


**Figure 2.** Typical configuration of the 3d $\lambda\phi^4$ model at $N = T = 45$: we show the NC plane, and dark/bright areas correspond to the two signs of $\phi$. The examples on the left (right) refer to low (high) $\lambda$, and the upper (lower) plots were obtained at weakly (strongly) negative $m^2$.

For $k = 0$ this is the magnetisation, and for finite $k$ it captures the possible dominance of a "stripe pattern", *e.g.* with $k$ parallel stripes, rotated in a suitable way (if two non-zero components of $\vec{p}$ are involved, the pattern is actually of a checker-board type).

Far from the transition lines, $M(k)$ indicates the phase unambiguously. The transition is identified best by varying $m^2 < 0$ at fixed $\lambda$, and searching for a peak in the connected two-point function

$$\langle M(k)^2\rangle_{\rm con} = \langle M(k)^2\rangle - \langle M(k)\rangle^2 \ . \qquad (2.7)$$

This provides accurate results for the critical values $m_c^2$. For all $\lambda$ values that we considered, this transition appears to be of second order.

On the other hand, the transition uniform/striped inside the ordered regime is rather hard to explore, as the uncertainty band in Figure 1 shows. Here we studied the thermal cycle, which reveals a clear hysteresis effect [11]; this is characteristic for a first order transition.

Next we considered the correlation functions close to the order/disorder transition, in the disordered regime (where finite size effects are harmless). Figure 3 refers to the spatial separation, and shows that the correlator

$$C(x_1, 0) = \langle \phi(\vec{0}; t)\, \phi(x_1, 0; t)\rangle \qquad (2.8)$$

has an unusual shape, both close to the uniform and close to the striped phase. The decay is fast, but not exponential; it is NC distorted.

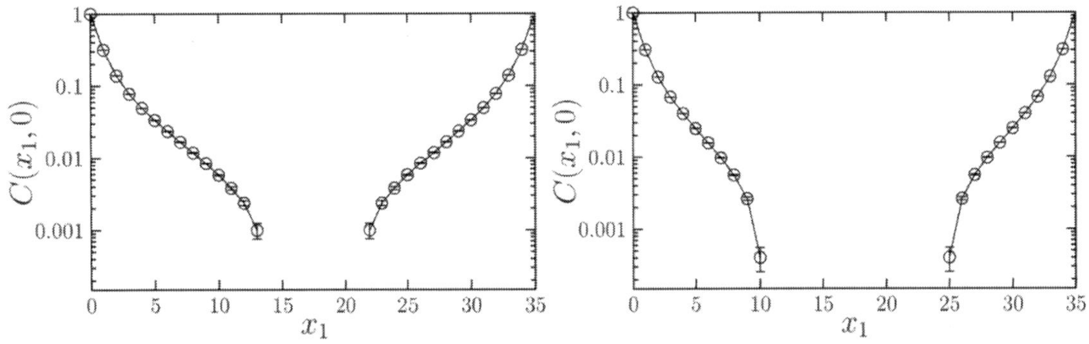


**Figure 3.** The correlation function $C(x_1, 0)$ for spatial separation, given in eq. (2.8). Both plots were obtained in the disordered phase, close to the ordering transition, at $N = T = 35$. On the left the parameters are $N^2\lambda = 70$, $N^2m^2 = -17.5$ (close to the uniform phase), and on the right $N^2\lambda = 3500$, $N^2m^2 = -140$ (close to the striped phase). In both cases, the decay is fast, but not exponential.

Nevertheless we can evaluate the energy based on the *temporal* correlation function. Here we first transform the spatial part of the configurations to momentum space, and we consider

$$G(\tau) = \langle \tilde{\phi}(\vec{p}, t)\, \tilde{\phi}(\vec{p}, t + \tau)\rangle \ . \qquad (2.9)$$

Now we do find an exponential decay — respectively a cosh function behaviour at finite $T$ — as Figure 4 illustrates for the example $\vec{p} = \vec{0}$.

This leads to the dispersion relation $E^2(\vec{p}^{\,2})$ shown in Figure 5, now for $N = T = 55$. In fact, the energy minimum is located at *non-zero* $\vec{p}$, which is a clear indication that we are close to the striped phase; decreasing $m^2$ then leads to the condensation of a corresponding striped pattern.

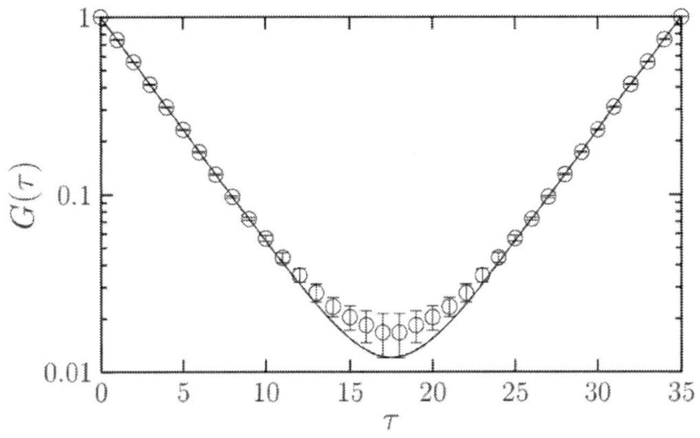


**Figure 4.** The temporal correlation function $G(\tau)$, given in eq. (2.9), measured at $N = T = 35$, $N^2\lambda = 350$, $N^2m^2 = -140$. The data agree very well with a cosh-fit.

It is evident that this pattern is non-uniform, but for the exact structure of the stripes various options are in close competition.

At very small momenta, the dispersion relation is consistent with the expected IR divergence.

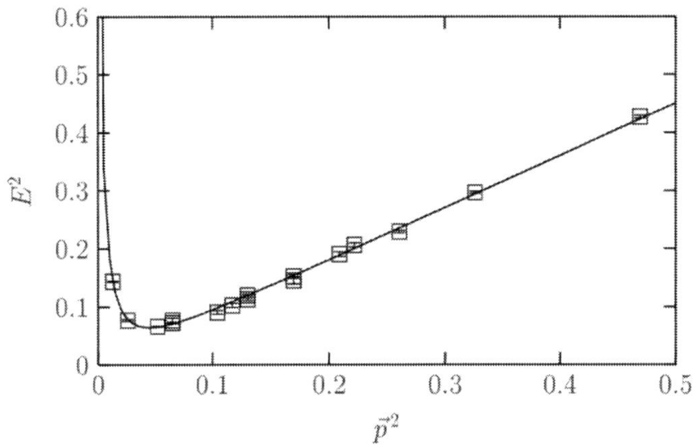


**Figure 5.** The dispersion relation $E^2(\vec{p}^2)$ determined at $N = T = 55$, $m^2 = -15$, $\lambda = 50$ (close to the striped phase). The energy takes its minimum at finite momentum, which leads to a striped pattern at somewhat lower $m^2$. A 4-parameter fit [11] is consistent with the trend to an IR divergence in infinite volume.

*2.3. Double Scaling Limit*

So far we have been dealing with lattice units. In order to take a continuum limit, we have to introduce a scale, *i.e.* we need a dimensional reference quantity. For this reason, we extrapolate the (broad) linear regime in the dispersion relation $E(|\vec{p}|)$ down to $\vec{p} = \vec{0}$. This linear extrapolation deviates from the dispersion at small momenta, but it defines an effective mass $M_{\rm eff}$, according to

$$E^2 = M_{\rm eff}^2 + \vec{p}^2 \; . \tag{2.10}$$

We are going to investigate the behaviour if $m^2$ approaches $m_c^2$ from above. Hence it is convenient to define

$$\Delta m^2 := m^2 - m_c^2 \, . \tag{2.11}$$

At fixed $\lambda$, and for $\Delta m^2 \gtrsim 0$, we observed the proportionality relation

$$M_{\rm eff}^2|_{\lambda=\rm const.} \propto \Delta m^2 \, , \tag{2.12}$$

as Figure 6 shows for $\lambda = 50$, as an example.

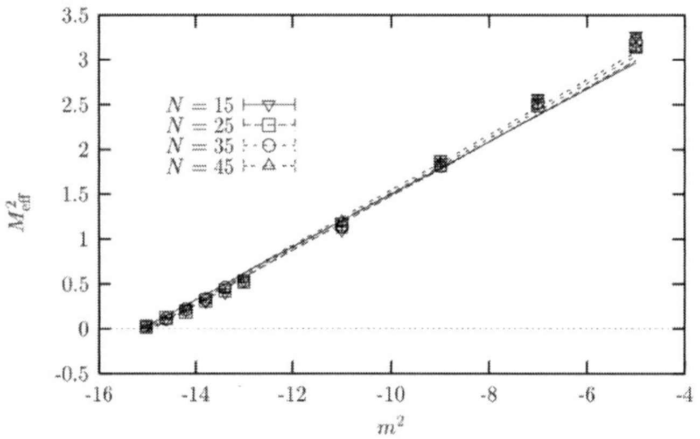


**Figure 6.** The effective mass squared, $M_{\rm eff}^2$, obtained by the extrapolation (2.10) of the dispersion relation. We see that $M_{\rm eff}^2$ depends linearly on $m^2$ resp. on $\Delta m^2$. The plot refers to $\lambda = 50$, where $M_{\rm eff}^2$ vanishes at $m_c^2 = -15.01(8)$.

Now we can take a continuum limit by keeping the effective mass in dimensional units, $M_{\rm eff}/a$, fixed. For simplicity we set this ratio to 1, hence $a = M_{\rm eff}^{-1}$ is the dimensional lattice spacing. Therefore, the DSL condition $\theta \propto Na^2 = \text{const.}$ is implemented by increasing $N$, and simultaneously decreasing $m^2$ so that it approaches $m_c^2$ from above, in such a manner that

$$N \, \Delta m^2 = \text{const.} \, . \tag{2.13}$$

Now we can re-consider the dispersion relation in dimensional units. The rest energy diverges like $E(|\vec{p}| \to 0)/a \propto \sqrt{N}$ [11], which confirms the UV/IR mixing also dimensionally. For a broad range of finite momenta $|\vec{p}|/a$, Figure 7 shows that the dispersion relation stabilises if we approach the DSL, and that the energy minimum is obtained around the dimensional momentum

$$\vec{p}^{\,2}/a^2 \lesssim 0.1 \ M_{\rm eff}^2 \, . \tag{2.14}$$

This tells us that the striped phase persists in the DSL, where stripes of finite width dominate. This observation implies the spontaneous breaking of translation and rotation symmetry in the striped phase of the 3d phase diagram.

That phenomenon leads to a tricky question: does the same happen also in $d = 2$ ? The next section will be devoted to this issue.

## 3. Does translation symmetry break in $d = 2$ ?

The Mermin-Wagner Theorem [15] tells us that usually a continuous global symmetry cannot break spontaneously in $d = 2$. At first sight this seems to imply that the striped phase cannot

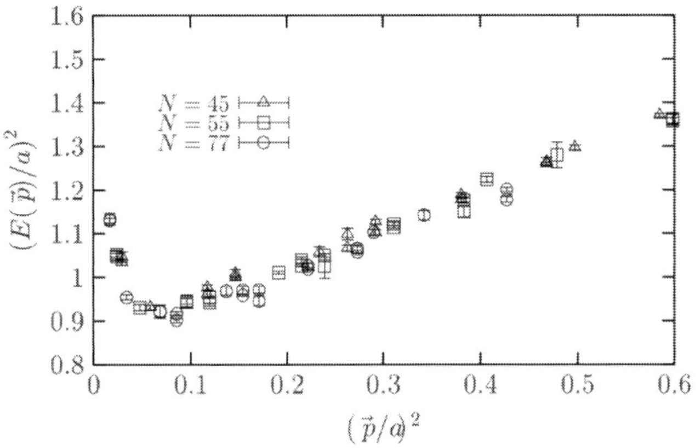


**Figure 7.** Dispersion relation in dimensional units, at $\lambda = 50$ and $N\,\Delta m^2 \simeq 100$. Both axes are given in units of $M_{\rm eff}^2$. We see that the dispersion stabilises, so it *is* possible to take a DSL next to the striped phase, with a finite width of the stripes to be formed at $m^2 < m_c^2$.

occur in the NC plane. However, the proof for this theorem is based on assumptions of standard quantum field theory, like locality — this does not hold on the NC plane.

Still, Gubser and Sondhi did not expect a striped phase in $d = 2$ [12]. They presented a consideration how the Mermin-Wagner Theorem could be extended even to NC field theory. They used an effective action approach of the Brazovskii-type, where the kinetic term is of quartic order in the momentum, which should make the exclusion of spontaneous symmetry breaking stronger. On the other hand, the effective action approach of Ref. [16] seems to affirm a striped phase.

From the numerical side, a striped phase in the NC $\lambda\phi^4$ model has been manifestly observed also in $d = 2$ [11, 17, 18]. However, this does not prove its existence in the DSL — it could also be an artifact of the lattice and of finite volume. The fate of this phase in the DSL has been investigated numerically only very recently [19].

The matrix model formulation corresponds to the description in Subsection 2.1, where the time direction collapses to one site. Figure 8 shows the phase diagram in $d = 2$. We see that the vertical axis needs a scaling factor $N^{3/2}$, which differs from the 3d case (cf. Figure 1). If $m^2$ is scaled in this way, we observe a convincing stabilisation of the order/disorder transition line for $N \geq 19$.

Figure 9 shows again typical configurations in the four sectors of this phase diagram, in analogy to the Figure 2, but now with a four-stripe pattern. The transition was again detected with the order parameter (2.6). Figure 10 gives examples how the uniform or striped order parameter rises for decreasing $m^2$, and how the corresponding connected correlator exhibits a peak at the transition.

Now we consider also here the correlation function. In $d = 3$ we focused on the correlation in time direction in order to extract the dispersion relation and to introduce the effective mass $M_{\rm eff}$, as a scale for the DSL. This is not available anymore in $d = 2$, so now we have to deal with the spatial correlation, and its unusual decay behaviour. Figure 11 shows an example in the disordered phase, but next to the the striped phase — for slightly lower $m^2$ a 4-stripe pattern will condense (like the example in Figure 9 on the bottom at the right).

Our concept is as follows: we decrease $m^2$ down towards $m_c^2$, and we increase the matrix size $N$ at the same time, such that the decay of the correlator stabilises down to the first dip. This replaces the usual reference to the exponential decay. The difference $\Delta m^2$, defined in eq. (2.11),

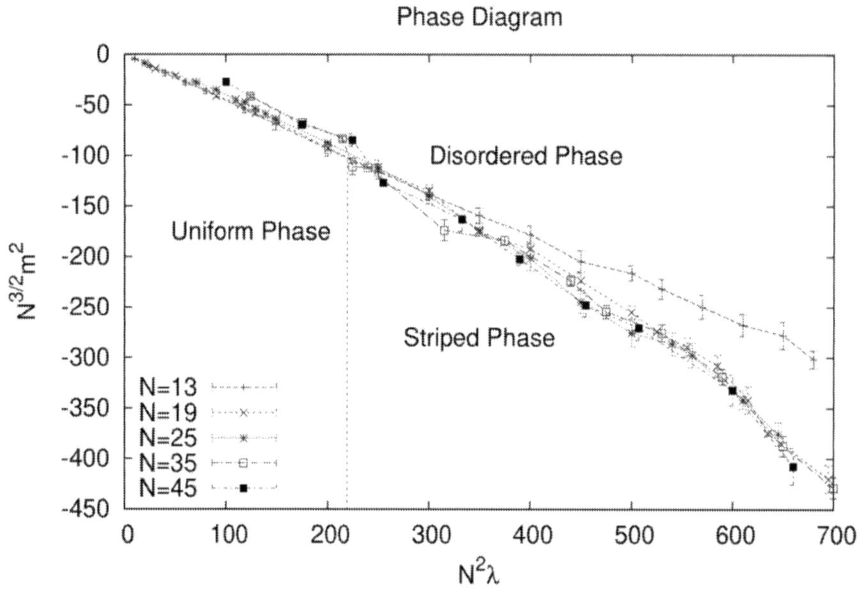


**Figure 8.** The phase diagram of the 2d NC $\lambda\phi^4$ model. In contrast to the 3d case, the vertical axis has to be chosen as $N^{3/2}m^2$. Then the transition line between disorder and the ordered phases stabilises at $N \geq 19$.

introduces a scale, which translates — with a suitable exponent — into the scale of the DSL,

$$a^2 \propto (\Delta m^2)^\sigma \ . \tag{3.1}$$

The exponent $\sigma$ has to be identified, then we can address the question whether or not it is possible to take a DSL and keep close to the striped phase. If this can (cannot) be done, this is evidence for the existence (absence) of the striped phase in the continuum and infinite volume limit of this model.

To tackle this question, we first choose a normalisation by setting the lattice spacing for $N = 35$ to $a = 1$. Thus the DSL converts a lattice distance $x$ into the dimensional distance

$$ax = \sqrt{\frac{35}{N}}\, x \ . \tag{3.2}$$

This distance should be compatible up to the first correlation dip for increasing $N$.

In analogy to the "dimensionless temperature", which is often used near a phase transition, $(T - T_c)/T_c$, we adjust the dimension by the suitable power of $m_c^2$, such that the DSL can be written as

$$Na^2 = N \frac{(\Delta m^2)^\sigma}{(m_c^2)^{1-\sigma}} \ . \tag{3.3}$$

Now let us consider two matrix sizes $N_1$ and $N_2$; we want to identify the mass shifts $\Delta m_1^2$ and $\Delta m_2^2$ which correspond to the same trajectory towards the DSL. We fix $\lambda$ to the same value, so that the dimensionless term $\lambda\theta$ remains constant. We fine-tune the mass shifts such that they lead to the same short-distance correlation decay. This is illustrated for three examples in

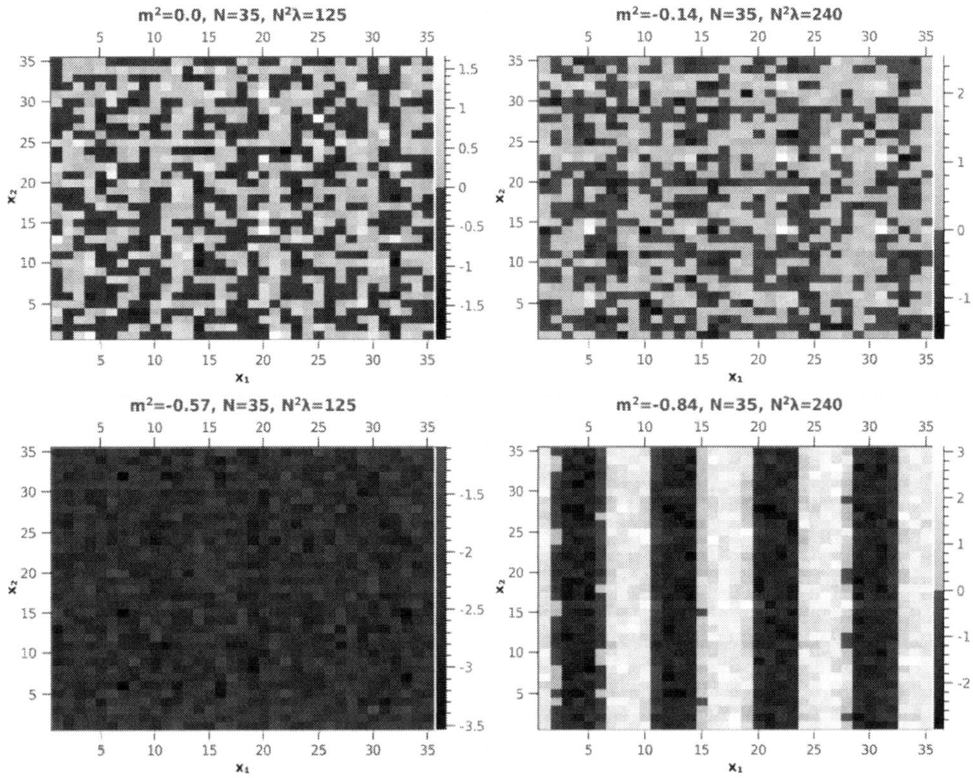


**Figure 9.** Typical configurations in four sectors of the phase diagram in Figure 8, analogous to Figure 2. The upper plots are in the disordered phase, next to the uniform and to the striped phase, respectively. The lower plots are examples for uniform and for striped ordering.

Figure 12. Once these values $\Delta m_i^2$ $(i = 1, 2)$, and the corresponding critical values $m_{c,i}^2$, are determined, we extract the exponent $\sigma$ from

$$\sigma = \frac{\ln(m_{1,c}^2/m_{2,c}^2)}{\ln(\Delta m_{1,c}^2/\Delta m_{2,c}^2) + \ln(m_{1,c}^2/m_{2,c}^2)} \ . \tag{3.4}$$

This has to be done for a variety of pairs $N_1$, $N_2$, and the crucial question is whether or not a stable $\sigma$-value is obtained.

| $\lambda$ | $N_1$ | $N_2$ | $\sigma$ |
|---|---|---|---|
| | 35 | 45 | 0.152(7) |
| 0.222 | 35 | 55 | 0.156(6) |
| | 45 | 55 | 0.161(11) |
| | 25 | 35 | 0.161(9) |
| 0.286 | 25 | 45 | 0.167(7) |
| | 35 | 45 | 0.178(23) |
| 0.4 | 25 | 35 | 0.147(13) |

**Table 1.** The $\sigma$-values obtained for various pairs of sizes $N_1$, $N_2$ after tuning $\Delta m^2$ such that the short-distance decay of the correlation function coincides.

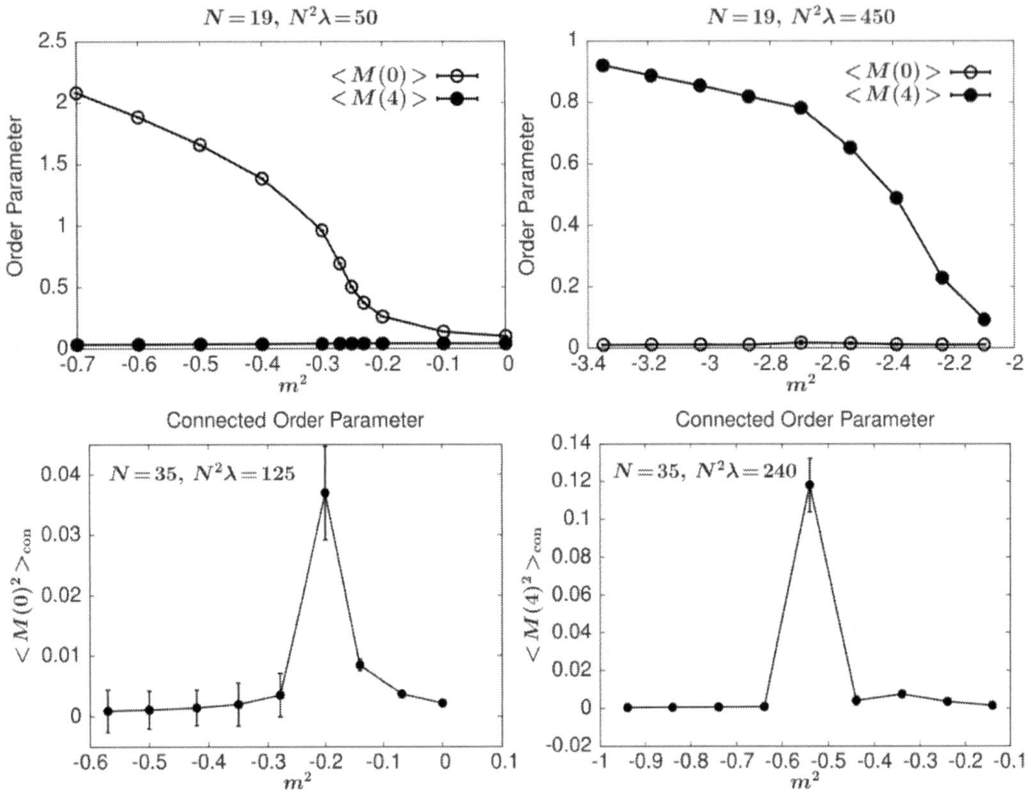


**Figure 10.** The order parameters $\langle M(0)\rangle$ and $\langle M(4)\rangle$, as defined in eq. (2.6), for the 2d NC $\lambda\phi^4$ model. We keep $N$ and $\lambda$ fixed and show the dependence on $m^2$. Above we see that for small (large) $\lambda$ and decreasing $m^2$ the disorder turns into a uniform (4-stripe) pattern. Below we show examples where the 2-point functions of these order parameters have a peak, which allows us to identify the critical value $m_c^2$.

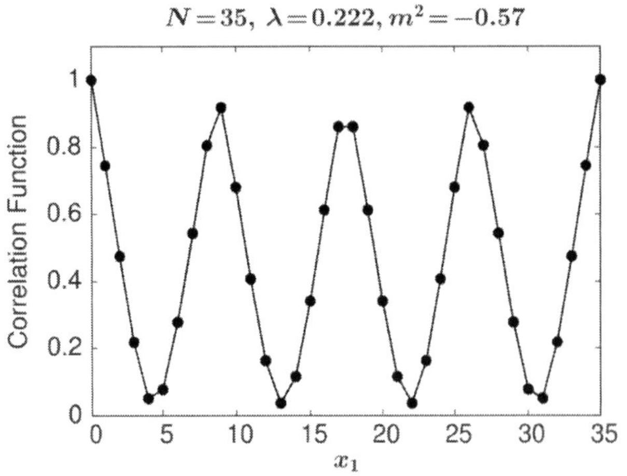


**Figure 11.** The correlation function $\langle\phi_{(0,0)}\phi_{(x_1,0)}\rangle$ near the striped phase, at $(N^{3/2}m^2, N^2\lambda) = (-118, 272)$. For somewhat lower $m^2$ a 4-stripe pattern condenses.

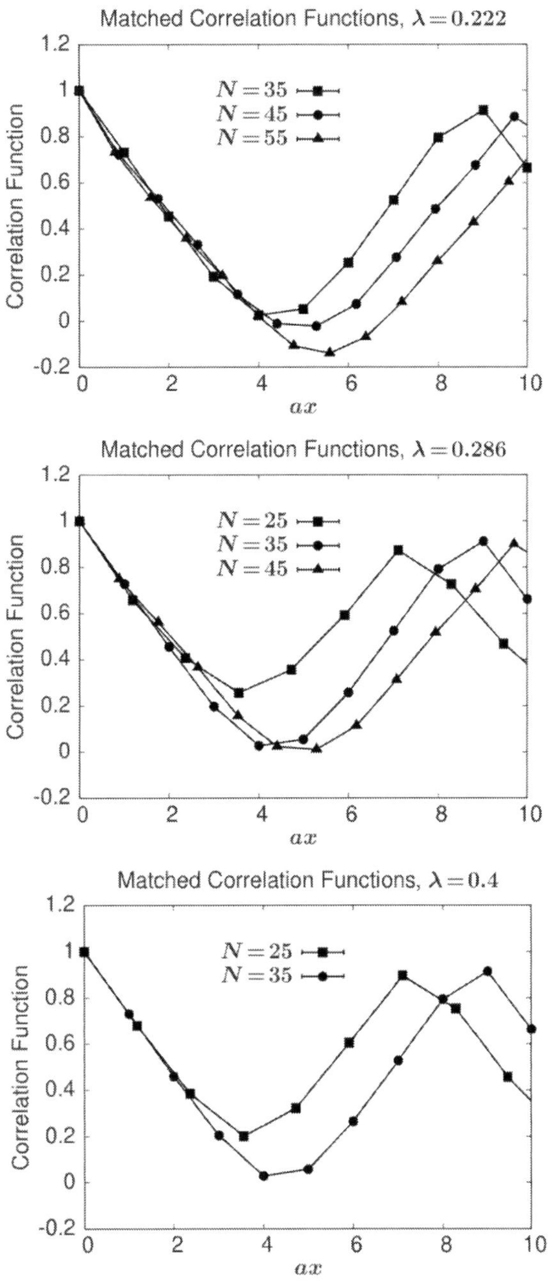


**Figure 12.** Three examples for "matched correlation functions": at fixed $\lambda$, but for different sizes $N$, the parameter $\Delta m^2 = m^2 - m_c^2$ is tuned such that the correlation decay down to the first dip has the same slope. Then the distance in physical units — as given in eq. (3.2) — agrees. In this way we identify $\Delta m^2$ values to be inserted in eq. (3.4), which determines the exponent $\sigma$.

There are practical constraints for these evaluations: $\lambda$ has to be large enough to be close to the striped (not uniform) phase for the smaller $N$ involved. On the other hand, for the larger $N$ the product $N^2\lambda$ should not be too large — otherwise we run to the far right in the

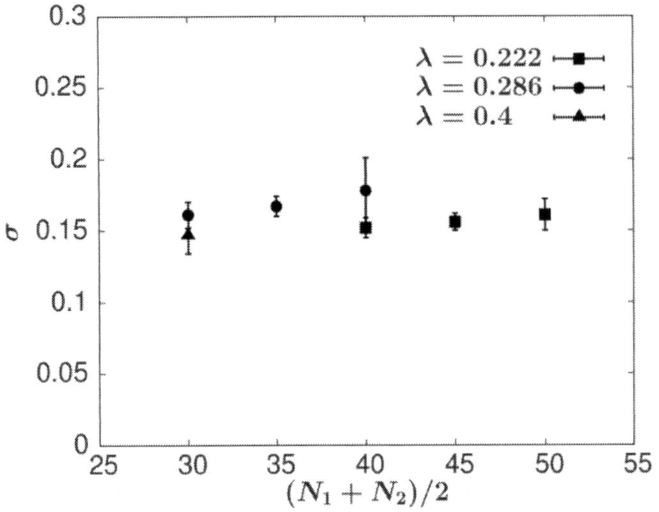


**Figure 13.** An illustration of the values given in Table 1. We see a clear trend to a plateau value of $\sigma = 0.16(1)$.

phase diagram in Figure 8, where the effective potential forms a landscape of many deep valleys (meta-stable local minima), so reliable simulations are more and more difficult to achieve (the trouble starts already with the thermalisation).

Still a numerically accessible window could be found, and we give our results for three $\lambda$ values in Table 1 and Figure 13. We considered the uncertainties which affect $\sigma$ (errors in $\Delta m_i^2$ and in $m_{c,i}^2$), but the precision is sufficient to confirm a clear trend towards a stable exponent

$$\sigma = 0.16(1) \ . \tag{3.5}$$

Thus we can indeed approach the DSL consistently, running to the right in the phase diagram in Figure 8, while staying in the vicinity of the striped phase. This implies that the latter does persist in the DSL. Hence in the NC world translation symmetry can indeed break spontaneously, even in $d = 2$.

## 4. Conclusions

We have studied the $\lambda\phi^4$ model in $d = 3$ [11] and in $d = 2$ [19]. In both cases, the spaces include a non-commutative plane. In order to explore this model beyond perturbation theory, we introduced a (fuzzy) lattice regularisation and mapped the theory onto a Hermitian matrix model, following Ref. [10]. This enables Monte Carlo simulations, which were performed with a Metropolis algorithm.[2]

For both dimensions we observed that a strongly negative bare mass parameter $m^2$ enforces some order.

- If $\lambda$ is small, so that non-commutativity effects are weak, this order is the uniform magnetisation, as in the commutative variant of this model.
- If $\lambda$ is large, so that non-commutativity effects are strongly amplified, this order is "striped". That phase does not occur in the commutative case.

[2] There have been related studies of the $\lambda\phi^4$ model on a *fuzzy sphere* instead of a NC plane [20]. Also in that case Monte Carlo simulations were performed after the mapping onto a Hermitian matrix model. However, the non-commutativity relation differs from the form (1.1) that we have discussed here.

In both dimensions we gave numerical evidence that the striped phase persists in the Double Scaling Limit, which extrapolates simultaneously to the continuum and to infinite volume, while keeping the non-commutativity parameter $\theta$ fixed. This indicates the spontaneous breaking of translation and rotation symmetry in this limit.

For the 2d case this might appear surprising due to the Mermin-Wagner Theorem. However, this theorem does not apply to non-commutative field theory, since it assumes locality and an IR regular behaviour.

## Acknowledgments

Part of the results presented here are based on our collaboration with Jun Nishimura. We thank him, as well as Antonio Bigarini and Jan Volkholz, for their contributions, and Urs Gerber for reading the manuscript.

This work was supported by the Mexican *Consejo Nacional de Ciencia y Tecnología* (CONACyT) through project 155905/10 "Física de Partículas por medio de Simulaciones Numéricas", and by the Spanish *MINECO* (grant SEV-2012-0249). The recent simulations were performed on the cluster of the Instituto de Ciencias Nucleares, UNAM.

[1] Snyder H S 1947 *Phys. Rev.* **71** 38
Yang C-N 1947 *Phys. Rev.* **72** 874
[2] Connes A 1995 *J. Math. Phys.* **36** 6194
[3] Girvin S M 1999 arXiv:cond-mat/9907002 [cond-mat.mes-hall]
[4] http://inspirehep.net/
[5] Seiberg N and Witten E 1999 *JHEP* **09** 032
[6] Doplicher S, Fredenhagen K and Roberts J E 1995 *Commun. Math. Phys.* **172** 187
[7] Filk T 1996 *Phys. Lett.* B **376** 53
[8] Szabo R J 2003 *Phys. Rept.* **378** 207
[9] Minwalla S, Van Raamsdonk M and Seiberg N 2000 *JHEP* **0002** 020
[10] Ambjørn J, Makeenko Y M, Nishimura J and Szabo R J 2000 *JHEP* **05** 023
[11] Bietenholz W, Hofheinz F and Nishimura J 2003 *Acta Phys. Polonica* B **34** 4711; 2004 *JHEP* **0406** 042
Hofheinz F 2004 *Fortsch. Phys.* **52** 391
[12] Gubser S S and Sondhi S L 2001 *Nucl. Phys.* B **605** 395
[13] Chen G-H and Wu Y-S 2002 *Nucl. Phys.* B **622** 189
[14] Castorina P and Zappalà D 2002 *Phys. Rev.* D **68** 065008
Mandanici G 2004 *Int. J. Mod. Phys.* A **19** 3541
Hernández J M, Ramírez C and Sánchez M 2013 *Phys. Rev.* D **87** 125012
[15] Mermin D and Wagner H 1966 *Phys. Rev. Lett.* **17** 113
[16] Castorina P and Zappalà D 2008 *Phys. Rev.* D **77** 027703
[17] Bietenholz W, Hofheinz F and Nishimura J 2003 *Nucl. Phys. (Proc. Suppl.)* **119** 941
[18] Ambjørn J and Catterall S 2002 *Phys. Lett.* B **549** 253
[19] Mejía-Díaz H 2013 *Diagrama de fase del modelo* $\lambda\phi^4$ *bidimensional no-conmutativo* B.Sc. thesis, Universidad Nacional Autónoma de México
Mejía-Díaz H, Bietenholz W and Panero M, in preparation
[20] Martin X 2004 *JHEP* **0404** 077
Panero M 2006 *SIGMA* **2** 081; 2007 *JHEP* **0705** 082
Medina J, Bietenholz W and O'Connor D 2008 *JHEP* 0804 (2008) 041
García Flores F, Martin X and O'Connor D 2009 *Int. J. Mod. Phys.* **A24** 3917
Ydri B 2014 arXiv:1401.1529 [hep-th]

# Top quark physics at the LHeC

**Antonio O. Bouzas and F. Larios**

Departamento de Física Aplicada, CINVESTAV-Mérida, A.P. 73, 97310 Mérida, Yucatán, México

E-mail: `larios@mda.cinvestav.mx`

**Abstract.** We study the DIS and photo-production modes of $t\bar{t}$ pairs at the proposed LHeC and its potential to probe the electromagnetic and weak dipole moments (MDM and EDM for $tt\gamma$) of the top quark. A framework of eight independent gauge invariant dimension six operators involving the top quark and the electroweak gauge bosons is used. Four of those operators modify the charged $tbW$ coupling which can be probed through the single (anti) top production mode as reported in the literature. One generates $tt\gamma(Z)$ as well as $tbW$ couplings, while other two do not generate $tbW$ but only $tt\gamma(Z)$. Our focus is on the MDM and EDM of the top quark for which the photo-production mode of $t\bar{t}$ can be an excellent probe. At the proposed electron energies of $E_e = 60$ and 140 GeV the LHeC could set constraints stronger than the indirect limits from $b \to s\gamma$ and the potential limits of the LHC through $t\bar{t}\gamma$ production[1].

## 1. Introduction

The Large Hadron Electron Collider (LHeC) is the proposal of a new electron beam with an energy $E_e = 60$ GeV, or possibly $E_e = 140$ GeV, to collide with one of the 7 TeV LHC proton beams at the high luminosity phase[2]. In this work we focus on top quark production and the potential of this machine to study the anomalous top-gauge boson couplings. In particular, for the case of the charged $tbW$ effective vertex a recent study has shown that the LHeC sensitivity will surpass the one achievable at the LHC[3]. As we shall see below, for the case of $t\bar{t}$ production, even though the rate is about one order of magnitude lower, the potential for measuring the $tt\gamma$ magnetic (MDM) and electric dipole (EDM) moments is also better than at the LHC14. The reason for this is that in $t\bar{t}$ photoproduction the highly energetic incoming photon couples only to the $t$ quark so that the cross section depends directly on the $tt\gamma$ vertex. In contrast, at the LHC the way to probe the $tt\gamma$ vertex is through $t\bar{t}\gamma$ production, and in this case the outgoing photon could come from other charged sources, like the top decay products. The DIS regime of $t\bar{t}$ production will also be able to probe the $ttZ$ coupling, albeit with less sensitivity. In the framework of the effective Lagrangian with $SU(2) \times U(1)$ gauge invariant operators, some of the $ttZ$ couplings are generated by the same operators that give rise to $tbW$ and $tt\gamma$. This correlation could be used to accomplish a complete and very sensitive analysis of $tbW$, $tt\gamma$ and $ttZ$ couplings at the LHeC.

## 2. Dimension-six $SU(2) \times U(1)$ effective operators

The minimal non-redundant set of dimension-six gauge-invariant operators that give rise to effective top quark vertices with the gauge bosons is [4]:

$$O_{\phi q}^{(3,ij)} = i\phi^\dagger \tau^I D_\mu \phi \, \bar{q}_{Li}\gamma^\mu \tau^I q_{Lj} \, , \qquad O_{uW}^{ij} = \bar{q}_{Li}\sigma^{\mu\nu}\tau^I u_{Rj} \, \tilde{\phi} \, W_{\mu\nu}^I \, ,$$

$$
\begin{aligned}
O_{\phi q}^{(1,ij)} &= i\phi^\dagger D_\mu \phi \, \bar{q}_{Li}\gamma^\mu q_{Lj} \,, & O_{dW}^{ij} &= \bar{q}_{Li}\sigma^{\mu\nu}\tau^I d_{Rj}\, \phi W_{\mu\nu}^I \,, \\
O_{\phi u}^{ij} &= i\phi^\dagger D_\mu \phi \, \bar{u}_{Ri}\gamma^\mu u_{Rj} \,, & O_{uB\phi}^{ij} &= \bar{q}_{Li}\sigma^{\mu\nu} u_{Rj}\, \tilde{\phi}\, B_{\mu\nu} \,, \\
O_{\phi\phi}^{ij} &= i\tilde{\phi}^\dagger D_\mu \phi \, \bar{u}_{Ri}\gamma^\mu d_{Rj} \,, & O_{uG\phi}^{ij} &= \bar{q}_{Li}\lambda^a\sigma^{\mu\nu} u_{Rj}\, \tilde{\phi}\, G_{\mu\nu}^a \,.
\end{aligned}
$$

Where, $W_{\mu\nu}^I = \partial_\mu W_\nu^I - \partial_\nu W_\mu^I + g\epsilon_{IJK} W_\mu^J W_\nu^K$ and $B_{\mu\nu} = \partial_\mu B_\nu - \partial_\nu B_\mu$ the $SU(2)_L$ and $U(1)_Y$ field strength tensors, respectively. In addition, for the operators on the left column $D_\mu = \partial_\mu - ig\frac{1}{2}\tau^a W_\mu^a - ig'\frac{1}{2}B_\mu$ is the Higgs covariant derivative.

We make the assumption:

$$
C_{\phi q}^{(3,33)} = -C_{\phi q}^{(1,33)} \equiv C_{\phi q} \,.
$$

We can write down the effective $tt\gamma$, $ttZ$ and $tbW$ couplings in terms of form factors:

$$
\begin{aligned}
\mathcal{L}_{t\bar{t}\gamma} &= \frac{g}{\sqrt{2}}\bar{t}\left(\gamma^\mu W_\mu^+ (F_1^L P_L + F_1^R P_R) - \frac{1}{2m_W}\sigma^{\mu\nu} W_{\mu\nu}^+ (F_2^L P_L + F_2^R P_R)\right) b \,, \\
&+ e\bar{t}\left(Q_t \gamma^\mu A_\mu + \frac{1}{4m_t}\sigma^{\mu\nu} F_{\mu\nu}(\kappa + i\tilde{\kappa}\gamma_5)\right) t \\
&+ \frac{g}{2c_W}\bar{t}\gamma^\mu Z_\mu \left((1 - \frac{4}{3}s_W^2 + F_{1Z}^L) P_L + (-\frac{4}{3}s_W^2 + F_{1Z}^R) P_R\right) t \\
&+ \frac{g}{2c_W}\bar{t}\left(\frac{1}{4m_t}\sigma^{\mu\nu} Z_{\mu\nu}(\kappa_Z + i\tilde{\kappa}_Z\gamma_5)\right) t
\end{aligned}
\tag{1}
$$

The relation between the form factors and the operator coefficients $C_x^r$ is given by:

$$
\begin{aligned}
F_1^L &= V_{tb} + \frac{v^2}{\Lambda^2} C_{\phi q} \,, & F_1^R &= \frac{1}{2}\frac{v^2}{\Lambda^2} C_{\phi\phi}^r \,, \\
F_2^L &= -\sqrt{2}\frac{v^2}{\Lambda^2} C_{tW}^r \,, & F_2^R &= -\sqrt{2}\frac{v^2}{\Lambda^2} C_{bW}^r \,, \\
F_{1Z}^L &= \frac{v^2}{\Lambda^2} C_{\phi q} \,, & F_{1Z}^R &= \frac{1}{2}\frac{v^2}{\Lambda^2} C_{\phi t} \,, \\
\kappa &= \frac{2\sqrt{2}}{e}\frac{v m_t}{\Lambda^2}\left(s_W C_{tW}^r + c_W C_{tB}^r\right) \,, & \kappa_Z &= \frac{4\sqrt{2}}{e}\frac{v m_t}{\Lambda^2} s_W c_W \left(c_W C_{tW}^r - s_W C_{tB}^r\right) \,.
\end{aligned}
\tag{2}
$$

The imaginary parts of the coefficients generate $CP$-odd interactions. Comparing with other definitions we obtain:

$$
\begin{aligned}
\kappa &= -F_{2V}^\gamma = \frac{2m_t}{e}\mu_t = Q_t a_t \,, \\
\tilde{\kappa} &= F_{2A}^\gamma = \frac{2m_t}{e} d_t \,.
\end{aligned}
\tag{3}
$$

Where $a_t = (g_t - 2)/2$ is the anomalous MDM in terms of the gyromagnetic factor $g_t$. Recent constraints coming from the branching ratio and a $CP$ asymmetry for $b \to s\gamma$ can be found in Ref. [5]: $-2.0 < \kappa < 0.3$ and $-0.5 < \tilde{\kappa} < 1.5$.

## 3. Top quark production at the LHeC

The most important top–production processes at the LHeC are single top, $t\bar{t}$, and associated $tW$ production. In Table 3 we show the values of the associated cross sections for three electron energies. As seen there, the main source of production is single top via the charged current $t$-channel (see Figure 1), whereas for the other modes, $t\bar{t}$ and $tW$, there is a lower though still sizeable production cross section. Given the advantage of an experimental environment cleaner than the LHC, we can envisage a good performance of this machine to do top quark physics.

Journal of Physics: Conference Series **651** (2015) 012004 doi:10.1088/1742-6596/651/1/012004

| Operator | Coefficient | $CP$-even $(O^{33}_x + O^{33\dagger}_x)$ | $CP$-odd $i(O^{33}_x - O^{33\dagger}_x)$ |
|---|---|---|---|
| $(tb)O^{(3,33)}_{\phi q}$ | $C_{\phi q}$ | $\frac{g}{\sqrt{2}}\phi_0^2\left(W^+_\mu \bar{t}_L\gamma^\mu b_L + h.c.\right)$ | $- - -$ |
| $O^{33}_{\phi\phi}$ | $C_{\phi\phi}$ | $\frac{g}{2\sqrt{2}}\phi_0^2\left(W^+_\mu \bar{t}_R\gamma^\mu b_R + h.c.\right)$ | $i\frac{g}{2\sqrt{2}}\phi_0^2\left(W^+_\mu \bar{t}_R\gamma^\mu b_R - h.c.\right)$ |
| $(tb)\ O^{33}_{uW}$ | $C_{tW}$ | $2\phi_0\left[D^-_{\mu\nu}\bar{b}_L\sigma^{\mu\nu}t_R + D^+_{\mu\nu}\bar{t}_R\sigma^{\mu\nu}b_L\right]$ | $i2\phi_0\left[D^-_{\mu\nu}\bar{b}_L\sigma^{\mu\nu}t_R - D^+_{\mu\nu}\bar{t}_R\sigma^{\mu\nu}b_L\right]$ |
| $O^{33}_{dW}$ | $C_{bW}$ | $2\phi_0\left[D^+_{\mu\nu}\bar{t}_L\sigma^{\mu\nu}b_R + D^-_{\mu\nu}\bar{b}_R\sigma^{\mu\nu}t_L\right]$ | $i2\phi_0\left[D^+_{\mu\nu}\bar{t}_L\sigma^{\mu\nu}b_R - D^-_{\mu\nu}\bar{b}_R\sigma^{\mu\nu}t_L\right]$ |
| $(tt)O^{(3,33)}_{\phi q}$ | $C_{\phi q}$ | $\frac{g}{2c_w}\phi_0^2\bar{t}_L\gamma^\mu t_L Z_\mu$ | $- - -$ |
| $O^{(1,33)}_{\phi q}$ | $-C_{\phi q}$ | $-\frac{g}{2c_w}\phi_0^2\bar{t}_L\gamma^\mu t_L Z_\mu$ | $- - -$ |
| $O^{33}_{\phi u}$ | $C_{\phi t}$ | $-\frac{g}{2c_w}\phi_0^2\bar{t}_R\gamma^\mu t_R Z_\mu$ | $- - -$ |
| $(tt)\ O^{33}_{uW}$ | $C_{tW}$ | $\sqrt{2}\phi_0 D^3_{\mu\nu}\bar{t}\sigma^{\mu\nu}t$ | $i\sqrt{2}\phi_0 D^3_{\mu\nu}\bar{t}\sigma^{\mu\nu}\gamma_5 t$ |
| $O^{33}_{uB\phi}$ | $C_{tB}$ | $\frac{1}{\sqrt{2}}\phi_0 B_{\mu\nu}\bar{t}\sigma^{\mu\nu}t$ | $i\frac{1}{\sqrt{2}}\phi_0 B_{\mu\nu}\bar{t}\sigma^{\mu\nu}\gamma_5 t$ |
| $O^{33}_{uG\phi}$ | $C^{33}_{uG\phi}$ | $\frac{1}{\sqrt{2}}\phi_0\bar{t}\sigma^{\mu\nu}\lambda^a G^a_{\mu\nu}t$ | $i\frac{1}{\sqrt{2}}\phi_0\bar{t}\sigma^{\mu\nu}\gamma_5\lambda^a G^a_{\mu\nu}t$ |

**Table 1.** Diagonal operators with $CP$-even and $CP$-odd parts written separately. For $O^{(1,33)}_{\phi q}$, $O^{(3,33)}_{\phi q}$ and $O^{33}_{dW}$ only the terms that involve the top quark are shown. We define $\phi_0 = v + h$, $D^\pm_{\mu\nu} = \partial_\mu W^\pm_\nu \pm igW^\pm_\mu W^3_\nu$ and $D^3_{\mu\nu} = \partial_\mu W^3_\nu - igW^+_\mu W^-_\nu$. The real (imaginary) part of each coefficient multiplies the $CP$-even (odd) part of the corresponding operator (the scale factor $\Lambda^{-2}$ is taken as 1 $\text{TeV}^{-2}$).

| Operator | Indirect | LHC (7,8 TeV) |
|---|---|---|
| $O^{(3,33)}_{\phi q}$ | $-0.35 < C_{\phi q} < 2.35$ | $-2.1 < C_{\phi q} < 6.7$ |
| $O^{33}_{\phi\phi}$ | $0.004 < C^r_{\phi\phi} < 0.056$ | $-6.6 < C^r_{\phi\phi} < 7.6$ |
| $O^{33}_{\phi u}$ | $-0.1 < C_{\phi t} < 3.7$ | |
| $O^{33}_{uW}$ | $-1.6 < C^r_{tW} < 0.8$ | $-1.0 < C^r_{tW} < 0.5$ |
| $O^{33}_{dW}$ | $-0.01 < C^r_{bW} < 0.004$ | $-1.7 < C^r_{bW} < 1.3$ |
| $O^{33}_{uB\phi}$ | $-6.0 < C^r_{tB} < 0.9$ | |
| $O^{33}_{uG\phi}$ | $-0.1 < C^r_{uG\phi} < 0.03$ | $-0.3 < C^r_{uG\phi} < 0.06$ |

**Table 2.** Current bounds on the coefficients (real part). The indirect bounds for the first four coefficients are taken from electroweak data, whereas for the last three: $C^r_{bW}$ , $C^r_{tB}$ and $C^r_{uG\phi}$, they are taken from $b \to s\gamma$ measurements. Direct bounds come from measurements on the $W$ helicity in top decays as well as single top production.

| Process | $E_e = 60\text{GeV}$ | $E_e = 140\text{GeV}$ | $E_e = 300\text{GeV}$ |
|---|---|---|---|
| $ep(b) \to \nu\bar{t}$ | 2.0 | 5.9 | 13.0 |
| $e(\gamma)p(g) \to t\bar{t}$ | 0.023 | 0.12 | 0.38 |
| $ep(g) \to et\bar{t}$ | 0.020 | 0.11 | 0.34 |
| $e(\gamma)p(b) \to \bar{t}W^+ + tW^-$ | 0.031 | 0.143 | 0.434 |
| $ep(b) \to e\bar{t}W^+ + etW^-$ | 0.021 | 0.099 | 0.30 |
| $\gamma p(g) \to t\bar{t}$ | 0.7 | 3.2 | 9.0 |

**Table 3.** The SM cross sections (pb) for single anti-top, $t\bar{t}$ and associated $tW^-$ (or $\bar{t}W^+$) production processes at the LHeC. The bottom row shows $t\bar{t}$ production at an LHeC–based $\gamma p$ collider.

Journal of Physics: Conference Series **651** (2015) 012004 doi:10.1088/1742-6596/651/1/012004

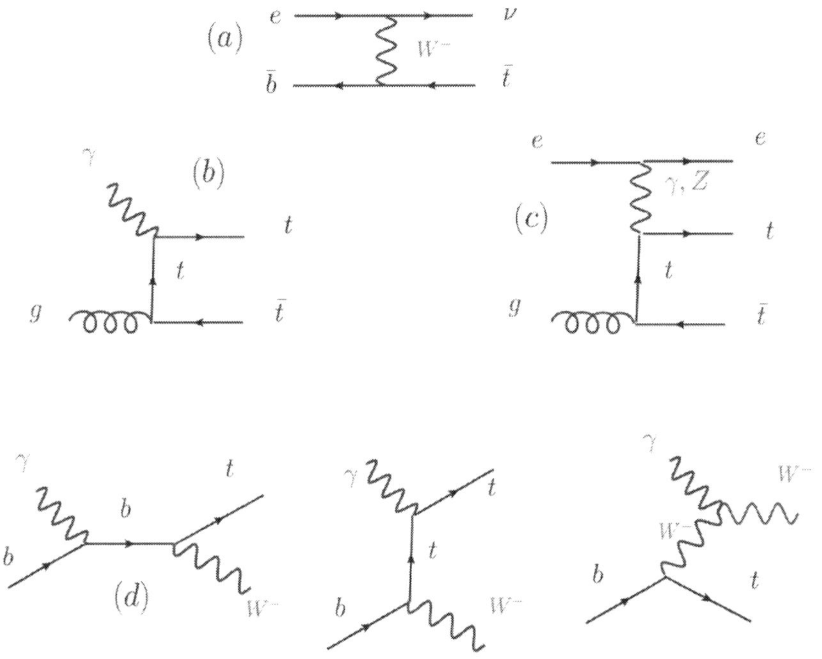


**Figure 1.** The dominant top quark production processes at the LHeC: (a) Single top production, (b) $t\bar{t}$ photoproduction, (c) $t\bar{t}$ DIS production and (d) $tW^-$ photoproduction.

## 4. Limits from $t\bar{t}$ photoproduction.

The branching fractions for $W$ decay are 21.32% for light leptonic decays $\ell\nu$ ($\ell = e$, $\mu$), 11.25% for $\tau\nu$ decays, and 67.6% for hadronic decays. Thus, for $t\bar{t}$ production followed by $bW$ decays we have the branching fractions given in Table 4. The dominant modes are the hadronic ($jjjj$) and the semileptonic ($\ell jj$).

| $\ell\ell$ | $\tau\tau$ | $jjjj$ | $\ell\tau$ | $\ell jj$ | $\tau jj$ |
|---|---|---|---|---|---|
| 4.55% | 1.27% | 45.70% | 4.80% | 28.82% | 15.2% |

**Table 4.** Approximate branching fractions for the decay of $t\bar{t}$ through $t \to bW$.

### *4.1. Semileptonic mode*

For the semileptonic mode the signal ($S$) and signal plus total irreducible background ($S + B$) in the SM are defined as

$$S:\ \gamma g \to t\bar{t} \to b\bar{b}jj\ell\nu\,, S + B:\ \gamma g \to b\bar{b}jj\ell\nu, \tag{4}$$

with $\ell = e^\pm, \mu^\pm$ and $j = u, \bar{u}, d, \bar{d}, s, \bar{s}, c, \bar{c}$. In the SM the signal process $S$ involves 16 Feynman diagrams, whereas $S + B$ involves 3704 diagrams in total, 2952 with one QCD vertex and five electroweak vertices like the signal process, and 752 with three QCD vertices and three electroweak vertices.

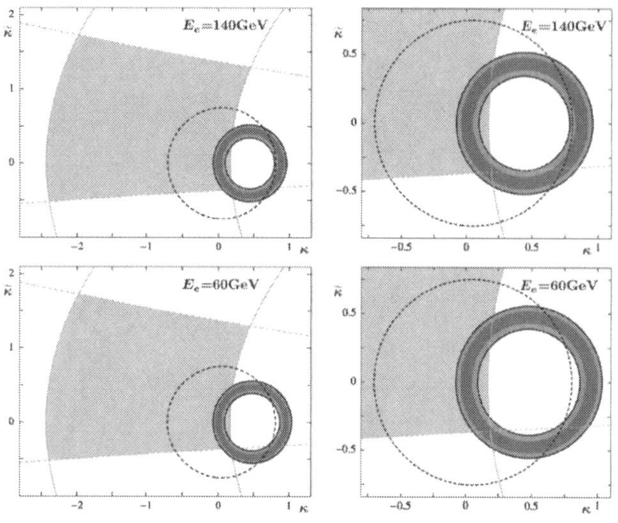


**Figure 2.** Bounds on the top quark dipole moments $\kappa$ and $\tilde{\kappa}$. Light-gray area: region allowed by the measurements of the branching ratio and $CP$ asymmetry of $B \to X_s\gamma$ [5]. Dashed line: region allowed by a hypothetical experimental result for $\sigma(pp \to t\bar{t}\gamma)$ with semileptonic final state at the LHC at $\sqrt{s} = 14$ TeV with $E_T^\gamma > 10$ GeV and 5% experimental uncertainty. Solid line: region allowed by a hypothetical measurement of $\sigma(\gamma p \to t\bar{t})$ with semileptonic final state, with the cuts $C_f$, (5), and 18% experimental uncertainty. Dark-gray area: same as previous, with 10% experimental error.

For the computation of the cross section we impose on the final-state momenta a set of appropriate phase-space cuts. We have considered several such sets defined as

$$\begin{aligned}
C_0 &: \; |\eta(j)| < 5, \; |\vec{p}_T(j)| > 1\,\text{GeV}, \; |\eta(\ell)| < 5, \; |\vec{p}_T(\ell)| > 1\,\text{GeV}, \\
C_1 &: \; \left\{ \begin{array}{l} |\eta(j)| < 5, \; |\vec{p}_T(j)| > 5\,\text{GeV}, \; |\eta(\ell)| < 5, \; |\vec{p}_T(\ell)| > 5\,\text{GeV}, \not{E}_T > 5\,\text{GeV}, \\ |\eta(b)| < 3, \; |\vec{p}_T(b)| > \left\{ \begin{array}{ll} 15\,\text{GeV} & (E_e = 60\,\text{GeV}) \\ 20\,\text{GeV} & (E_e = 140\,\text{GeV}) \end{array} \right. , \end{array} \right. \\
C_f &: \; C_1, \; \Delta R(x) > 0.4 \; (x = b\bar{b}, \; \ell\ell, \; \ell b, \; bj, \; jj),
\end{aligned} \tag{5}$$

where $b$ stands for $b$ or $\bar{b}$, $j$ refers to the light jets, $\ell$ to the charged leptons, and $\Delta R = \sqrt{(\Delta\eta)^2 + (\Delta\phi)^2}$ is the distance in the $\eta$-$\phi$ plane. The kinematic variables $\eta$, $\phi$, correspond to the laboratory frame. The cuts $C_0$ are a minimal set needed to render the scattering amplitude for background processes free from infrared instabilities, due to the emission of massless leptons and quarks. We use $C_0$ only for reference. We have tested the cuts in the different kinematic variables one by one to assess their efficiency to reduce the ratio $\epsilon_\sigma = (\sigma(S+B)-\sigma(S))/\sigma(S+B)$. We have found that only the cuts in $b$ and $\bar{b}$ lead to a significant enhancement of the signal. In the set $C_1$ we use a standard centrality cut for $\eta(b)$ and choose the cut in $|\vec{p}_T(b)|$ so that $|\epsilon_\sigma|$15%. The cuts on leptons and light jets do not seem effective at improving the signal-to-background ratio, so in $C_1$ we keep them as loose as realistically possible. Finally, in the set $C_f$, which is the one used in our computations, we add standard isolation cuts for the $b$ and light-quark jets, and the charged leptons, as idealized analogues of the ones required in actual experimental measurements. In Table 5 the effects of those cuts on the signal and total cross sections, computed in the SM, are summarized. Whereas a $\sim$ 15% background is sufficiently small for our purposes, further enhancement of the signal is in principle possible by imposing

Journal of Physics: Conference Series **651** (2015) 012004 doi:10.1088/1742-6596/651/1/012004

| | $E_e = 60$ GeV | | | $E_e = 140$ GeV | | |
|---|---|---|---|---|---|---|
| | $\sigma(S)$[fb] | $\sigma(S+B)$[fb] | $\epsilon_\sigma$ | $\sigma(S)$[fb] | $\sigma(S+B)$[fb] | $\epsilon_\sigma$ |
| $\emptyset$ | 5.91 | | | 30.94 | | |
| $C_0$ | 5.84 | 9.21 | 36.6% | 30.92 | 47.09 | 34.3% |
| $C_1$ | 4.50 | 5.26 | 14.4% | 25.59 | 29.87 | 14.3% |
| $C_f$ | 3.85 | 4.50 | 14.4% | 22.25 | 25.90 | 14.1% |

**Table 5.** The effect of the cuts defined in (5) on the SM signal and total semi–leptonic cross sections. $\emptyset$ refers to no cuts.

additional cuts, for instance, on the invariant mass of the hadronic decay products. Let us assume a cut of the form

$$\left| m_t - \sqrt{(p_b + p_q + p_{q'})^2} \right| < W, \tag{6}$$

where $p_b$ stands for the four-momentum of either one of the two $b$-tagged jets and $p_q$, $p_{q'}$ for those of the non-$b$ jets. Then, at $E_e = 140$ GeV and $W = 30$ GeV, with the cut (6) in addition to $C_f$, we get $\sigma(S) = 22.06$ fb and $\sigma(S+B) = 24.90$ fb, corresponding to $\epsilon_\sigma = 11$ %, which constitutes a slight improvement on $C_f$. An even larger enhancement of the signal would be obtained in the ideal case in which the missing momentum carried by the neutrino could be fully reconstructed. In that case, imposing the cuts $C_f$ together with (6) and the analogous cut on the leptonic decay products leads to $\sigma(S) = 21.79$ fb and $\sigma(S+B) = 23.30$ fb, yielding $\epsilon_\sigma = 6$ % which is less than one half of the background level in table 5.

With an integrated luminosity of 100 fb$^{-1}$ and the cross sections from Table 5, at $E_e = 60$ GeV we expect $\sim 385$ photoproduction events. Taking into account a $b$-tagging efficiency of 60% per $b$-jet, we are left with about 140 events corresponding to a statistical error of 8.4%. Similarly, at $E_e = 140$ GeV the expected statistical error is 3.5%.

One important source of systematic errors lies in the SM reducible background to the signal process $S$ in (4), given by processes of the form $e^-(\gamma)p \to jjj\ell\nu$ (where j stands for a gluon or a quark or antiquark of the first two generations), or $e^-(\gamma)p \to bjj\ell\nu$ (where $b$ refers to $b$ or $\bar{b}$). The former class of processes involves two $b$-mistaggings, and their cross-section is smaller than that of the signal by about two orders of magnitude which, multiplied by the probability of two mistaggings, results in a negligible contribution. We take the $b$-mistagging probability to be 1/10 for $c$, and 1/100 for lighter partons. The second class of processes, involving a single $b$-mistagging, comprises 7408 Feynman diagrams. The overwhelmingly dominant contribution to the cross section, however, originates in diagrams containing two resonant intermediate propagators.

Thus, the reducible background is essentially given by the processes

$$\gamma b \to tW \to bgc\bar{s}\ell\nu, \gamma b \to tW \to bgu\bar{d}\ell\nu, \tag{7}$$

where the quark symbolds stand for either those quarks or their antiquarks, and $\ell$ for $e^\pm$, $\mu^\pm$. All possible quark and lepton flavor combinations in the final state result in 204 diagrams for the charmed process, and as many diagrams for the charmless final state, with a cross section of 10.5 fb each at $E_e = 140$ GeV, and 2.04 fb each at $E_e = 60$ GeV. We have explicitly separated the charmed and charmless final states in (7) due to the different mistagging probabilities for the $c$ and lighter partons. For each of the processes in (7) we have to ascertain the fraction of events in which some or none of the three non-$b$ jets pass the cuts for $b$-jets (so they can therefore potentially be mistagged), and how many of them, and whether those jets passing the cuts are $c$ or lighter. For brevity, we skip the combinatorial analysis and the results for the partial cross sections for each case and just state the results. At $E_e = 140$ GeV the cross section for events with a single $b$-mistagging is 1.15 fb, or 5.16% of $\sigma(S)$ as given in Table 5, and at $E_e = 60$ GeV it is 0.25 fb, or 6.5% of $\sigma(S)$.

| $\epsilon = 10\%$ | min | max | $\epsilon = 18\%$ | min | max |
|---|---|---|---|---|---|
| $C^r_{tW}$ | $-0.28$ | 0.32 | $C^r_{tW}$ | $-0.48$ | 0.62 |
| $C^i_{tW}$ | $-1.02$ | 1.02 | $C^i_{tW}$ | $-1.37$ | 1.37 |
| $C^r_{tB}$ | $-0.15$ | 0.17 | $C^r_{tB}$ | $-0.26$ | 0.33 |
| $C^i_{tB}$ | $-0.65$ | 0.65 | $C^i_{tB}$ | $-0.87$ | 0.87 |

**Table 6.** The bounds obtained from $t\bar{t}$ photoproduction including irreducible background, with the set of cuts $C_f$ from (5).

Adding the statistical and mistagging errors discussed above in quadrature we obtain an error of 10.6% at $E_e = 60$ GeV and 6.2% at $E_e = 140$ GeV. Allowing for other unspecified sources of systematical error, we consider total experimental errors of 18% and 10% at $E_e = 60$ and 140 GeV, respectively, as plausible estimates.

In order to assess more accurately the effects on our results of the irreducible background processes passing the cuts, we repeated a small part of the analysis of the previous section including background effects. We considered only the semileptonic mode in PHP, $e^-(\gamma)p(g) \to b\bar{b}jj\ell\nu$, including all possible insertions of the anomalous operators $O^{33}_{uW\phi}$ and $O^{33}_{uB\phi}$. The resulting amplitude consists of 5136 Feynman diagrams, excluding those with internal Higgs lines as done for calculation with the signal process. The results obtained considering one coupling at a time are displayed in Table 6.

*4.2. Limits from single top production*

As it has been well known that anomalous $tbW$ couplings can change the $W$ boson helicity in the top decay process [6] various kinematical asymmetries of the top decay products that directly depend on the $W$ helicity were considered by [3]. Assuming an uncertainty of 2% in the experimental measurements they obtain constraints that are approximately as follows: $-0.05 < C_{\phi q} < 0.05$, $-1.6 < C^r_{\phi\phi} < 2.6$, $-0.04 < C^r_{tW} < 0.04$ and $-0.4 < C^r_{bW} < 0.8$. Notice that the constraints on $C^r_{\phi\phi}$ and $C^r_{bW}$ are much weaker, this is because those operators are related to right-handed bottom quarks and there is a negligible interference with the SM amplitude. On the other hand, if we assume that the single top cross section is measured with 2% (essentially systematic) error that is assumed for the asymmetries in Ref. [3] we obtain (based only on the cross section) $-0.34 < C_{\phi q} < 0.33$, $|C_{\phi\phi}| < 2.8$, $-0.7 < C^r_{tW} < 0.9$ and $|C^r_{bW}| < 1.1$. The bounds for $C_{\phi q}$ and $C^r_{tW}$ obtained from the variation of $\sigma(ep \to \nu\bar{t})$ are about one order of magnitude weaker than the bounds obtained by analyzing the $W$ boson helicity in the decay of $\bar{t}$. On the other hand, the bounds for $C^r_{\phi\phi}$ and $C^r_{bW}$ (that involve $b_R$) are about the same order of magnitude.

Our focus is on the potential to probe the MDM and the EDM of the top quark through the $t\bar{t}$ photoproduction process. The sensitivity changes very little going from $E_e = 60$ GeV to $E_e = 140$ GeV; it only depends on the accuracy achieved in measuring the production cross section (which can be much better at 140 GeV due to the larger event sample). We assume two possible values of the experimental error $\Delta\sigma/\sigma = 10\%, 18\%$ and derive allowed regions for the MDM $\kappa = 2m_t\mu_t/e$ and the EDM $\tilde{\kappa} = 2m_t d_t/e$ as shown in Figure 2. In both cases, the measurement of the $t\bar{t}$ photoproduction at the LHeC could greatly improve the limits imposed by the indirect constraints from $b \to s\gamma$ and even the limits imposed by a future measurement of $t\bar{t}\gamma$ production at the LHC (14 TeV). Specifically, measuring $\sigma(\gamma p \to t\bar{t})$ with 10% (18%) error would yield the bounds $|\kappa| < 0.05(0.09)$ and $|\tilde{\kappa}| < 0.20(0.28)$.

**Acknowledgments**

We thank RedFAE, Conacyt and SNI for support.

Journal of Physics: Conference Series **651** (2015) 012004 doi:10.1088/1742-6596/651/1/012004

## References

[1] A. O. Bouzas and F. Larios, *Phys. Rev.* D**88**, 094007 (2013).
[2] J. L. Abelleira Fernandez *et al.* [LHeC Study Group Collaboration], *J. Phys.* G**39**, 075001 (2012). See [arXiv:1206.2913v2 [physics.acc-ph]].
[3] S. Dutta, A. Goyal, M. Kumar and B. Mellado, arXiv:1307.1688 [hep-ph].
[4] J. A. Aguilar-Saavedra, *Nucl. Phys.* B**812** (2009) 181.
[5] A. O. Bouzas and F. Larios, *Phys. Rev.* D**87**, 074015 (2013).
[6] C. -R. Chen, F. Larios and C. -P. Yuan, *Phys. Lett.* B**631**, 126 (2005); J. A. Aguilar-Saavedra, N. F. Castro and A. Onofre, *Phys. Rev.* D**83**, 117301 (2011); J. A. Aguilar-Saavedra and J. Bernabeu, *Nucl. Phys.* B**840**, 349 (2010).

# Quality Certification 4 (QC4) for RE4 in the Upgrade in CMS

**S. Carpinteyro Bernardino on behalf of CMS colaboration.**
Facultad de Ciencias Física Matemáticas, Benemérita Universidad Autónoma de Puebla.
E-mail: bernardinocarpinteyro@cern.ch

**Abstract.** The CMS muon system consist of Drif Tubes (DTs) in barrel region, Cathode Strip Chambers (CSCs) in the end-cap region and Resistive Plate Chambers (RPCs) in both regions. The RPCs with their excellent time resolution were chosen as dedicated muon trigger detector for the CMS experiment. After the Long Shutdown 1 (LS1) the LHC luminosity should reach $10^{34}cm^{-2}s^{-1}$ thus it is necessary to improve the level 1 trigger efficiency of the CMS detector for this high luminosity, to achieve that a 4th RPCs layer (RE4) in both end-caps of the CMS detector and all the components must be installed and completely validated. Based on the valuable experience (Barrel & Endcap) are defined several steps of Quality Certification (QC): QC1 (Components), QC2 (Gaps), QC3 (Chambers), QC4 (chambers & Super Modules), QC5 (Commissioning at P5). In this work is presented some results of QC4.

## 1. Introduction

After the Long Shutdown 1 (LSD1) the LHC luminosity should reach $10^{34}cm^{-2}s^{-1}$ [1]. Therefore during the LSD1 the Compact Muon Solenoid (CMS) collaboration is upgrading several sub-systems in the detector. In the end-caps, CMS is using Cathode Strip Chambers (CSCs) as muon tracking and trigger detectors, while Resistive Plate Chambers (RPCs) serve as dedicated trigger detectors, and improve the muon reconstruction [2]. In particular, the instrumentation of the muon system, will be extendeded in both end-caps to ensure efficient muon triggering and reconstruction in that region at high luminosities. The RPCs are widely employed for the muon trigger systems thanks to their fast time resolution (~1ns), suitable space resolution (~1cm) besides its low production cost. The RPC developed by R. Santonico in the early 80s, are parallel plate gas detectors, made of highly resistive plates such as glass or bakelite coated with graphite to provide the anode and cathode and are operated either in avalanche or streamer mode with a typical combination of gas mixture [3, 4]. They are designed to create an avalanche and then also configured to arrest the avalanche from spreading out. The fourth end-cap layer is being installed during the long shutdown (2013-2014), then LHC will operate with its full designed luminosity. A common Quality Control (QC) protocol for the chamber and Super Modules (SM) production has been carefully prepared for each level of the construction [5]. Some results of individual components need to satisfy detailed specifications are presented in this work.

## 2. Region End-cap 4 (RE4) Upgrade

As the instantaneous luminosity of the LHC continues to increase over time during Phase 1, the CMS experiment has to ensure that the detector can operate in a stable and efficient manner for the increasingly high particle rates. In particular, the muon system should be able to keep its trigger rate and efficiency under control. CMS will among other things extend its RPC system with a fourth layer in both endcaps [1, 2]. The RPC upgrade is essentially driven by the impact of the instantaneous peak luminosity on the trigger system [3] with the RPC trigger requiring hits in at least three stations, the endcap systems with only three stations does not have the necessary redundancy to control the trigger rate at the increased luminosity which causes the observed drop in efficiency for the endcaps with only three stations, while preserving high trigger efficiency. Adding the fourth layers in the endcaps, enabling a 3-out-of-4 trigger logic in those regions, will bring the RPC end-cap performance to a similar level as in the barrel region. In figure 1 the RPC Level-1 Trigger Efficiencies performance is simulated for the present system with three stations and the four station after the upgrade.

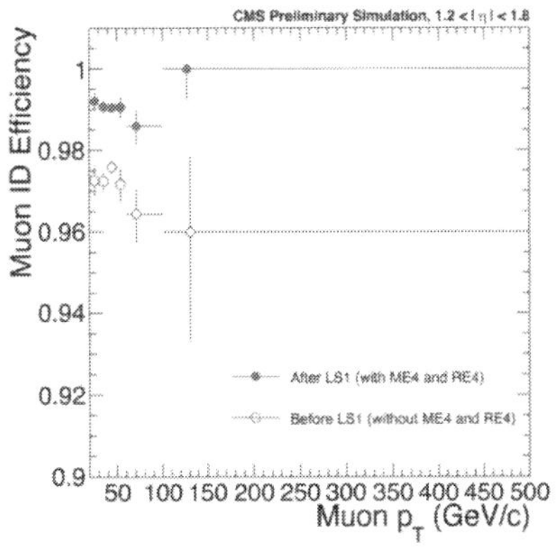


**Figure 1.** Simulated efficiencies with and without RE4 and ME4.

## 3. Resistive Plate Chambers (RPC)

Resistive Plate Chambers are widely employed for the muon trigger systems at the Large Hadron Collider (LHC) thanks to their fast time resolution, suitable space resolution and low cost. The basic design of a RPC consist of two parallel resistive plates such as glass or Bakelite (phenol-formaldehyde polymers) about 3 mm thick, separated by a gas gap whose width is maintained by spacers [3, 4]. The gas is kept at atmospheric pressure and DC voltage is applied to the plates via a layer of graphite paint on their external surface. The graphite electrodes are covered with an isolating layer that is etched with a thin copper layer that works as a pickup pad to detect any movement of charge within the chamber.

As the voltage on the anode is increased from zero, various modes of operation of chambers are encountered; Recombination mode where the field is not enough for the primary electrons to drift to the anode and they are lost by recombination to the positive ions. The second region is ionization mode where the field is enough for the electrons to reach the anode, but still not enough to multiply, nevertheless all of them are collected. The next region is called the proportional mode region; here above the critical or threshold voltage multiplication starts and the detector signal is proportional to the original ionization, thus the chamber can be used for energy measurement.

Increasing the voltage further one encounters the region of limited proportionality where multiplication reaches saturation. Thus gain no longer increases whith the voltage but avalanche transits into Streamer. In this region of operation there is a strong probability of photoemission from avalanches where secondary avalanches merge with original avalanche creating a streamer. This mode of operation has advantage of very large gains, hence simpler electronics is needed. Nevertheless, the gas mixture requires a strong quencher as one of its components. Increasing voltage further still more one reaches the Geiger or discharge mode which is dominated by tremendous amount of photoemission where the full length of anode is arrested by avalanche. To operate in this mode the HV needs to be pulsed and very strong quenchers are needed.

## 4. Quality Control (QC)

A full central control on the production at the level of components & chambers [5]. Based on the valuable experience (Barrel & Endcap) has defined several steps for the QC:

- QC1: Components, RPC-Technical Coordination.
- QC2: Gaps, Kodel+Assembly sites.
- QC3: Chambers, Assembly sites: CERN, Mumbai, Ghent.
- QC4: Chambers & Super Modules, 904 CERN.
- QC5: Commissioning at P5, 904 CERN & P5.

For each step; Quality Protocols to assure manufacture process reliability are established. QC protocols to select the good objects which are inside the technical specifications ranges are applied. Documentation protocols: RPC construction Data Base (DB) are used [6, 7].

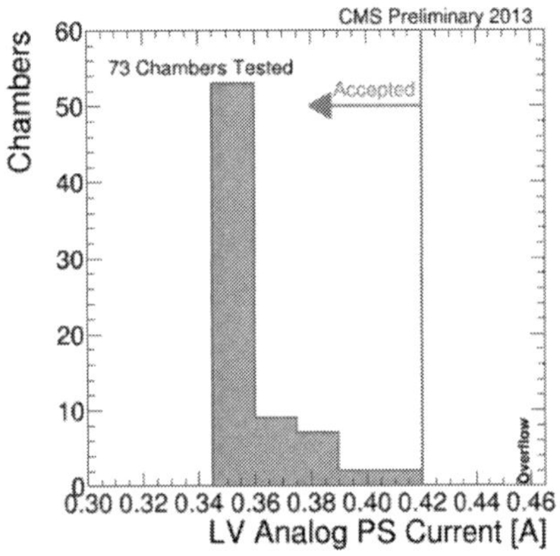


**Figure 2.** Signal current for analogic signal sent to Front-End Board (FEBs).

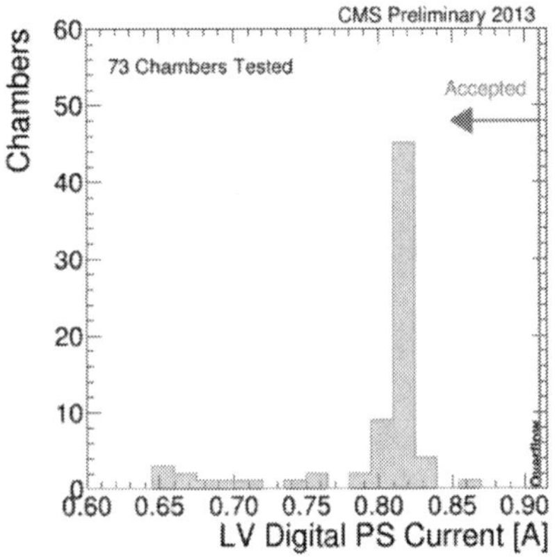


**Figure 3.** Signal current for digital signal sent to Front-End Boart (FEBs).

## 5. Quality Certification level 4 (QC4)

Quality Certification level 4 is divided in 3 parts; QC4-1, QC4-2 and QC4-3. The goal of the first part of the certification QC4-1 is to validate that RPC parameters are still in range. QC4-1 have the stages: Connectivity test, Electrical test, Leak test and HV Long Stability test is the only test which is carried out in the QC4-2 phase. If all the QC4-1 & QC4-2 test are successful the chambers can be used for the assembly of super modules in QC4-3.

Connectivity test starts with a visual inspection, then the connectivity with the use of a frequency meter check for bad or disconnected strips and for no dead or too noise chambers. The frequency meter send a signal directly to strips and then is returned to check that they are working well.

In Electrical test an analogical signal and digital signal are sent to Front-End Board (FEBs). For analogic signal the current must be less than 0.42mA and for digital signal the current must be less than 0.91mA. The plot in figure 2 summarize the value for analogic current response of chambers are acepted if $i$-analog $<$0.422mA. For this test all the chambers are in range. The

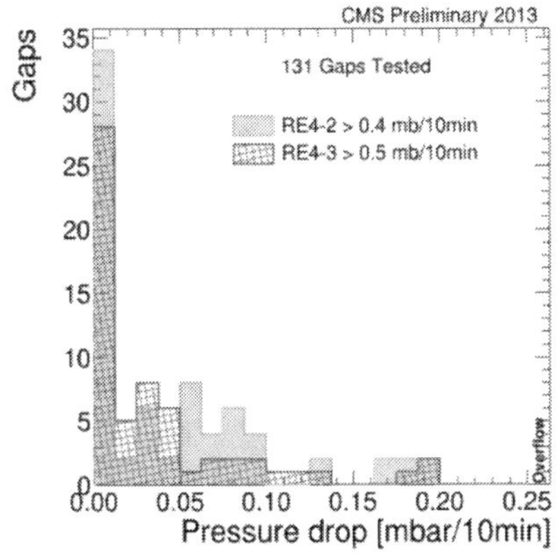


**Figure 4.** Pressure drop for RE4-2 chambers and RE4-3 chambers.

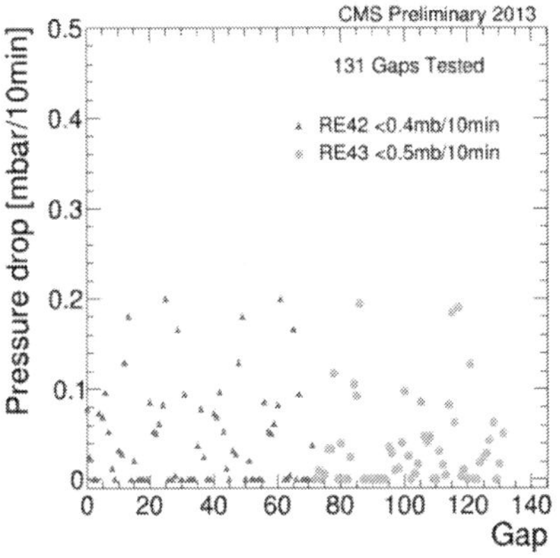


**Figure 5.** Distribution of pressure drop for chambers.

plot in figure 3 summarize the value for $i$-digital current response of chambers are acepted if $i$-digi <0.422mA. For this test all the chambers are in range. for the analog low voltage (LV) power supply (PS) and around 0.25A for the digital PS. Adding the Distribution Board, every chamber should draw about 0.36A from the analog PS and 0.75A from the digital PS plus 50-100 mA for the distribution board. In the second step in the Electrical Test the threshold in the FEBs is checked that the value is 215mV, in case the threshold is adjusted.

In Gas Leak there are four kind of RPC gas leaks and they are classified in; Large gas leak, normal gas Leak, Acceptable gas leak, and no gas leak.

The pressure drops is measured for each chamber in a time interval of 10 minutes in stable conditions. These measurements were taken with a tool provided by CERN gas group called Gas Leak Box. For all the chambers the gas leak are in range therefore all them are accepted for this test, the results are showed in figures 4 and 5.

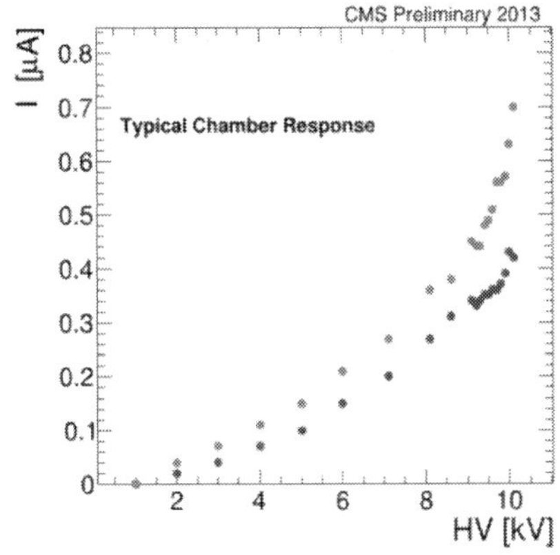


**Figure 6.** Typical chamber response in HV Scan test.

The three HV jupiter connectors are changed by a tripolar connector, therefore all RPC chambers need to undergo HV Scan test. The test is started by applying 1kV to each gap to detect broken connections and electrical short circuits. Then the high voltage is raised by 1kV every 10 minutes up to 8kV. Then to 9kV by 0.5kV, at this voltage gas avalanches begin to develop. Finally the high voltage is raised to 10kV in steps of 0.1kV. In figure 7 is showed an example of how an operational chamber should responds to the HV Scan test.

The main accent of QC4 fall down on the Long Stability test. During a long period the chamber current will be constantly monitored at 2 HV values. One cycle of 12 hours monitoring should consist of 10 hours of current monitoring. The chambers are alternately kept at 9.7kV for 10 hours, and at 6kV for 2 hours to observe the behaviour of the ohmic dark currents of the gas gaps, and this for a period of 4 weeks. If the dark current has a rising trend or is too high, the chamber is rejected and not used for further assembly. This cycle should be repeated more than 5 times for the 4 weeks. In this stage the chambers are cooling with CMS mix gas at 1atm pressure.

The value of the HV is corrected according to the pressure of the CMS mix gas; when the pressure of the gas increase then the HV is increased, when the pressure of the gas decrease then the HV is decreased.

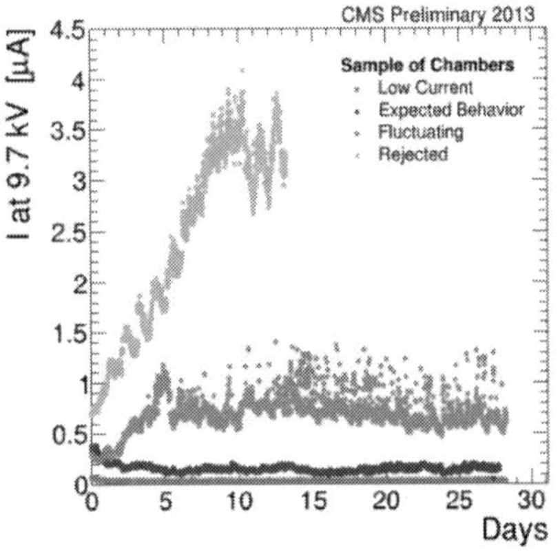


**Figure 7.** Diferent typical behavior for chambers under Long Stability test.

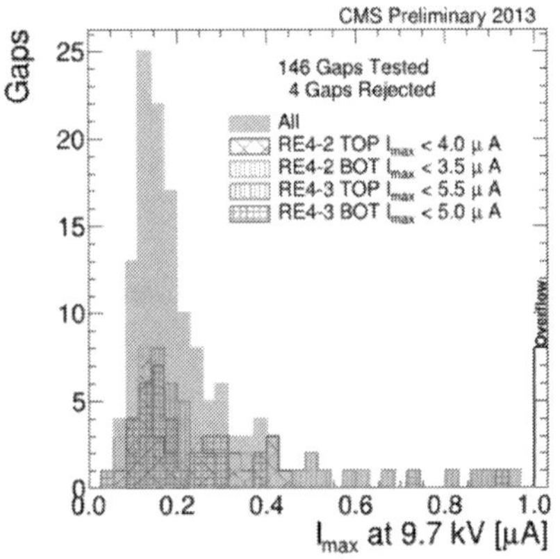


**Figure 8.** Distribution of number of gaps for current increment.

Is defined four main behaviour of the chambers under Long Stability test, see figure 7 where high resistivity is showed in reed for this case the current in the chambers is very small but different of zero. The expected behaviour in blue when the current in the chambers is into a well defined interval. The called Nervous behavior is showed in purple for this the current in the chambers is dispersed in a wide interval. For last we have the behavior for a reject chamber in green, in this case the current is always increasing indicating that something is wrong in the chamber.

In the plot in figure 8 the curren increment in the gaps when they are under Long Stability test at 9.7kV is showed and all them are in range.

## 6. General Chamber and Super Module Status

When all the QC4-1 & QC4-2 tests are successful for the chambers, this can be used for the assembly of Super Modules, where one chamber type 2 is coupled to a one chamber type 3 to form a Super Module assembly. Currently the 36 SM needed for the positive side of end-cap system have been delivered and the firts end-cap upgrade has been installed successfully. The production for the SM for the negative end-cap is ongoing and therefore the QC4 test is carry on for that chambers.

## Acknowledgments

Acknowledgments I would like to thank the Eplanet program for the partial support gotten during my stay at CERN.

Journal of Physics: Conference Series **651** (2015) 012005 doi:10.1088/1742-6596/651/1/012005

## References

[1] CMS collaboration, The upgrade of the CMS RPC system during the first LHC long shutdown, 2012 JINST 8 T02002.
[2] CMS Collaboration, The CMS experiment at the CERN LHC, JINST 3 (2008) S08004.doi:10.1088/1748-0221/3/08/S08004.
[3] L.M. Pant, Resistive Plate Chambers for the RE4 upgrade of the CMS endcap system, 2012 JINST 7 P10025.
[4] V. K. Bhandari, Production of resistive plate chambers for CMS Experiment. EHEP Group, Departament of Physics, Panjav University Chandigarh, April, 2012
[5] https://apex.cern.ch/pls/htmldbc_msr/f?p=106.
[6] https://twiki.cern.ch/twiki/bin/view/Main/QC4Software.
[7] https://twiki.cern.ch/twiki/bin/viewauth/CMS/RPCOffLineAnalysis.

# The pion distribution amplitude from SDE-BSE

**J J Cobos-Martínez**

Departamento de Física, Universidad de Sonora, Boulevard Luis Encinas J. y Rosales, Colonia Centro, Hermosillo, Sonora 83000, Mexico

E-mail: `j.j.cobos.martinez@gmail.com`

**Abstract.** A brief exposition of the Schwinger-Dyson–Bethe-Salpeter equations of Quantum Chromodynamics and their application to hadron physics is given. Results for the rainbow-ladder trucantion scheme are presented. The Pion distribution amplitude is calculated in the SDE-BSE approach to hadron physics employing a novel method of computation [28]. The SDE-BSE is a well founded continuum approach to nonperturbative hadron physics that unifies a range of hadron observables.

## 1. Introduction

Strong interaction phenomena poses a wealth of fundamental questions with profound significance of our understanding of Nature and the structure of matter of which we and our Universe are composed [34]. The field of hadron physics is the study of strong interacting matter in all of its manifestations and the understanding of its properties and interactions in terms of the underlying fundamental theory, Quantum Chromodynamics (QCD) [34]. QCD is the theory of quarks, gluons, and their interactions; it is a self-contained part of the Standar Model of particle physics whose only input parameters are the masses of the quarks and the coupling constant between these and the gluons; it is a consistent quantum field theory with a simple and elegant Lagrangian, based entirely on the invariance under the local non-Abelian $SU(3)$ colour gauge group, and renormalisability. However, the Lagrangin written on the blackboard does not by itself explain the data on strongly interacting matter, and it is not clear how the plethora of observed bound states, the hadrons, and their properties arise from the fundamental quark and gluon fields, and the parameters of QCD.

At large momentum transfer, due to the property of asymptotic freedom of QCD, one could use the familiar perturbative language, the Feynman diagrams and the like, to describe hadron interactions. At these small dinstances, hadrons and their interactions are described as bound states of weakly interacting quarks and gluons. This description, however, starts to break down at energy scales of around 1-2 GeV, and it is surely inadequate at length scales corresponding to the size of the nucleon. At such lenght scales the strong coupling constant is large enough to invalidate perturbation theory and ones has to employ different methods to deal with what is called strong QCD.

In order to go from the nonperturbative quarks and gluons to the study of hadron physics, we need special tools such QCD correlation functions, a bridge between theory and experiment. At present, our choices for the nonperturbative calculation of theses correlation functions are lattice-QCD and the Schwinger-Dyson equations, both of which have their own advantages and drawbacks. Lattice-QCD Monte Carlo methods, based on the discretisation of spacetime,

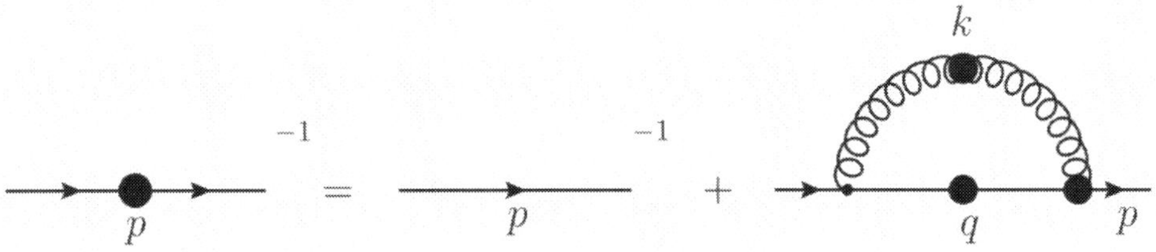


**Figure 1.** Quark Schwinger-Dyson Equation; filled circles indicate fully dressed objects.

include all the nonperturbative physics, and therefore are the only *ab initio* calculation method available so far. However, the simulations suffer from limitations at small momenta due to finite volume effects and one has to rely on extrapolation methods to obtain the infinite volume limit. Furthermore, calculations including quarks use unphysically large values for the quark masses, and extrapolations to the physical values are required. On the other hand, with the Schwinger-Dyson equations, which are the quantum equations of motion for the correlation functions of the theory, we can work in either Minkowski or Euclidean space for any value of the quark masses. However, here we are working with an infinite number of coupled integral equations, and in order to obtain a closed system of equations we must introduce ansatze for the higher correlation functions that are not expliclitly solved for, introducing a model dependence. However, this model dependence is limited to infrared momenta. Nevertheless, these two methods are entirely complementary in their strengths and weaknesses.

In this short talk we give a brief exposition of the Schwinger-Dyson and Bethe-Salpeter equations approach to hadron physics. The first few sections present generalities of this system of equations and of the rainbow-ladder truncation scheme. The main result is given in the last section where a novel method to obtain the pion distribution amplitude is presented [28]. An application of this distribution amplitude to the asymptotic form of the pion electromagnetic form factor is also shown. For a more comprehensive and up to date presentation see [35] and references therein.

## 2. The Schwinger-Dyson and Bethe-Salpeter Equations

The Schwinger-Dyson equations (SDEs) are the equations of motion of a quantum qield heory (QFT). They are derived from the full generating functional of the quantum theory. The starting point for the derivation of the SDEs (also called $n$-point functions) is the observation that the functional integral of a total derivative is zero:

$$\int \mathcal{D}\phi \frac{\delta}{\delta\phi} = 0. \tag{1}$$

In the Euclidean space formulation of QCD, with $\{\gamma_\mu, \gamma_\nu\} = 2\delta_{\mu\nu}$, $\gamma_\mu^\dagger = \gamma_\mu$, and $a \cdot b = \sum_{i=1}^{4} a_i b_i$, the renormalised SDE for the dressed quark propagator (a 2-point function) for a particular quark flavour is

$$S^{-1}(p) = Z_2 i\gamma \cdot p + Z_4 m(\mu) + Z_1 \int \frac{d^4q}{(2\pi)^4} g^2 D_{\mu\nu}(p-q) \frac{\lambda^a}{2} \gamma_\mu S(q) \Gamma^a_\nu(p,q), \tag{2}$$

where $g$ is the renormalisation-point-dependent coupling constant, $D_{\mu\nu}$ is the renormalised dressed-gluon propagator, $\Gamma^a_\nu$ is the renormalised dressed-quark-gluon vertex, and $m(\mu)$ the renormalised current-quark mass; $Z_1(\mu, \Lambda)$ and $Z_2(\mu, \Lambda)$ are the quark-gluon vertex and the quark wave function renormalisation constants, respectively, which depend on the renormalisation point $\mu$ and regularisation mass scale $\Lambda$. This equation is depicted in Fig 1.

The dressed-quark propagator, as well as the dressed-gluon propagator, and dressed-quark-gluon vertex, depend on the renormalisation point at which they are defined; however obvervables do not depend on this.

Since both $D_{\mu\nu}$ and $\Gamma^a_\nu$ satisfy their own SDE, which in turn are coupled to higher $n$-point functions and so on *ad infinitum*, the quark SDE, Fig 1, explicitly shows that the SDEs form a coupled infinite set of nonlinear integral equations. A tractable problem is defined once a truncation scheme has been specified.

The general form of the renormalised dressed-quark propagator, obtained as the solution of Eq. (2), is given in terms of two Lorentz-scalar dressing functions, written in various forms as

$$S^{-1}(p) = i\gamma \cdot pA(p^2,\mu^2) + B(p^2,\mu^2) \tag{3}$$
$$= Z^{-1}(p^2,\mu^2)\left(i\gamma \cdot p + M(p^2)\right), \tag{4}$$

which are equivalent. In the last form, $Z(p^2, \mu^2)$ is known as the wave-function renormalisation, and $M(p^2)$ is the dressed quark mass function, which is independent of the renormalisation point if the quark propagator is renormalised multiplicatively.

The solution of the quark SDE Eq. (2) is subject to the renormalisation condition, for $\mu$ large and spacelike,

$$S^{-1}(p)\Big|_{p^2=\mu^2} = i\gamma \cdot p + m(\mu), \tag{5}$$

where $m(\mu)$ is the flavour-dependent renormalised current-quark mass.

The best known truncation scheme of the SDEs is the weak coupling expansion, which reproduces every diagram in perturbation theory. It is a systematic-improvable truncation scheme, and a essential tool for the investigation of large momentum transfer phenomena because QCD is asymptotically free. However, it excludes the possibility of obtaining information about the low-energy regime relevant for the hadron structure and reactions, and the phenomena of dynamical chiral symmetry breaking (DCSB) and confinement, which are all essentially nonperturbative.

As an example, consider the chiral limit in perturbative QCD, defined by $Z_2 m_{\rm bare} \equiv 0$ for all $\Lambda \gg \mu$. In this case, the theory is chirally symmetric, and a perturbative evaluation of the quark propagator dressing functions, Eq. (3), gives [4, 7, 8]

$$B(p^2) = m\left[1 - \frac{\alpha}{\pi}\ln\left(\frac{p^2}{m^2}\right) + \cdots\right], \tag{6}$$

where the ellipsis denote higher order terms in $\alpha$. However, it is always true that at any order in perturbation theory

$$\lim_{m\to 0} B(p^2) = 0, \tag{7}$$

and hence dynamical chiral symmetry breaking is impossible in perturbation theory, and the quark SDE cannot generate a mass gap.

### 3. The Meson Bethe-Salpeter Equation

Meson bound states, whose flavour structure is given by a dressed quark-antiquark ($a\bar{b}$) pair, are described by the Bethe-Salpeter equation (BSE), depicted in Fig 2,

$$[\Gamma_H(p;P)]_{tu} = \int \frac{d^4q}{(2\pi)^4} [K(p,q;P)]_{tu;rs} \left[S^a(q_+)\Gamma_H(q;P)S^b(q_-)\right]_{sr}, \tag{8}$$

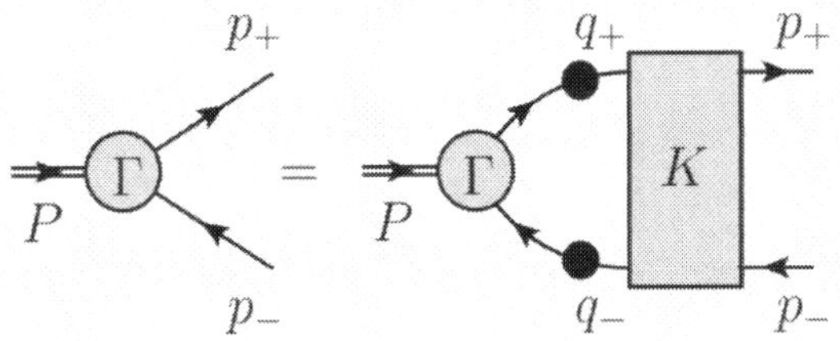


**Figure 2.** Bethe-Salpeter Equation: Γ is the fully-amputated quark-meson vertex or Bethe-Salpeter Amplitude; $K$ is the fully-amputated, two-particle irreducible, scattering kernel; filled dots on the quark lines indicate quark propagators are fully dressed.

where $H = (a\bar{b})$ indicates the flavour structure. Here, $\Gamma_H(p;P)$ is the meson Bethe-Salpeter amplitude (BSA) describing the bound state, $S^f(q_\pm)$ is the propagator for a dressed quark, and $K(p,q;P)_{tu;rs}$ is the quark-antiquark scattering kernel. Latin indices indicate the colour, flavour, and Dirac structure. Poincaré covariance and momentum conservation entail $q_+ = q + \eta P$, $q_+ = q - (1-\eta)P$, and similarly for $p_\pm$, with $P = p_+ - p_-$. The parameter $\eta \in [0,1]$ describes the meson momentum sharing between the quark-antiquark pair. Observables, however, do not depend on this.

The Bethe-Salpeter equation, Eq. (8) is a homogeneous eigenvalue equation that admits solutions only for discrete values of the meson momenta $P^2 = -m_H^2$, where $m_H$ is the mass of the meson under consideration. In a particular channel, the lowest mass solution corresponds to the physical ground state.

In the pseudoscalar channel ($J^P = 0^-$), the lowest mass solutions are the pion and kaon mesons, with flavour structure $u\bar{d}$ and $u\bar{s}$, respectively. The general form of the BSA in this channel is given by

$$\Gamma_H(p;P) = T^H\gamma_5\left[\mathrm{i}E_H(p;P) + \gamma\cdot PF_H(p;P) + \gamma\cdot p(p\cdot P)G_H(p;P) + \sigma_{\mu\nu}p_\mu P_\nu H_H(p;P)\right], \quad (9)$$

with $H = \pi$, $K$. For mesons that are eigenstates of the charge conjugation operation $C$, such as the $\pi^0$, there is an additional constraint on the Bethe-Salpeter amplitude to obtain a specified $C$-parity (In the isospin symmetric limit ($m_u = m_d$) we are working in, the mass and dressing functions of $\pi^\pm$ mesons will be equal to those of $\pi^0$). In Eq. (9), all the elements of the Lorentz-Dirac basis employed are even under $C$, and thus the only remaining quantity that can produce a definite $C$-parity is $p\cdot P$, which is odd under $C$. Therefore, a $C = +1\,(-1)$ solution will have dressing functions $E_H(p;P)$, etc that are even (odd) in $p\cdot P$. For mesons that are not eigenstates of $C$, each dressing function will contain both even an odd terms in $p\cdot P$. When using a Chebyshev expansion of $E_H(p;P)$ etc this means that for $C = +1\,(-1)$ we will only require even (odd) Chebyshev polynomials, and thus the odd (even) Chebyshev coefficients will vanish.
The pseudoscalar leptonic decay constant is calculated using

$$f_H P_\mu = Z_2\int\frac{\mathrm{d}^4q}{(2\pi)^4}\mathrm{Tr}\left[\frac{(T^H)^t}{2}\gamma_5\gamma_\mu S^a(q_+)\Gamma_H(q;P)S^b(q_-)\right], \text{ at } P^2 = -m_H^2, \quad (10)$$

where $H = \pi$, $K$, and the trace is over Dirac, colour, and flavour indices. In Eq. (10), $\Gamma_H(q;P)$ is the normalised BSA. [1]

In Eq. (8), $K(p,q;P)_{tu;rs}$ is the fully-amputated, two-particle irreducible, quark-antiquark scattering kernel. It is a four-point Schwinger function, obtained formally as the sum of a countable infinity of skeleton diagrams [9]. The complexity of $K(p,q;P)_{tu;rs}$ is one of the reasons why quantitative SDE and BSE studies employ modelling of $D_{\mu\nu}(k)$ and $\Gamma^a_\nu(k,p)$ [1], because $K(p,q;P)_{tu;rs}$ also appears in the SDE satisfied by $\Gamma^a_\nu(k,p)$. However, the lack of a full understanding of the interaction between quarks, through the complete knowledge of $K(p,q;P)_{tu;rs}$, does not prevent us from obtaining general results in hadron physics [2].

## 4. The Rainbow-Ladder Trucation and the Maris-Tandy Model

The kernel in the quark SDE, Eq. (2), is formed from the product of the dressed-gluon propagator and the dressed-quark-gluon-vertex, and therefore the structure of these quantities is largely responsible for the spectral properties of the dressed-quark propagator. However, in developing a truncation scheme for the quark SDE it is insufficient to concentrate only on these, as the SDE system forms an infinite tower. In the low-energy region of the strong interactions, dynamical chiral symmetry breaking is the dominant nonperturbative effect characterising the light hadron spectrum, in particular that of the octet of pseudoscalar mesons [10]. It is thus imperative that the implemented truncation scheme does not break chirally symmetry, and the pattern by which it is broken, if it is to be a viable approach.

In QCD, chiral symmetry and its breaking pattern are expressed through the axial-vector Ward-Takahashi identity [1, 2, 3]

$$\int \frac{d^4q}{(2\pi)^4} K_{tu;rs}(k,q;P)\left[\gamma_5 T^H S(q_-) + S(q_+)\gamma_5 T^H\right]_{sr} = \left[\Sigma(k_+)\gamma_5 T^H + \gamma_5 T^H \Sigma(k_-)\right]_{tu}, \quad (11)$$

thus constrainig the content of the quark-antiquark scattering kernel $K(p,q;P)_{tu;rs}$ if an essential symmetry of the strong interactions, and its breaking pattern, is to be preserved.

From a practical point of view, Eq. (11) provides a way of obtaining the quark-antiquark scattering kernel if we can solve this constraint, given an expression for the quark self-energy. However, this is not always possible, see for example [11], and we must find an alternative way of preserving the chiral symmetry properties of the strong interactions. In principle, one may construct a quark-antiquark scattering kernel satisfying Eq. (11) from a functional derivative of the quark self-energy with respect to the quark propagator [12], within the framework of the effective action formalism for composite operators developed in [6].

Fortunately, for the rainbow truncation of the quark self-energy, Eq. (13), the axial-vector Ward-Takahashi identity can be easily satisfied. The quark-antiquark scattering kernel that is consistent with the rainbow truncation of the quark self-energy, in the sense that the axial-vector Ward-Takahashi identity, Eq. (11), is satisfied, is given by

$$K(p,q;P)_{tu;rs} = -\mathcal{G}(k^2) D^{\text{free}}_{\mu\nu}(k) \left[\frac{\lambda^a}{2}\gamma_\mu\right]_{ts} \left[\frac{\lambda^a}{2}\gamma_\nu\right]_{ru}, \quad (12)$$

where $k = p - q$, and $\mathcal{G}(k^2)$ is the effective coupling of Eq. (14). This is the so-called ladder truncation of the BSE. This equation corresponds to a single effective dressed-gluon exchange, and its solution corresponds to re-summing this gluon rung, thus providing the (infinite) ladder.

Despite their complexity, recent progress has been made in unveiling the nonperturbative structure of $D_{\mu\nu}$ and $\Gamma^a_\nu$ using their SDE [13, 14], as well the lattice formulation of QCD [15, 16]. Ultimately, the detailed infrared behaviour of these quantities should not materially affect the observable consequences of the quark SDE and meson BSE, as long as the infrared enhancement in the quark SDE implements the appropriate amount of dynamical chiral symmetry breaking and, explains the (pseudo)Goldstone character of the pion [5].

In modelling the quark SDE kernel we follow [1, 3], and make the following approximation in the self-energy, the so-called rainbow approximation,

$$Z_1 \int \frac{d^4q}{(2\pi)^4} g^2 D_{\mu\nu}(k) \frac{\lambda^a}{2} \gamma_\mu S(q) \Gamma^a_\nu(k,p) \to \int \frac{d^4q}{(2\pi)^4} \mathcal{G}(k^2) D^{\rm free}_{\mu\nu}(k) \frac{\lambda^a}{2} \gamma_\mu S(q) \frac{\lambda^a}{2} \gamma_\nu, \tag{13}$$

where the phenomenological "effective" coupling $\mathcal{G}(k^2)$ contains information about the behaviour of the product $G(k^2, \mu^2)F^1(k, p, \mu)$, where $F^1(k, p, \mu)$ is the form factor associated with the $\gamma_\nu$ structure in the dressed-quark-gluon vertex $\Gamma^a_\nu$, and $G(k^2, \mu^2)$ that in the dressed-gluon propagator. The model is completely defined once a form for $\mathcal{G}(k^2)$ has been specified. Note that we have actually made the replacement $\Gamma^a_\nu(k, p) \to (\lambda^a/2)\gamma_\nu$, and absorbed $Z_1 g^2$ into the effective coupling $\mathcal{G}(k^2)$. The function $\mathcal{G}(k^2)$ can be interpreted as an effective gluon dressing function. The solution of the resulting equation resumms an effectively-dressed gluon rainbow.

In principle, constraints on the form of $\mathcal{G}(k^2)$ come from the SDE satisfied by dressed-gluon propagator and the dressed-quark-gluon vertex. However, we know that the behaviour of the QCD running coupling $\alpha(k^2)$ in the ultraviolet, that is $k^2 >$ 2-3 GeV$^2$, is well described by perturbation theory [17], and therefore the model dependence is restricted to the infrared region. On the other hand, the effective interaction in the quark SDE should exhibit sufficient infrared enhancement capable of triggering dynamical chiral symmetry breaking, through a nonzero quark condensate, and the generation of a momentum-dependent quark mass dressing function [18] that connects the current-quark mass to a constituent-like quark mass.

Various models for the effective interaction $\mathcal{G}(k^2)$ combining the ultraviolet behaviour known from perturbative QCD and an ansatz for the infrared part have been designed in the past. These have been applied to different detailed studies of dynamical chiral symmetry breaking, hadron structure and reactions [1, 3, 19, 20, 21].

In choosing a form for the effective coupling we follow reference [1, 3], and write the effective gluon dressing function $\mathcal{G}(k^2)$ as

$$\frac{\mathcal{G}(k^2)}{k^2} = 4\pi^2 \frac{D}{\omega^6} k^2 \exp\left(-\frac{k^2}{\omega^2}\right) + 4\pi \frac{\gamma_m \pi}{\frac{1}{2}\ln\left[\tau + \left(1 + \frac{k^2}{\Lambda_{\rm QCD}}\right)^2\right]} \mathcal{F}(k^2), \tag{14}$$

where $\mathcal{F}(k^2) = \left(1 - \exp\left(-\frac{k^2}{4m_t^2}\right)\right)/k^2$, and fixed parameters $m_t = 0.5$ GeV, $\tau = e^2 - 1$, $N_f = 4$, $\gamma_m = 12/(33 - 2N_f)$, $\Lambda^{N_f}_{\rm QCD} = 0.234$ GeV. The remaining parameters, $\omega$ and $D$, are phenomenological parameters fitted, together with the renormalised quark masses $m_u = m_d$ and $m_s$, to pion and kaon observables.

The detailed study of [3] found that pseudoscalar and vector meson ground state properties are practically insensitive to variations in the model parameters, with $\omega =$ 0.3-0.5 GeV and $(\omega D)^{1/3} = 0.72$ GeV, as long as the integrated strength of the effective interaction is strong enough to generate an acceptable amount of chiral symmetry breaking, as required by the chiral quark condensate. However, this is not true for excited states made up of light quarks, where the long range part of the effective quark-quark interaction plays a significant role [23, 22, 24].

In Fig 3 we present numerical solutions to the mass dressing function $M(p^2) = B(p^2, \mu^2)/A(p^2, \mu^2)$, renormalised at $\mu = 19$ GeV, where various values of the renormalised current-quark mass are considered, while $\omega = 0.4$ GeV and $D = 0.93$ GeV$^2$ are kept fixed. It is evident that DCSB is a reality in the rainbow truncation with the effective coupling given by Eq. (14). At ultraviolet momenta, the magnitude of the mass function is determined from the renormalised current-quark mass. In the infrared, however, and specially for light-quarks, $M(p^2)$ is significantly enhanced. For light quarks, this enhancement is orders of magnitude larger than the mass present in the Lagrangian. Fig 3 also shows that the evolution from the current-quark mass to a constituent-like quark mass occurs at the scale of $\approx 1$ GeV$^2$, as required

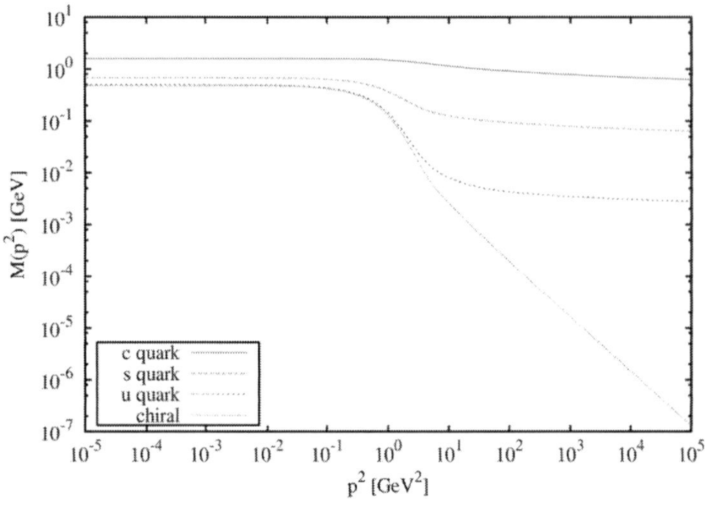


**Figure 3.** Quark mass function in the rainbow-ladder truncation with the Maris-Tandy dressing function for various values of the renormalised current-quark mass, with $\omega = 0.4$ GeV and $D = 0.93$ GeV$^2$.

from hadron phenomenology. Furthermore, this effect is particularly important for light quarks, that is the evolution and magnitude of their mass function in the infrared is dominated by the nonperturbative effect of DCSB. The domain in which the effect of DCSB is relevant decreases as the renormalised current-quark mass increases.

The behaviour of the quark mass function and wave function renormalisation of Fig 3 has also been confirmed semi-quantitatively in lattice simulations of QCD [36, 37, 38]. Agreement for a range of quark masses requires the effective interaction to be flavour dependent, and dressing the quark-gluon-vertex ensures this dependence, as pointed out in [39].

## 5. The Pion Parton Distribution Amplitude

The theoretical interest in the leading twist light-cone distribution amplitude (DA) of hadrons is due to their relevance in the description of exclusive reactions [25, 26, 27] from the fundamental accepted theory of the strong interactions, namely Quantum Chromodynamics. In terms of the Bethe-Salpeter wave functions these distributions are defined by keeping track of the momentum fraction $x$ and integrating out the dependence on the transverse momentum $k_\perp$:

$$\phi(x,\mu) \sim \int_{k_\perp^2<\mu^2} d^2k_\perp \phi(x,k_\perp). \tag{15}$$

The DAs describe probability amplitudes to find the hadron in question in a state with minimum number of Fock constituents and at small transverse separation, which provides a ultraviolet cutoff. The dependence on the ultraviolet cutoff scale $\mu$ is given by the Efremov-Radyushkin-Brodsky-Lepage evolution equations [25, 27] and can be calculated in perturbative QCD, while the distribution amplitude at a certain low scale provides the necessary nonperturbative input for the rigorous QCD treatment of exclusive reactions with large momentum transfer [25, 26, 27].

In terms of dressed-quark propagators and the pion Bethe-Salpeter amplitude, the pion distribution amplitude is given by [28]

$$f_\pi \phi_\pi(x) = Z_2 N_c \text{Tr} \int \frac{d^4 q}{(2\pi)^4} \delta(n \cdot q_+ - xn \cdot P) \gamma_5 \gamma \cdot n S(q_+) \Gamma_\pi(q; P) S(q_- P), \tag{16}$$

where $f_\pi$ is the pion decay constant; $n$ is a light-like vector, $n^2 = 0$; $P$ is the pion's four momentum, $P^2 = -m_\pi^2$ and $n \cdot P = -m_\pi$, with $m_\pi$ being the pion's mass; $\Gamma_\pi$ is the pion BSA; and $S$ the dressed quark propagator. Due to Poincaré covarince, observables are independent of the momentum sharing parameter $\eta \in [0, 1]$, with $q_+ = q + \eta P$ and $q_- = q - (1 - \eta)P$. Using Eq. (16), the integer moments of the distribution $\phi_\pi(x)$, $\langle x^m \rangle = \int_0^1 dx x^m \phi_\pi(x)$, are given by [28]

$$f_\pi (n \cdot P)^{m+1} = Z_2 N_c \text{Tr} \int \frac{d^4 q}{(2\pi)^4} (n \cdot q_+)^m \gamma_5 \gamma \cdot n S(q_+) \Gamma_\pi(q; P) S(q_- P). \tag{17}$$

The quark propagator $S$ and pion BSA $\Gamma_\pi$ are obtained as numerical solutions of their respective equations in the Rainbow-Ladder truncation using the Maris-Tandy interaction, which preserves the one-loop renormalisation group behaviour of QCD and guarantees that the quark mass function $M(p^2)$ is independent of the renormalisation point.

The computation of the moments in Eq. (17) is difficult with numerical solutions so we employ algebraic parametrisations of these that serve as interpolations in evaluating the moments. For the quark propagator a convenient representation is in terms of complex conjugate poles, a feature consistent with confinement . It is found that 2 complex conjugate poles is adequate [28]

$$S(p) = \sum_{l=1}^{2} \left( \frac{z_l}{i\gamma \cdot p + m_l} + \frac{z_l^*}{i\gamma \cdot p + m_l^*} \right). \tag{18}$$

The parameters $z_l$ and $m_l$ are numerically determined to provide the best fit to numerical solutions of the quark SDE. The quality of the fit is very good, see [28]. For the pion BSA, each scalar dressing function is expressed via a Nakanishi-like representation, with parameters fitted to that function's first two Chebyshev moments [28]

Using these representations it is now straigthforward to compute arbitrarily many moments using Feynman integral techniques and a subsequent numerical integrations over the Feynman parameters introduced. Once all needed moments are calculated (tipycally 50) a novel reconstruction procedure is implemented to obtain the distribution amplitude $\phi_\pi(x)$ [28]. In this procedure, the distribution amplitude is expanded in terms of the orthonormal set of Gegenbauer polynomials of order $\alpha$, instead of the typical expansion in terms of Gegenbauer polynomials of order $\alpha = 3/2$; that is with $\alpha$ fixed. With $\alpha_- = \alpha - 1/2$ we write

$$\phi_\pi(x, \mu) = x^{\alpha_-}(1 - x)^{\alpha_-} \left[ 1 + \sum_{j=2,4,\ldots} a_j^\alpha(\mu) C_j^\alpha(2x - 1) \right]. \tag{19}$$

The parameter $\alpha$ and expasion coefficients $a_j^\alpha(\mu)$ are determined from a best fit to the numerical moments Eq. (17). Once these are obtained one may project out the result onto an $\alpha = 3/2$ basis, as dictated by the solution of the ERBL evolution equation. However, this requires many nonzero coefficient $\{a_j^{3/2}\}$, and introduce spurious oscillations that are typical of Fourier-like approximations to a simple function.

The plot in Fig 4 shows the RL result with the Maris-Tandy interaction, with $D\omega = (0.87 \text{ GeV})^3$, and $\omega = 0.3$. It is described by

$$\phi_\pi^{\text{RL}}(x, \mu) = 1.74 x^{\alpha_-^{RL}}(1 - x)^{\alpha_-^{\text{RL}}} \left[ 1 + a_2^{\alpha^{\text{RL}}}(\mu) C_j^{\alpha^{\text{RL}}}(2x - 1) \right], \tag{20}$$

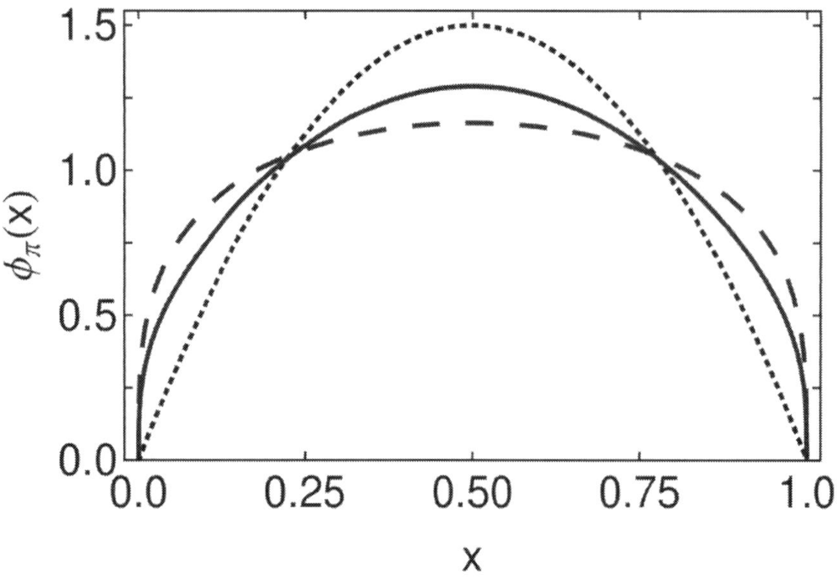


**Figure 4.** Computed distribution amplitude at $\mu = 2$ GeV. Curves: dashed, rainbow-ladder; solid: DCSB-improved kernel (see [28] and references therein); and dotted, asymptotic distribution.

with $\alpha^{\rm RL} = 0.79$, and $a_2^{\alpha^{\rm RL}} = 0.0029$. Projected onto a $\alpha = 3/2$ basis this corresponds to $a_2^{3/2} = 0.23, \ldots, a_{14}^{3/2} = 0.022$, etc. That $j \geq 14$ is required before $a_j^{3/2} < 0.1a_2^{3/2}$ highlights the merit of reconstruction via Gegenbauer polynomials of order $\alpha$ at any reasonable scale $\mu$. The merit is still greater still if one oly has access to a single nontrivial moment [30], as in lattice QCD [31]. For the moments $\langle(2x-1)^2\rangle$ lattice-QCD gives $\langle(2x-1)^2\rangle^{\text{lattice-QCD}} = 0.27 \pm 0.04$ [31], while the RL results gives $\langle(2x-1)^2\rangle^{\rm RL} = 0.28$, and the asymptotic distribution gives $\langle(2x-1)^2\rangle^{\rm asy} = 0.2$. Also shown in Fig 4 is the asymptotic QCD distribution amplitude $\phi_\pi^{\rm asy}(x) = 6x(1-x)$.

Various qualitative signficant results can be read from Fig 4. The most important being that DCSB is expressed in the pion PDA through a marked broadening with respect to $\phi_\pi^{\rm asy}$. This may be because the PDA has been computed at low renormalisation scale in the chiral limit, whereat the quark mass function owes entirely to DCSB; and on the domain $0 < p^2 < \mu^2$, the nonperturbative interactions responsible for DCSB produce significant structure in the dressed-quark's self-energy. The PDA is an integral of the pion's Bethe-Salpeter wave-function, whose pointwise behaviour is rigurously connected with that of the quark self-energy. Hence, the structure of the pion PDA at the hadronic scale is pure expression of DCSB. As the scale is removed to extremely large values, phase space growth diminishes the impact of nonperturbative DCSB interactions so that the PDA relaxes to its asymptotic form [28].

As a first application [29] of the calculated pion PDA, consider the perturbative QCD (pQCD) prediction for the asymptotic pion electromagnetic form factor $F_\pi(Q^2)$ [25, 26, 27]

$$Q^2 F_\pi(Q^2) \overset{Q^2\to\infty}{=} 16\pi\alpha_S(Q^2) f_\pi^2 \left|\frac{1}{3}\int dx \frac{1}{x}\phi_\pi(x)\right|^2 \tag{21}$$

$$= 16\pi\alpha_S(Q^2) f_\pi^2, \tag{22}$$

where in the $f_\pi = 92.2$ MeV and $\alpha_S(Q^2)$ is the strong running coupling constant at the scale $Q^2$, where $Q$ is the four-momentum transfer in factorised electro-pion scattering. To get Eq. (22) we have used the asymptotic pion PDA, $\phi_\pi(x) = \phi_\pi^{\rm asy}(x)$. From the experimental data on $F_\pi(Q^2)$ [32, 33] there is a very weak suggestion that the condition that $Q^2 F_\pi(Q^2) =$ constant is being approached. Furthermore the value at this should happen is not predicted by pQCD. Since the data is several times larger than the pQCD prediction, nonperturbative effects are still dominant. For example, for the largest experimental point [32, 33], $Q^2 = 2.45$ GeV$^2$, $Q^2 F_\pi(Q^2) = 0.41$ GeV$^2$. Compare this with a larger $Q^2$ using the pQCD analysis. For $Q^2 = 4$ GeV$^2$ we have $Q^2 F_\pi(Q^2) = 0.15$ GeV$^2$. These observations beg the question: Has $\phi_\pi(x)$ really reached its asymptotic form at currently accessible energies? At leading-log accuracy the Gegenbauer expansion coefficients evolve with scale according to [25, 27], with $\mu_1 = 2$ GeV,

$$a_n^{3/2}(\mu) = a_n^{3/2}(\mu_1)\left[\frac{\alpha_S(\mu_1)}{\alpha_S(\mu)}\right]^{\gamma_n^0/\beta_0}, \quad \alpha_S(Q^2) = \frac{4\pi}{\beta_0}\ln\left(\frac{Q^2}{\Lambda_{\rm QCD^2}}\right), \tag{23}$$

with

$$\beta_0 = 11 - (2/3)N_f, \quad \gamma_n^{(0)} = \frac{4}{3}\left[3 + \frac{2}{(n+1)(n+2)} - 4\sum_{k=1}^{n+1}\frac{1}{k}\right]. \tag{24}$$

Now comes the surprising result, see [29]: Using $N_f = 4$ quark flavours, and $\Lambda_{\rm QCD} = 0.234$ GeV it is necessary to evolve to $\mu = 100$ GeV before $a_2^{3/2}$ falls to 50% of its value. This means that the asymptotic value $\phi_\pi^{\rm asy}(x)$ lies at very large momenta and the magnitude of $Q^2 F_\pi(Q^2)$ reflect the scale of dynamical chiral symmetry breaking. The analysis in shows that hard contrubutions to the pion form factor dominate for $Q^2 \geq 8$ GeV$^2$. Therefore, in evaluating $Q^2 F_\pi(Q^2)$, $\phi_\pi(x)$ has not yet reached its asymptotic value. Using the asymptotic form for $\phi_\pi(x)$ we have

$$\left|\frac{1}{3}\int dx \frac{1}{x}\phi_\pi^{\rm asy}(x)\right|^2 = 1. \tag{25}$$

On the other hand, using the RL result for $\phi_\pi(x)$ we have

$$\left|\frac{1}{3}\int dx \frac{1}{x}\phi_\pi^{\rm RL}(x)\right|^2 = 3.2. \tag{26}$$

This means that the pQCD analysis result has to be multiplied by a factor of 3.2 and the asymptotic analysis of various models has to be compared to this new result [29], since the asymptotic domain given by pQCD lies at very very large momenta .

## 6. Conclusions

The SDE-BSE is a well founded continuum approach to nonperturbative hadron physics. Indeed, they are a natural framework for the exploration of strong QCD since they provide access to infrared as well as ultraviolet momenta, thus giving a clear connection with processes that are well understood because QCD is asymptotically free. Moreover, the SDE are the generating tool for perturbation theory.

However, the SDE-BSE form an infinite tower of coupled n-point functions that must be truncated in order to define a tractable problem. That is, we have to make an ansatz for the n-point functions whose SDE are not explicitly solved for, thereby introducing a *model dependence* that is difficult to quantify. Furthermore, drawing a connection between QCD (in the form of the SDE for the nonperturbative quarks and gluons) and hadron observables (for example in the form of the meson BSE) is difficult, and that is why modelling remains a keystone is the SDE-BSE approach to hadron physics. Phenomenological input is proving useful in designing an effective quark-antiquark interaction, and quantitative comparisons between the SDE-BSE and lattice-QCD studies are today complementing the (SDE-BSE) approach.

## References

[1] Maris P and Roberts C D 1997 *Phys. Rev.* C **56** 3369
[2] Maris P, Roberts C D and Tandy P C 1998 *Phys.Lett.* B **420** 267
[3] Maris P and Tandy P C 1999 *Phys. Rev.* C **60** 055214
[4] Cornwall J M 1982 *Phys. Rev.* D **26** 1453
[5] Maris P Raya A, Roberts C D,Schmidt S M 2003 *Eur. Phys. J.* A **18** 231
[6] Cornwall J M, Jackiw R and Tomboulis E 1974 *Phys. Rev.* D **10** 2428
[7] Pagels H 1979 *Phys. Rev.* D **19** 3080
[8] Miransky V A 1993 *Dynamical symmetry breaking in quantum field theories* (Singapore: World Scientific) p 533
[9] Bjorken J D and Drell S 1965 *Relativistic Quantum Fields* First ed (New York: McGraw-Hill)
[10] Weise W 2005 The QCD vacuum and its hadronic excitations *Preprint* arXiv:nucl-th/0504087
[11] Fischer C S, Nickel D and Wambach J 2007 *Phys. Rev.* D **76** 094009
[12] Munczek H J 1995 *Phys. Rev.* D **52** 4736
[13] Alkofer R, Fischer C S Llanes-Estrada F J and Schwenzer K 2009 *Annals Phys.* **324** 106
[14] Fischer C S 2003 Non-perturbative propagators, running coupling and dynamical mass generation in ghost - antighost symmetric gauges in QCD *Preprint* arxiv:hep-ph/0304233
[15] Skullerud J, Bowman P O, Kizilersu A, Leinweber D B and Williams A G 2005 *Nucl. Phys. Proc. Suppl.* **141** 244
[16] Cucchieri A and Mendes T 2007 *PoS LAT2007* 297
[17] Bethke S 2009 *Eur. Phys. J.* C **64** 689
[18] Roberts C D and Williams A G 1994 *Prog. Part. Nucl. Phys.* **33** 477
[19] Jain P and Munczek H J 1991 *Phys. Rev.* D **44** 1873
[20] Munczek H J and Jain P 1992 *Phys. Rev.* D **46** 438
[21] Alkofer R, Watson P and Weigel H 2002 *Phys. Rev.* D **65** 094026
[22] Holl A, Krassnigg A, Maris P, Roberts C D and Wright S V 2005 *Phys. Rev.* C **71** 065204
[23] Holl A, Krassnigg A and Roberts C D 2004 *Phys. Rev.* C **70** 042203
[24] Krassnigg A 2009 *Phys. Rev.* D **80** 114010
[25] Lepage G P and Brodsky S J 1980 *Phys. Rev.* D **22** 2157
[26] Farrar G R and Jackson D R 1979 *Phys. Rev. Lett.* **43** 246
[27] Efremov A V and Radyushkin A V 1980 *Phys. Lett.* B **94** 245
[28] Chang L, Cloet I C, Cobos-Martinez J J, Roberts C D, Schmidt S M and Tandy P C 2013 *Phys.Rev.Lett.* **110** 132001 (*Preprint* arXiv:1301.0324 [nucl-th])
[29] Chang L, Cloet I C, Roberts C D, Schmidt S M and Tandy P C 2013 Pion electromagnetic form factor at spacelike momenta *Preprint* arXiv:1307.0026
[30] Cloet I C, Chang L, Roberts C D, Schmidt S M and Tandy P C 2013 *Phys.Rev.Lett.* **111** 092001 (*Preprint* arXiv:1306.2645)
[31] Braun V M, *et al* 2006 *Phys.Rev.* D **74** 074501
[32] Huber G M, *et al* 2008 *Phys. Rev.* C **78** 045203
[33] Blok H P *et al* 2008 *Phys. Rev.* C **78** 045202
[34] Capstick S, *et al* 2000 Key issues in hadronic physics *Preprint* arXiv:hep-ph/0012238
[35] Cloet I C and Roberts C D 2013 Explanation and Prediction of Observables using *Preprint* arXiv:1310.2651
[36] Skullerud J, Leinweber D B and Williams A G 2001 *Phys. Rev.* D **64** 074508
[37] Bowman P O, Heller U M, Leinweber D B and Williams A G 2003 *Nucl. Phys. Proc. Suppl* **119** 323
[38] Bowman P O, Heller U M, Leinweber D B, Williams A G and Zhang J 2004 *Nucl. Phys. Proc. Suppl.* **128** 23
[39] Bhagwat M S, Pichowsky M A, Roberts C D and Tandy P C 2003" *Phys. Rev.* C **68** 015203

# Phenomenology of hadron structure — why low energy physics matters

**A. Courtoy**
IFPA, Inst. de Physique, Université de Liège, Belgium
INFN-LNF, Frascati, Italy

E-mail: aurore.courtoy@ulg.ac.be

**Abstract.** The description of the internal structure of hadrons is one of the main goal of QCD. At moderate energy scales, the hadronic representation succeeds to the partonic description, rendering challenging the description of the dynamics of scattering processes and hadronic structure. The information on the hadron structure is embodied in the long distance contributions which are defined as Parton Distribution Functions (PDFs). PDFs are a key framework for connecting the low and high-energy regimes, in that the knowledge on non-perturbative QCD carries important consequences at the high-energy level. We here review recent progress in the description of the proton, from complementary approaches such as fits of PDFs, phenomenological analyses and experimental predictions in view of the JeffersonLab upgrade and applications for high-energy colliders.

## 1. Introduction

QCD describes interactions as the exchange of a quantum degree of freedom called "color". The theory of strong interactions has the interesting property of asymptotic freedom. As the running coupling constant $\alpha_s(Q^2)$ becomes smaller, the perturbative treatment of QCD allows for the explanation of the hadronic phenomena, whose basic ingredients are the Parton Distributions.

On the other hand, hadrons that are actually observed in nature carry no color — they are color singlets. So far, physicists have failed to describe colorless hadrons within QCD due to the property of confinement: At low energy (typically below $\sim 1$ GeV), there is no justification for a perturbative treatment of QCD. We do not longer have a description of the relevant low energy observables from QCD. In this regime, non-perturbative approaches come into play.

In these proceedings, we will describe how, in these schemes, parameters are fixed by phenomenology. The two regimes of the strong interactions form an undivided whole, though the transition of the relevant degrees of freedom is not fully understood yet. As such, low energy parameters —as accounting for the physical phenomena— will guide high-energy observables. Among the many interesting implications of non-perturbative physics for perturbative QCD and high-energy observables, we will consider 2 main directions. The first is related to the PDFs. The quark degrees of freedom transition into hadronic degrees of freedom at a scale intimately related to the initial scale chosen for the PDF fits. We here discuss various approaches of the determination of the transition scale in Sections 3 & 4. The first steps toward "soft" evolution are discussed in Section 4. The phenomenological role of the running coupling constant in the infrared is highlighted in Section 5. This analysis induced a need to reconsider the PDFs at large values of Bjorken-$x$, close to the exclusive limit. Interestingly, the large-$x$ PDF carry important consequences at the high-energy level as is shown in Section 6. The second direction we want to explore, in Section 7, relates hadronic matrix elements to observables of physics Beyond the Standard Model. In particular, the scalar and tensor charges have been shown to be of great interest for precision measurements of new physics.

## 2. Parton Distribution Functions at High Energy

A way of connecting the perturbative and non-perturbative worlds has traditionally been through the study of Parton Distribution Functions (PDFs): Deep Inelastic processes are such that they enable us to look with a good resolution inside the hadron and allow us to resolve the very short distances, *i.e.* small configurations of quarks and gluons. At short distances, this part of the process is described through perturbative QCD. A resolution of such short distances is obtained with the help of non-strongly interacting probes. Such a probe, typically a photon, is provided by hard reactions. In that scheme, the PDFs reflect how the target reacts to the probe, or how the quarks and gluons are distributed inside the target. The insight into the structure of hadrons is reached at that stage: the large virtuality of the photon, $Q^2$, involved in such processes allows for the factorization of the hard (perturbative) and soft (non-perturbative) contributions in their amplitudes —in an OPE style. They are matrix elements of a bilocal current on the light-cone,

$$P^\mu q(x) \;=\; \int \frac{d\tau}{4\pi}\, e^{ix\tau}\, \langle PS|\, \bar{q}(0)\gamma^\mu q(\tau n)\, |PS\rangle \quad , \tag{1}$$

with $n^\mu = (1,0,0,-1)/(\sqrt{2}\Lambda)$ a light-like vector. The distribution functions are non-perturbative objects as they describe the large distance behavior of hadrons, a regime where confinement starts to matter. The virtuality of the photon introduces *the factorization scale*, *i.e.* PDFs explicitly depend on $Q^2$. This $Q^2$-evolution is dictated by the Dokshitzer-Gribov-Lipatov-Altarelli-Parisi (DGLAP) equations,

$$Q^2 \frac{\partial}{\partial Q^2} q(x,Q^2) \;=\; \frac{\alpha_s(Q^2)}{2\pi} \int_x^1 \frac{d\xi}{\xi}\, P\left(\frac{x}{\xi}, \alpha_s(Q^2)\right) q(\xi, Q^2) \quad , \tag{2}$$

here at leading order in $\alpha_s$— and where $P(x, Q^2)$ are the splitting functions.

At leading order (*leading-twist*), there are three types of PDFs: the ordinary number density, the helicity, and the transversity. The former, $q(x)$, is the well-known unpolarized PDF and is accessible through, *e.g.*, inclusive deep inelastic scattering (DIS), *i.e.* in the Bjorken regime. On the other hand, $g_1(x)$, the helicity PDF, is less constrained. The experimental knowledge on $h_1(x)$, the transversity PDF, is sparse as it is a chiral-odd quantity, not accessible through fully inclusive processes.

Collinear PDFs are universal and come into play in many processes, *e.g.* Drell-Yan processes, proton-proton collisions, $\cdots$ and are therefore of the utmost importance, especially for precision measurements.

Unpolarized PDFs have been extensively studied from first principles, in models and fits. The global fits combine various data set with different energy range. The precision obtained in the various fits, for light flavor and valence quarks, is high ; gluon and sea quarks as well as low and large-$x$ regions can incontestably be improved. For example, in DIS, the hard scattering part of the process can be presently described using splitting functions up to next-to-next-to-leading order (NNLO) [1, 2], and Wilson coefficients functions up to N$^3$LO [3]. PDF parametrizations along with their uncertainties have been obtained applying this framework up to NNLO, by a number of collaborations (see review in Ref. [4]). We cite MSTW, CTEQ, NNPDF, HERAFitter, GJR, ABM.[1]

The parameterizations present statistical error on the PDF itself as well as on the value of $\alpha_s(M_Z^2)$. The propagation of these uncertainties for the LHC phenomenology has been extensively studied, see *e.g.* [4]. A representative example is the Higgs production through gluon-gluon fusion: around half ($\sim 7\%$) of the uncertainty on the cross section comes from uncertainty on PDF and $\alpha_s(M_Z^2)$. Those numbers vary from one parameterization to the other.[2]

In the standard PDF fitting approaches, an *arbitrary* initial scale for the evolution equations, $Q_0^2 > 1$ GeV$^2$, is chosen: In parton distribution analyses the $x$-dependence of the PDFs is thus extracted at a particular scale $Q_0^2$, usually referred to as the *input scale*. A first systematic study of the effects of the choice of the input scale in global determinations of parton distributions and QCD parameters has been presented in Ref. [6], introducing the concept of *procedural bias* in PDF analyses. The first consequence of the choice of the input scale in a parton distribution analysis is that it affects the determination the strong coupling constant $\alpha_s(M_Z^2)$. In fact, the latter is determined together with the parton distributions and is substantially correlated with the gluon distribution which drives the QCD evolution. Besides, at low scales, there is an ambiguity related to the hadronic representation. The evolution equations modify the PDFs' radiative behavior in opposition to the low-energy valence behavior. The author of Ref. [6] uses a dynamical GJR parton distribution that optimally determines the input scale [7]. The latter is used as a guideline for the corresponding degrees of freedom —for $Q_0^2 < 1$ GeV$^2$ the PDF should tend to a valence-like behavior. $Q_0^2$ turns out to be of the order of 0.55GeV$^2$ (with $\Lambda_{\rm NLO}^{n_f=3} \sim 303$MeV), *i.e.*, in the region where non-perturbative inputs cannot be neglected, in a top-down approach.

In the next Sections, we will see how this non-perturbative input can come into play.

## 3. Hadronic Scale

It is still a challenge to describe consistently the dynamics of scattering processes and hadronic structure at moderate energy scales. Because at such a moderate scale the hadronic representation gives way to the partonic description, it is called *the hadronic scale*. The hadronic scale is peculiar to each hadronic representation and should be related to the *input scale* defined in the previous Section.

### *3.1. Determination of $Q_0$: Standard Approach*

From a bottom-up point of view, the evaluation of PDFs is guided by a standard scheme, set up in valuable litterature of the 90s [8–10]. This scheme runs in 3 main steps. First, we either build models consistent with QCD in a moderate energy range, typically the hadronic scale; or we use effective theories of QCD for the description of hadrons at the same energy range. Second, PDFs are evaluated in these models, giving a description of the Bjorken-$x$ dependence of the distribution. Third, the scale dependence of these distributions is studied. The last step allows to bring the moderate energy description of hadrons to the *factorization scale*, thanks to the QCD evolution equations (2). Here we are interested in the matching of non-perturbative models to perturbative QCD, using experimental data.

The hadronic scale is defined at a point where the partonic content of the model, defined through the second moment of the parton distribution, is known. For instance, the CTEQ

[1] A useful link to compare PDF sets can be found in Ref. [5] or the proceedings of the PDF4LHC working groups.
[2] Fig. 7 of Ref. [4] is very illustrative.

parameterization gives [3]

$$\left\langle (u_v + d_v)(Q^2 = 10\text{GeV}^2) \right\rangle_{n=2} = 0.36 \quad , \tag{3}$$

with $q_v$ the valence quark distributions and with $\langle q_v(Q^2) \rangle_n = \int_0^1 dx\, x^{n-1}\, q_v(x, Q^2)$. Scenarios for the hadronic representation have to be chosen. In an extreme scenario, *i.e.*, when we assume that the partons are pure valence quarks, the second Mellin moment is evolved downward until

$$\left\langle (u_v + d_v)(\mu_0^2) \right\rangle_{n=2} = 1 \quad . \tag{4}$$

The hadronic scale is found to be $\mu_0^2 \sim 0.1$ GeV$^2$.

This standard procedure to fix the hadronic (non-perturbative) scale pushes perturbative QCD to its limit. In effect, the hadronic scale turns out to be of a few hundred MeV$^2$, where the strong coupling constant has already started approaching its Landau pole. As it will be shown hereafter, the N$^m$LO evolution converges very fast, what justifies the perturbative approach. Consequently, the behaviour of the strong coupling constant plays a central role in the QCD evolution of parton densities. We here extend the standard procedure with the non-perturbative generalization of the QCD running coupling.

### *3.2. $Q_0$ from Non-Perturbative Physics*

We call perturbative evolution the renormalization group equations (RGE). The running of the coupling constant is driven by the RGE. In QCD, $\alpha_s$ is defined by renormalization conditions imposed at a large momentum scale where the coupling is small. The running coupling constant is dimensionless, but through dimensional transmutation, the strength of the interaction may be described by a dimensionful parameter. QCD scale, $\Lambda_{\text{QCD}}$, is then defined as the energy scale where the interaction strength reaches the value 1.

At N$^m$LO the scale dependence of the coupling constant is given by

$$\frac{d\,a(Q^2)}{d(\ln\, Q^2)} = \beta_{\text{N}^m\text{LO}}(\alpha_s) = \sum_{k=0}^{m} a^{k+2} \beta_k,$$

where $a = \alpha_s/4\pi$. We show here the solution to $k = 2$, *i.e.*, NLO.[4] The evolution equations for the coupling constant can be integrated out exactly leading to

$$\begin{aligned}
\ln(Q^2/\Lambda_{\text{LO}}^2) &= \frac{1}{\beta_0 a_{\text{LO}}} \quad , \\
\ln(Q^2/\Lambda_{\text{NLO}}^2) &= \frac{1}{\beta_0 a_{\text{NLO}}} + \frac{b_1}{\beta_0} \ln(\beta_0 a_{\text{NLO}}) - \frac{b_1}{\beta_0} \ln(1 + b_1 a_{\text{NLO}}) \quad ,
\end{aligned} \tag{5}$$

where $b_k = \beta_k/\beta_0$. These equations, except the first, do not admit closed form solution for the coupling constant, and we have solved them numerically. We show their solution, for the same value of $\Lambda = 250$ MeV, in Fig. 1.

We see in Fig. 1 (left panel) that the NLO and NNLO solutions agree quite well even at very low values of $Q^2$. They agree even better if we change the value of $\Lambda$ for the NNLO slightly, confirming the fast convergence of the expansion. This analysis concludes, that even close to the Landau pole, the convergence of the perturbative expansion is quite rapid, specially if we use a

[3] MSTW gives a similar result.
[4] $\beta_0 = 11 - \frac{2}{3}\, n_f \quad , \beta_1 = 102 - \frac{38}{3}\, n_f$, where $n_f$ stands for the number of effectively massless quark flavors and $\beta_k$ denote the coefficients of the usual four-dimensional $\overline{MS}$ beta function of QCD.

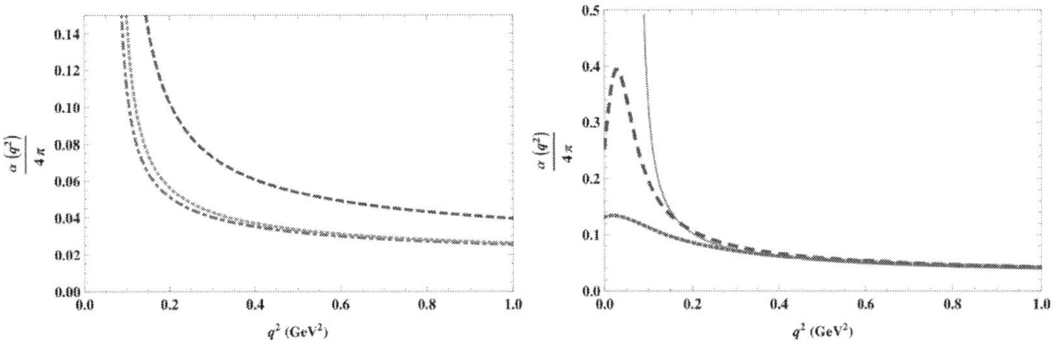


**Figure 1.** The running of the coupling constant. *Left panel*: The short dashed curve corresponds to the LO solution, the medium dotted-dashed curve to NLO solution and the tiny dashed curve to the NNLO solution ($\Lambda$ = 250 MeV). *Right panel*: The running of the effective coupling. The dotted and dashed curves represent the non-perturbative evolution with the parameters $m_0$ = 0.3GeV and, respectively, for the medium dashed blue curve $\rho = 1.53$, for the short dashed purple curve $\rho = 2.2$. The solid curve shows the NLO evolution with $\Lambda$ = 250 MeV.

different value of $\Lambda$ to describe the different orders, a feature which comes out from the fitting procedures. *This fast convergence ensures that perturbative evolution can still be used at rather low scales.* However, when entering the non-perturbative regime, other mechanisms take place that influence the QCD evolution. That is what we will call here non-perturbative evolution.

It is well established by now that the QCD running coupling (effective charge) freezes in the deep infrared. This non-perturbative property can be understood from various non-perturbative approaches [11–15], *e.g.* from the point of view of the dynamical gluon mass generation [11].[5] At the level of the Schwinger-Dyson equations the generation of such a mass is associated with the existence of infrared finite solutions for the gluon propagator, i.e. solutions with $\Delta^{-1}(0) > 0$. Such solutions may be fitted by "massive" propagators of the form $\Delta^{-1}(Q^2) = Q^2 + m^2(Q^2)$; $m^2(Q^2)$ is not "hard", but depends non-trivially on the momentum transfer $Q^2$. One physically motivated possibility, which we shall use in here, is the so called logarithmic mass running, which is defined by

$$m^2(Q^2) = m_0^2 \left[ \ln\left(\frac{Q^2 + \rho m_0^2}{\Lambda^2}\right) \Big/ \ln\left(\frac{\rho m_0^2}{\Lambda^2}\right)\right]^{-1-\gamma} . \tag{6}$$

Note that when $Q^2 \to 0$ one has $m^2(0) = m_0^2$. Even though in principle we do not have any theoretical constraint that would put an upper bound to the value of $m_0$, phenomenological estimates place it in the range $m_0 \sim \Lambda - 2\Lambda$ [16, 17]. The other parameters were fixed at $\rho \sim 1-4$, $\gamma = 1/11$ [11, 13, 18]. The non-perturbative generalization of $\alpha_s(Q^2)$ the QCD running coupling, comes in the form

$$a_{\rm NP}(Q^2) = \left[\beta_0 \ln\left(\frac{Q^2 + \rho m^2(Q^2)}{\Lambda^2}\right)\right]^{-1} , \tag{7}$$

where we use the same notation as before and NP stands for Non-Perturbative. Note that its zero gluon mass limit leads to the LO perturbative coupling constant momentum dependence. The $m^2(Q^2)$ in the argument of the logarithm tames the Landau pole, and $a(Q^2)$ freezes at a finite

[5] Even though the gluon is massless at the level of the fundamental QCD Lagrangian, and remains massless to all order in perturbation theory, the non-perturbative QCD dynamics generate an effective, momentum-dependent mass, without affecting the local $SU(3)_c$ invariance, which remains intact.

value in the IR, namely $a^{-1}(0) = \beta_0 \ln(\rho m^2(0)/\Lambda^2)$ [11, 19, 20] as can be seen on the right panel of Fig. 1. As shown in Fig. 2, the coupling constant in the perturbative and non-perturbative approaches are close in size for reasonable values of the parameters from very low $Q^2$ onward ( $Q^2 > 0.1$ GeV$^2$). This result supports the perturbative approach used up to now in model calculations, since it shows, that despite the vicinity of the Landau pole to the hadronic scale, the perturbative expansion is quite convergent and agrees with the non-perturbative results for a wide range of parameters.

Ref. [21] justifies the perturbative evolution approach by comparing it to the non-perturbative momentum dependence as determined by the phenomenon of the freezing of the coupling constant, and to analyze the consequences of introducing an effective gluon mass.

## 4. Non-perturbative QCD and the Hadron Scale

The perturbative and non-perturbative approaches can be inferred from the point of view of the hadronic models. We use, as an example, the original bag model, in its most naive description, consisting of a cavity of perturbative vacuum surrounded by non-perturbative vacuum. The bag model is designed to describe fundamentally static properties, but in QCD all matrix elements must have a scale associated to them as a result of the RGE of the theory. A fundamental step in the development of the use of hadron models for the description of properties at high momentum scales was the assertion that all calculations done in a model should have a RGE scale associated to it [22]. The momentum distribution inside the hadron is only related to the hadronic scale and not to the momentum governing the RG equation. Thus a model calculation only gives a boundary condition for the RG evolution as can be seen for example in the LO evolution equation for the moments of the valence quark distribution

$$\langle q_v(Q^2)\rangle_n = \langle q_v(\mu_0^2)\rangle_n \left(\frac{\alpha_s(Q^2)}{\alpha_s(\mu_0^2)}\right)^{d^n_{NS}}, \tag{8}$$

where $d^n_{NS}$ are the anomalous dimensions of the Non Singlet distributions. Inside, the dynamics described by the model is unaffected by the evolution procedure, and the model provides only the expectation value, $\langle q_v(\mu_0^2)\rangle_n$, which is associated with the hadronic scale. The hadronic scale is related to the maximum wavelength at which the structure begins to be unveiled. This explanation goes over to non-perturbative evolution. The non-perturbative solution of the Dyson–Schwinger equations results in the appearence of an infrared cut-off in the form of a gluon mass which determines the finiteness of the coupling constant in the infrared. The crucial statement is that the gluon mass does not affect the dynamics inside the bag, where perturbative physics is operative and therefore our gluons inside will behave as massless. However, this mass will affect the evolution as we have seen in the case of the coupling constant. The generalization of the coupling constant results to the structure function imply that the LO evolution Eq. (8) simply changes by incorporating the non-perturbative coupling constant evolution Eq. (7).

The non-perturbative results, using the same parameters as before, are quite close to those of the perturbative scheme and therefore we are confident that the latter is a very good approximate description. We note however, that the corresponding hadronic scale, for the sets of parameters chosen, turns out to be slightly smaller than in the perturbative case ($\mu_0^2 \sim 0.1$ GeV$^2$), even for small gluon mass $m_0 \sim 0.3$ GeV and small $\rho \sim 1$. One could reach a pure valence scenario at higher $Q^2$ by forcing the parameters but at the price of generating a singularity in the coupling constant in the infrared associated with the specific logarithmic form of the parametrization. We feel that this strong parametrization dependence and the singularity are non physical since the fineteness of the coupling constant in the infrared is a wishful outcome of the non-perturbative analysis. In this sense, the non-perturbative approach seems to favor a scenario where at the

Journal of Physics: Conference Series **651** (2015) 012007 doi:10.1088/1742-6596/651/1/012007

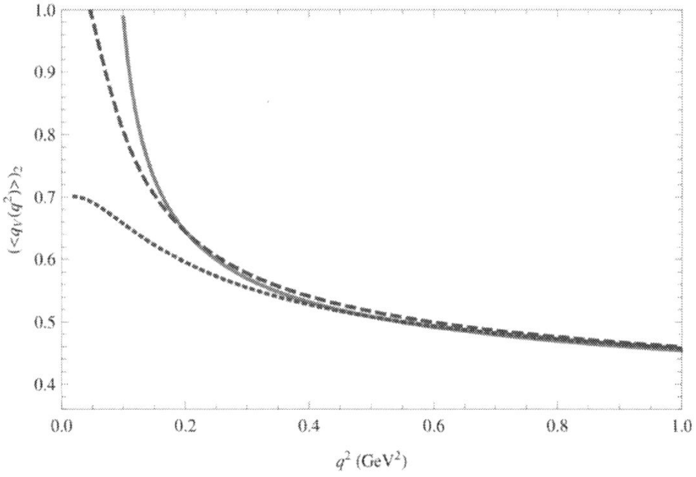


**Figure 2.** Left: The running of the effective coupling. The dotted and dashed curves represent the non-perturbative evolution with the parameters used above. The solid curve shows the NNLO evolution with $\Lambda = 250$ MeV. Right: The evolution of the second moment of the valence quark distribution. The solid curve represents the perturbative LO approximation.

hadronic scale we have not only valence quarks but also gluons and sea quarks [23, 24]. We mean by this statement that to get a scenario with only valence quarks we are forced to very low gluon masses and very small values $\rho$, while a non trivial scenario allows more freedom in the choice of parameters.

## 5. The Hadronic Scale from Perturbative Approaches

Although the perturbative stage of a hard collision is distinct from the non-perturbative regime characterizing the hadron structure, early experimental observations suggest that, in specific kinematical regimes, both *the perturbative and non-perturbative stages arise almost ubiquitously*, in the sense that the non-perturbative description follows the perturbative one. In this Section, we discuss an example of perturbative approach where the transition from perturbative to non-perturbative QCD is clearly identified.

There exists, in DIS processes, a dual description between low-energy and high-energy behavior of a same observable, *i.e.* the unpolarized structure functions. Bloom and Gilman observed a connection between the structure function $\nu W_2(\nu, Q^2)$ in the nucleon resonance region and that in the deep inelastic continuum [25, 26]. The resonance structure function was found to be equivalent to the deep inelastic one, when averaged over the same range in the scaling variable. This concept is known as *parton-hadron duality*: the resonances are not a separate entity but are an intrinsic part of the scaling behavior of $\nu W_2$. The meaning of duality is more intriguing when the equality between resonances and scaling happens at a same scale. It can be understood as a natural continuation of the perturbative to the non-perturbative representation.

Bloom–Gilman duality implies a one-to-one correspondence between the behavior of the structure function, $F_2$, for unpolarized electron proton scattering in the resonance region, and in the perturbative QCD regulated scaling region. In DIS, the relevant kinematical variables are the Bjorken scaling variable, $x = Q^2/2M\nu$ with $M$ being the proton mass and $\nu$ the energy transfer in the lab system, the four-momentum transfer, $Q^2$, and the invariant mass for the proton, $P$, for the virtual photon, $q$, and for the system, $W^2 = (P+q)^2 = Q^2\,(1-x)\,/x + M^2$.

For large values of Bjorken $x \geq 0.5$, and $Q^2$ in the multi-GeV$^2$ region, the cross section is dominated by resonance formation, *i.e.* $W^2 \leq 5$ GeV$^2$. While it is impossible to reconstruct the detailed structure of the proton resonances, these remarkably follow the pQCD predictions when averaged over the resonance region.

To answer the question of the nature of a dual description, an option is to focus on purely perturbative analysis from perturbative QCD evolution. Although Bloom–Gilman duality has been known for years, quantitative analyses could be attempted only more recently, having at disposal the extensive, high precision data from Jefferson Lab [27, 28]. Perturbative QCD-based studies [29–31], have been presented that include higher-twist contributions or, more generally, the evidence for non-perturbative inserts, which are required to achieve a fully quantitative fit of PDFs, especially at large-$x$. In Ref. [32], we discuss the Bloom–Gilman duality from a purely pertubative point of view, by analysing the scaling behavior of the resonances at the same low-$Q^2$, high-$x$ values as the $F_2$ data from JLab. Our study leads to an analysis of the role of the running coupling constant in the infrared region in tuning the experimental data.

A quantitative definition of *global duality* is accomplished by comparing limited intervals defined according to the experimental data. Hence, we analyze the scaling results as a theoretical counterpart, or an output of perturbative QCD, in the same kinematical intervals and at the same scale $Q^2$ as the data for $F_2$. It is easily realized that the ratio,

$$R^{\rm exp/th}(Q^2) \equiv \frac{\int_{x_{\rm min}}^{x_{\rm max}} dx\, F_2^{\rm exp}(x,Q^2)}{\int_{x_{\rm min}}^{x_{\rm max}} dx\, F_2^{\rm th}(x,Q^2)} = 1 \quad , \tag{9}$$

if duality is fulfilled.[6]

Duality is violated (the ratio (9) is not 1) when considering the fully perturbative expression, and is still violated after corrections by the target mass terms. One possible explanation for the apparent violation of duality is the lack of accuracy in the Parton Distribution Functions (PDF) parametrizations at large-$x$.[7] Therefore, the behavior of the nucleon structure functions in the resonance region needs to be addressed in detail in order to be able to discuss theoretical predictions in the limit $x \to 1$. In such a limit, terms containing powers of $\ln(1-z)$, $z$ being the longitudinal variable in the evolution equations, that are present in the Wilson coefficient functions $B^q_{\rm NS}(z)$ become large and have to be resummed, *i.e.* Large-$x$ Resummation (LxR). Resummation was first introduced by linking this issue to the definition of the correct kinematical variable that determines the phase space for real gluon emission at large $x$. This was found to be $\widetilde{W}^2 = Q^2(1-z)/z$, instead of $Q^2$ [36]. As a result, the argument of the strong coupling constant becomes $z$-dependent [37],

$$\alpha_s(Q^2) \to \alpha_s\left(Q^2 \frac{(1-z)}{z}\right) \quad . \tag{10}$$

In this procedure, however, an ambiguity is introduced, related to the need of continuing the value of $\alpha_s$ for low values of its argument, *i.e.* for $z \to 1$. In Ref. [32], we have reinterpretated $\alpha_s$ for values of the scale in the infrared region. To do so, we investigated the effect induced by changing the argument of $\alpha_s$ on the behavior of the $\ln(1-z)$-terms in the convolution with the coefficient function $B_{\rm NS}$:

$$F_2^{NS}(x,Q^2) = xq(x,Q^2) + \frac{\alpha_s}{4\pi}\sum_q \int_x^1 dz\, B^q_{\rm NS}(z)\, \frac{x}{z}\, q\left(\frac{x}{z},Q^2\right) \quad , \tag{11}$$

[6] In the analysis of Ref. [32], we use, for $F_2^{\rm exp}$, the data from JLab (Hall C, E94110) [28] reanalyzed (binning in $Q^2$ and $x$) as explained in [33] as well as the SLAC data [34].

[7] In our analysis, we use the MSTW08 set at NLO as initial parametrization [35]. We have checked that there were no significant discrepancies when using other sets.

We resum those terms as

$$\ln(1-z) \;=\; \frac{1}{\alpha_{s,\mathrm{LO}}(Q^2)}\int^{Q^2} d\ln Q^2\,\left[\alpha_{s,\mathrm{LO}}(Q^2(1-z))-\alpha_{s,\mathrm{LO}}(Q^2)\right]\equiv \ln_{\mathrm{LxR}}\quad, \tag{12}$$

including the complete $z$ dependence of $\alpha_{s,\mathrm{LO}}(\tilde{W}^2)$ to all logarithms. Using the 'resummed' $F_2^{\mathrm{theo}}$ in Eq. (9), the ratio $R$ decreases substantially, even reaching values lower than 1. It is a consequence of the change of the argument of the running coupling constant. At fixed $Q^2$, under integration over $x < z < 1$, the scale $Q^2 \times (1-z)/z$ is shifted and can reach low values, where the running of the coupling constant starts blowing up. At that stage, our analysis requires non-perturbative information.

In the light of quark-hadron duality, it is necessary to prevent the evolution from enhancing the scaling contribution over the resonances. We define the limit from which non-perturbative effects have to be accounted for by setting a maximum value for the longitudinal momentum fraction, $z_{max}$. Two distinct regions can be studied: the "running" behavior in $x < z < z_{max}$ and the "steady" behavior $z_{max} < z < 1$. Our definition of the maximum value for the argument of the running coupling follows from the realization of duality in the resonance region. The value $z_{max}$ is reached at

$$R^{\mathrm{exp/th}}(z_{\mathrm{max}},Q^2) \;=\; \frac{\displaystyle\int_{x_{\mathrm{min}}}^{x_{\mathrm{max}}} dx\, F_2^{\mathrm{exp}}(x,Q^2)}{\displaystyle\int_{x_{\mathrm{min}}}^{x_{\mathrm{max}}} dx\, F_2^{NS,\mathrm{Resum}}(x,z_{\mathrm{max}},Q^2)} = \frac{I^{\mathrm{exp}}}{I^{\mathrm{Resum}}} = 1\quad . \tag{13}$$

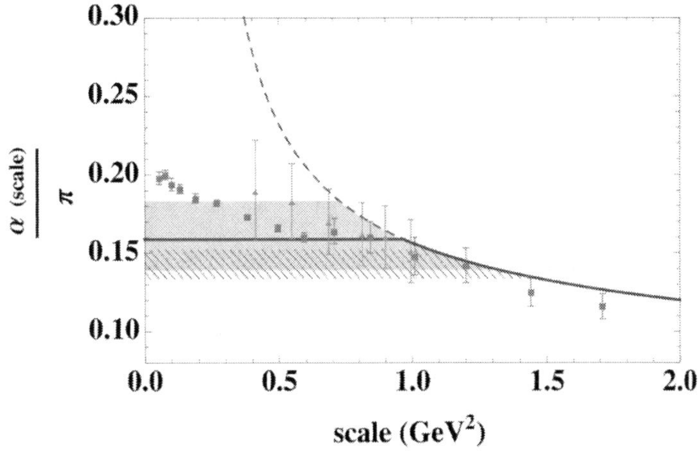


**Figure 3.** Extraction of $\alpha_s$. See text.

The direct consequence of Eq. (13) is that duality is realized, within our assumptions, by allowing $\alpha_s$ to run from a minimal scale only. From that minimal scale downward, the coupling constant does not run, it is frozen. This feature is illustrated on Fig. 3. We show the behavior of $\alpha_{s,\mathrm{NLO}}(\mathrm{scale})$ in the $\overline{\mathrm{MS}}$ scheme and for the same value of $\Lambda^{\mathrm{NLO}}_{\overline{\mathrm{MS}},\mathrm{MSTW}} = 0.402$ GeV used throughout our analysis. The theoretical errorband correspond to the extreme values of

$$\alpha_{s,\mathrm{NLO}}\left(Q_i^2\frac{(1-z_{max,i})}{z_{max,i}}\right)\quad , \tag{14}$$

Journal of Physics: Conference Series **651** (2015) 012007 doi:10.1088/1742-6596/651/1/012007

$i$ corresponds to the data points. The determination of the transition scale $Q_0^2$ is probably the main result of our analysis. Of course, we expect the transition from non-perturbative to perturbative to occur at one unique scale. The discrepancy between the 10 values we have obtained has to be understood as the resulting error propagation. The grey area represents the approximate frozen value of the coupling constant,

$$0.13 \leq \frac{\alpha_{s,\mathrm{NLO}}(\mathrm{scale} \to 0\mathrm{GeV}^2)}{\pi} \leq 0.18 \quad . \tag{15}$$

The solid blue curve represents the (mean value of the) coupling constant obtained from our analysis using inclusive electron scattering data at large $x$. The blue dashed curve represents the exact NLO solution for the running coupling constant in $\overline{\mathrm{MS}}$ scheme. The grey area represents the region where the freezing occurs for JLab data, while the hatched area corresponds the freezing region determined from SLAC data. This error band represents the theoretical uncertainty in our analysis.

In the figure we also report values from the extraction using polarized $eP$ scattering data in Ref. [38–40]. These values represent the first extraction of an effective coupling in the IR region that was obtained by analyzing the data relevant for the study of the GDH sum rule. To extract the coupling constant, the $\overline{\mathrm{MS}}$ expression of the Bjorken sum rule up to the 5th order in alpha (calculated in the $\overline{\mathrm{MS}}$ scheme) was used. The red squares correspond to $\alpha_s$ extracted from Hall B CLAS EG1b, with statistical uncertainties; the orange triangles corresponds to Hall A E94010 / CLAS EG1a data, the uncertainty here contains both statistics and systematics. The agreement with our analysis, which is totally independent, is impressive. We notice, and it is probably one of the most important result of our analysis, that the transition from perturbative to non-perturbative QCD seems to occur around 1 GeV$^2$.

At that stage, a comparison with fully non-perturbative effective charges and "modified pQCD" is noteworthy. It is shown in Fig. 4. The grey areas are as in Fig. 3 with $\Lambda^{\mathrm{NLO}}_{\overline{\mathrm{MS}},\mathrm{MSTW}} =$ 0.402 GeV ; the dashed blue curve is the exact NLO solution with the same $\Lambda$. The dotted-dashed orange curve corresponds to the result of Ref. [12], using the version ($b$) of their fit with $a = b = 1$. The latter analysis was performed in the MOM renormalization scheme. Though the $\beta$ function does not depend on the scheme up to 2 loops, the definition of $\Lambda$ varies from scheme to scheme. The comparison of the results is made possible using the relation, [41]

$$\Lambda_{\overline{\mathrm{MS}}} = \frac{\Lambda_{\mathrm{MOM}}}{3.334} \quad , \tag{16}$$

leading to the value of $\Lambda^{\mathrm{Ref.\ [12]}}_{\overline{\mathrm{MS}}} = (0.71/3.334)$ GeV$\sim$ 0.21GeV. The value of $\alpha(0)$ is fixed to $8.915/N_c$. The red curves are variations of the effective charge of Ref. [11], in Eq. (7) with a logarithmic running for the gluon mass described by Eq. (6) where $(m_0^2, \rho, \Lambda)$ are parameters to be fixed. The solid red curve corresponds to the set $(m_0^2 = 0.3\mathrm{GeV}^2, \rho = 1.7, \Lambda = 0.25\mathrm{GeV})$, the dashed red curve to $(m_0^2 = 0.5\mathrm{GeV}^2, \rho = 2., \Lambda = 0.25\mathrm{GeV})$. This result is also obtained in the MOM scheme, the value of $\Lambda$ turns out to be similar in both Fischer *et al.* and Cornwall's approaches. The cyan curves correspond to two scenarios of the effective charges of Ref. [13]. Their numerical solution is fitted by a functional form similar to Eq. (7). The 2 sets of parameters, corresponding to $m_0 = 500$MeV (dashed-dotted curve) and 600 MeV (medium dashed curve), are then driven by the shape of the numerical solution. They are plotted here with the same $\Lambda^{n_f=0}_{\mathrm{MOM}} = 300$ MeV as in the publication, but for $n_f = 3$ for sake of comparison. Further investigation on comparison of schemes is needed. The short dashed green curves corresponds to Shirkov's analytic perturbative QCD to LO [14] with $\Lambda^{\mathrm{LO}}_{\overline{\mathrm{MS}},\mathrm{MSTW}}$. The value of $\alpha(0)$ is fixed to $4\pi/\beta_0$. Finally, the pink curve is the freezing value of Ref. [15].

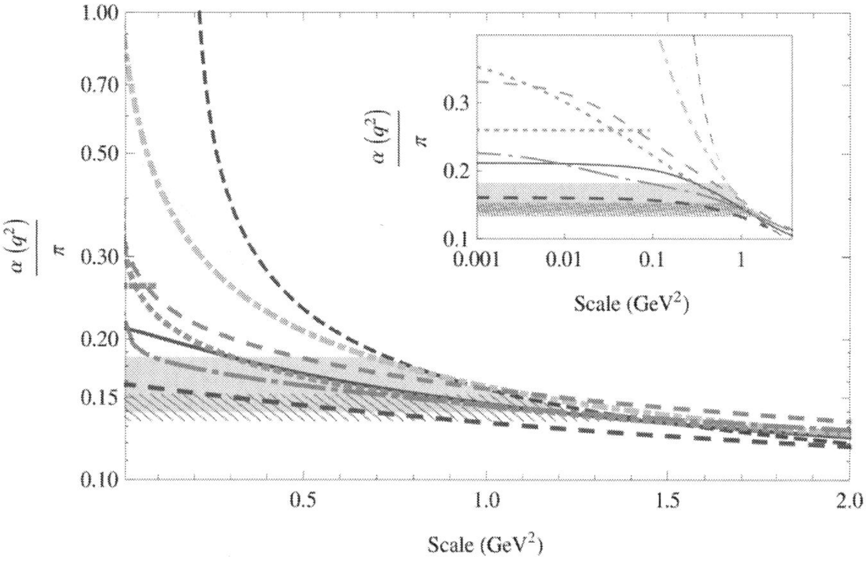


**Figure 4.** Comparison of the effective coupling constant from Ref. [42]. See text.

Notice that the extracted value for $\alpha_s(Q^2 < 1\text{GeV}^2)$ is only constrained by the integral in the resummed version of Eq. (11): no conclusion can be drawn on its value at $Q^2 = 0\text{GeV}^2$. While it is not possible to conclude on the value of $\alpha_s(0)$, we notice that it is possible to find sets of parameters for which the transition from perturbative to non-perturbative QCD occurs around 1 GeV$^2$.

## 6. PDFS at Large-$x$

As discussed in the previous Section, the apparent violation of duality could be explained by the lack of accuracy in the Parton Distribution Functions (PDF) parametrizations at large-$x$. The remedy is obviously a better understanding and description of the large-$x$ PDFs. The large-$x$ resummation proposed in Ref. [32] could be implemented at the PDF level, such that, in a global fit procedure, the large-$x$ region would account for non-perturbative effects, leading to a "cleaner" functional form for $f_1(x)$. An improved treatment of nuclear effects, especially in the resonance region, has already been carried out by the CJ (CTEQ-JLab) collaboration [43]. Independent and complementary advancements are needed as uncertainties at large-$x$ highly matter for, *e.g.*, search of New Physics' particles.

The impact of PDF uncertainties at large-$x$ on heavy boson production has been studied in Ref. [44], using the CJ PDF set[8]. Hadron-hadron collisions involve at least two interacting partons, with momentum fractions $x_1$ and $x_2$, respectively. At fixed center of mass energy $\sqrt{s}$ and boson rapidity $2y = \ln\left(\frac{E+p_z}{E-p_z}\right)$ where $E$ and $p_z$ are the boson energy and longitudinal momentum in the hadron center of mass frame, the parton momentum fractions are given (at leading order in the strong coupling constant) by

$$x_{1,2} = \frac{M}{\sqrt{s}} \exp(\pm y) \quad , \tag{17}$$

[8] The new version is cited but the 2011 version was used.

Journal of Physics: Conference Series **651** (2015) 012007 doi:10.1088/1742-6596/651/1/012007

where $M$ is the mass of the produced boson. At low rapidities the cross sections are relatively insensitive to uncertainties in the large-$x$ PDFs. At larger rapidities, however, there is far greater sensitivity to the large-$x$ behavior, leading to $\approx 15\%$ uncertainty in the differential cross section for $y_Z = 4$ at the LHC, and for $y_Z = 2.8$ at the Tevatron, which correspond to parton fractions of $x \approx 0.7$. As for the elusive $W', Z'$ bosons, it is clear now that the increasing of their mass $M$ will directly increase the relevant PDF's $x$ values, so that higher mass bosons will more readily sample the high-$x$ region where the nuclear uncertainties are more prominent. The cross sections will also decrease rapidly with increasing boson mass, so that the effects of the large-$x$ PDF uncertainties will become more significant as the mass increases. The effect on the exclusion limits from, *e.g.*, ATLAS [45] is important.

## 7. Tensor and Scalar Charges

As we have already stated in these proceedings, the hadronic structure carries important information for high energy observables. Hadronic observables are often related to the manifestation of fundamental processes at the quark level. We here discuss the structural scalar and tensor currents.

### *7.1. Relation to New Physics*

Non-standard electroweak couplings, studied in the decay of ultracold neutrons [46], are proposed that are related to hadronic matrix elements. Beyond the well-known weak interactions of the Standard Model, new physics'coupling could be probed in neutron $\beta$-decay. The latter are related to the isovector scalar and axial-vector hadronic matrix elements [47],

$$\langle p(P_p)|\, \bar{u}d \,|n(P_n)\rangle = g_S(\Delta^2)\, \overline{u}_p(P_p) u_n(P_n) \quad , \tag{18a}$$

$$\langle p(P_p)|\, \bar{u}\sigma_{\mu\nu}d \,|n(P_n)\rangle = g_T(\Delta^2)\, \overline{u}_p(P_p)\sigma_{\mu\nu} u_n(P_n) + \ldots \quad , \tag{18b}$$

with $\Delta = P_n - P_p$ and the ellipsis refer to higher order terms. The —isoscalar and isovector— scalar charges and the —isovector— tensor charge correspond to the form factors for $q^2 = 0$, *i.e.*

$$\langle 1 \rangle_{\sigma u - \sigma d} = g_S(0) \quad , \tag{19a}$$

$$\langle 1 \rangle_{\sigma u + \sigma d} = \sigma(0) = \sigma_{\pi N} \quad , \tag{19b}$$

$$\langle 1 \rangle_{\delta u - \delta d} = g_T(0) \quad . \tag{19c}$$

Furthermore an analysis of the uncertainties in the spin-independent and spin-dependent elastic scattering cross sections of supersymmetric dark matter particles on protons and neutrons [48] concludes that the largest single uncertainty comes from the spin-independent scattering matrix element $\langle N|\, \bar{q}q \,|N\rangle$ linked to the $\sigma_{\pi N}$ term.

For each quark flavor, the scalar and axial charges are related to the following forward matrix elements, respectively,

$$\begin{aligned} \langle 1 \rangle_{\sigma u} &\equiv \int_0^1 dx \left[ e^q(x,Q^2) + e^{\bar{q}}(x,Q^2) \right] \quad , \\ \Rightarrow M\, e^q(x) &= \int \frac{d\xi^-}{4\pi} e^{ixP^+\xi^-} \left\langle PS \right| \bar{q}(0)\, q(\xi) \left| PS \right\rangle \Big|_{\xi^+=\vec{\xi}=0} \quad ; \end{aligned} \tag{20}$$

$$\begin{aligned} \langle 1 \rangle_{\delta q}(Q^2) &\equiv \delta q(Q^2) = \int_0^1 dx \left[ h_1^q(x,Q^2) - h_1^{\bar{q}}(x,Q^2) \right] \quad , \\ \Rightarrow h_1^q(x) &= \int \frac{d\xi^-}{4\pi} e^{ixP^+\xi^-} \left\langle PS_\perp \right| \bar{q}(0)\, i\sigma^{\perp +}\gamma_5 q(\xi) \left| PS_\perp \right\rangle \Big|_{\xi^+=\vec{\xi}=0} \quad . \end{aligned} \tag{21}$$

Notice that, contrary to the axial charge, there is no proper sum rule associated to the tensor "charge" (the name in itself is inaccurate). There are non-vanishing anomalous dimension associated to it and the tensor charge therefore evolves with the hard scale $Q^2$ [49]. It has been calculated on the lattice [46, 50, 51] and in various models [52–56], and it turns out not to be small. On the other hand, the $\sigma_{\pi N}$ is renormalization point invariant. The sum rule is however related to a singularity and is not of practical use. While the axial charge is a charge-even operator, from Eq. (21), it is evident that the tensor charge is odd under charge conjugation and, therefore, it does not receive contributions from $q\bar{q}$ pairs in the sea and is dominated by valence contributions.

### *7.2. Determination of Tensor and Scalar Charges through PDFs*

The main experimental access to the tensor and scalar [9] charges is provided thanks to the sum rules Eqs. (20, 21). The knowledge on the PDFs $h_1(x)$ and $e(x)$, up to theoretical limitations, will bring some light on the values of $g_{S/T}$.

Let's start with the transversity PDF. A comprehensive review of the properties of the transversity distribution function can be found in Ref. [57]. Transversity $h_1$, as leading-twist collinear PDF, enjoys the same status as $q$ and $g_1$ [49, 58]. The distribution of transversely polarized quarks $q^\uparrow$ in a transversely polarized nucleon $p^\uparrow$ (integrated over transverse momentum) can be written as

$$f_{q^\uparrow/p^\uparrow}(x) = q(x) + \mathbf{S}\cdot\mathbf{S}_q\, h_1^q(x) \quad , \tag{22}$$

in which $\mathbf{S}$ is the nucleon spin and $\mathbf{S_q}$ the quark spin. Therefore, transversity can be interpreted as the difference between the probability[10] of finding a parton (with flavor $q$ and momentum fraction $x$) with transverse spin parallel and anti-parallel to that of the transversely polarized nucleon. The only first principle based property on the transversity distribution is the Soffer inequality. Because a probability must be positive, we get the important Soffer inequality [59],

$$2|h_1^q(x,Q^2)| \le q(x,Q^2) + g_1^q(x,Q^2)\,, \tag{23}$$

which is true at all $Q^2$ [60, 61]. An analogous relation holds for antiquark distributions.

In spin-$\frac{1}{2}$ hadrons there is no gluonic function analogous to transversity. The most important consequence is that $h_1^q$ for a quark with flavor $q$ does not mix with gluons in its evolution and it behaves as a non-singlet quantity; this has been verified up to NLO, where chiral-odd evolution kernels have been studied so far [61–63].

[9] The $\sigma$-term can be accessed through pion-pion scattering for values of $\Delta^2 \neq 0$.
[10] The probabilistic interpretation is valid in the light-cone gauge.

There are two complementary extractions of the transversity distributions from semi-inclusive processes: the TMD parametrization (also known as Torino fit) [64, 65] and the collinear extraction (also known as Pavia fit) [66, 67]. The former is based on the TMD framework in which the chiral-odd partner of $h_1(x, k_\perp)$ is the Collins fragmentation function ; the latter is based on a collinear framework, involving the chiral-odd dihadron fragmentation function, *i.e.* $H_1^{\sphericalangle}$. Parameterizations for the dihadron FFs have been obtained independently [68]. One of the main differences lie in that the collinear extraction does not require the use of a fitting functional form: it is a point-by-point extraction. However, for practical reasons, a statistical study of the transversity PDF has been performed as well. So far, both approaches has found compatible results in the range in $x$ where data exist. However, recent progress on TMD evolution are expected to affect the Torino fit. It is important to notice that the parameterizations are biased by the choice of the fitting functional form. The behavior of the best-fit parametrization is largely unconstrained outside the range of data, leading to confusing results at low and large-$x$ values. This is nicely illustrated by the collinear —Pavia— transversity collaboration, on Fig. 5, where 2 different functional forms, with a equally good $\chi^2/d.o.f.$, have been used.

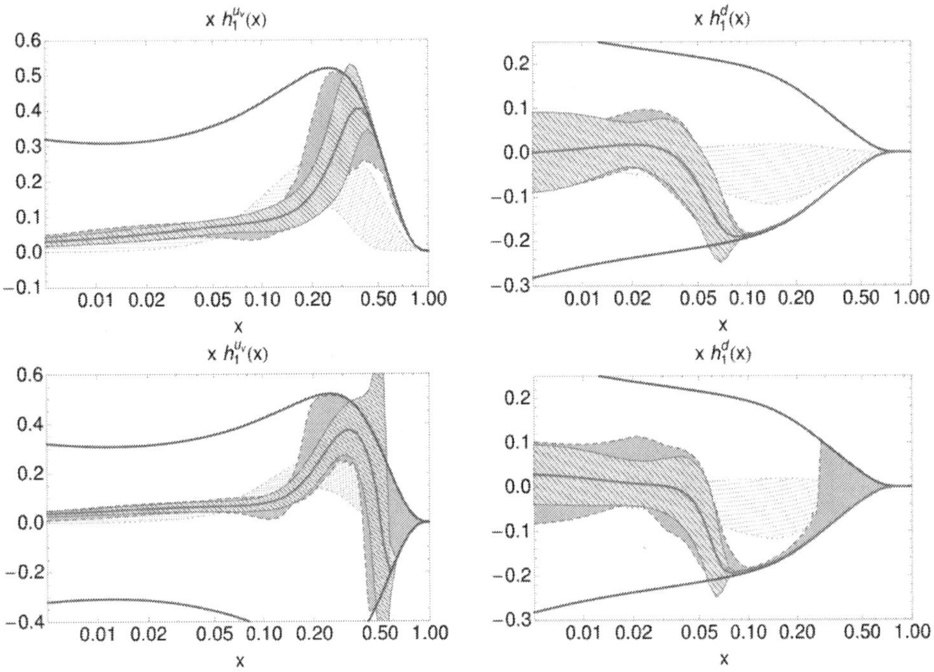


**Figure 5.** Fit of valence collinear transversities at $Q^2 = 2.4\text{GeV}^2$. For $u$ and $d$ distributions, respectively, left and right columns. The two upper plots correspond to the fits using a flexible parametrization, while the two lower plots correspond to a fit using an extra-flexible parameterization, see Ref. [67]. The red bands represent the standard fit within $1\sigma$ errors, while the green bands stand for a $1\sigma$ Monte Carlo based analysis ($n \times 64\%$ fit of $n$ data replica at $1\sigma$). The light blue bands correspond to the Torino13 [65] and the full blue curves to the Soffer bound.

Another independent access to the tensor charge has recently been published using exclusive processes [69]. Through yet another "sum rule", the tensor charge corresponds to the first Mellin moment of the chiral-odd $H_T$ generalized parton distribution.

In Fig. 6, we show some estimates of the tensor charge for up and down quarks separately. In

Fig. 7, we illustrate the strong dependence on the functional form. More data, especially in the low and large $x$ regions, are needed to constrain the fits. Hopefully, more data from PHENIX are soon to be released. Worth to mention too are the proposals at JLab —both for CLAS12 and SoLID [70, 71].

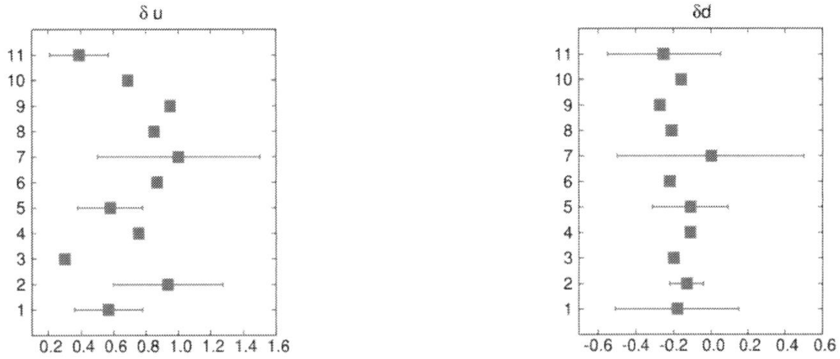


**Figure 6.** The tensor charge $\delta q$ for $u$ (left) and $d$ (right) quarks, computed using the transversity distributions from TMD collaboration [65] (standard) **(11)**. The gray areas correspond to the statistical uncertainty bands in that extraction. The results of the model calculations in **(10)** the quark-diquark model [52], **(9)** the chiral quark-soliton model [53], **(8)** lattice QCD [72], **(7)** sum rules [54], **(6)** a constituent quark model [55], **(5)** spin-flavour SU(6) symmetry [56] and **(4)** Dyson-Schwinger with dressed constituents [73]. A recent calculation in Dyson-Schwinger including axial-vector diquarks with contact interaction **(3)** [74]. **(2)** comes from the chiral-odd GPD $H_T$ sum rule, with the GPD fits of Ref. [69]. The number **(1)** corresponds to the standard rigid version of the fit via DiFF [67], see Fig. 7 for comparison of the 2 fits' results.

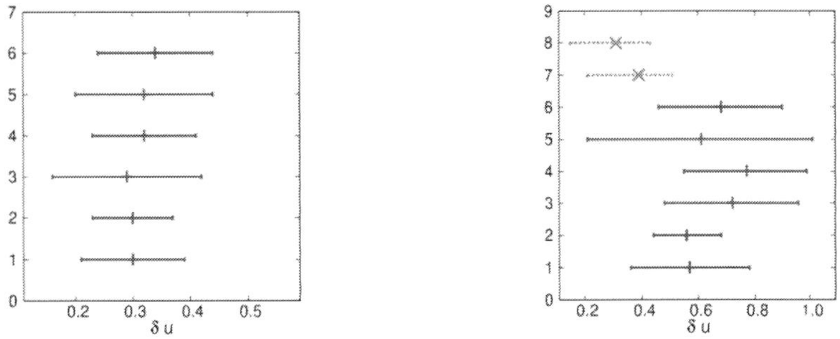


**Figure 7.** *Left panel*: tensor charge integrated over data range for the u-quark. *Right panel*: tensor charge integrated on the theoretical support $x \in [0, 1]$. Respectively for, from 1 to 8, standard rigid, Monte Carlo rigid, standard flexible, Monte Carlo flexible, standard extra-flexible, Monte Carlo extra-flexible of the collinear fit [67] and the fit for $A_0$ and $A_{12}$ asymmetries at Belle combined with single-hadron SIDIS of the TMD fit collaboration [65].

As for the scalar charges, they are, though in a more elusive way, related to the twist-3 PDF $e(x)$. See Ref. [75] for a review of the chiral-odd twist-3 PDF. QCD equations of motion allow to decompose the chiral-odd twist-3 distributions into 3 terms

$$e^q(x) = e^q_{\rm loc}(x) + e^q_{\rm tw\text{-}3}(x) + e^q_{\rm mass}(x) \quad . \tag{24}$$

The first term comes from the local operator and is related to the pion-nucleon sigma term,

$$\begin{aligned} e^q_{\rm loc}(x) &= \frac{1}{2M}\int \frac{d\lambda}{2\pi} e^{i\lambda x}\langle P|\bar{\psi}_q(0)\psi_q(0)|P\rangle \quad , \\ &= \frac{\delta(x)}{2M}\langle P|\bar{\psi}_q(0)\psi_q(0)|P\rangle \quad ; \end{aligned} \tag{25}$$

the second term is a genuine twist-3 contribution, *i.e.* pure quark-gluon interaction term ; while the last term is related to the current quark mass.

Owing to Eq. (19b), $\sigma_{\pi N}$ is related to $e(x=0)$, due to the delta-function singularity. Furthermore, being a subleading contribution, this twist-3 PDF is hardly known. It is however accessible in single- [76] and two-hadron semi-inclusive DIS off unpolarized targets, through the Beam Spin Asymmetry $A_{LU}$ at CLAS.[11] Just like in the case of transversity extraction, the chiral-odd partner of $e(x)$ is the chiral-odd dihadron fragmentation function, *i.e.* $H_1^{\sphericalangle}$. The analysis of the soon-to-be-released data is ongoing. [77].

The bounds on $g_S, g_T$ constrain the new effective couplings $\tilde{C}_S, \tilde{C}_T$ through the matching conditions from our quark-level effective theory —with coupling $\epsilon_S, \epsilon_T$— to a nucleon-level effective theory [46]

$$\begin{aligned} \tilde{C}_S &= g_S\epsilon_S \quad , \\ \tilde{C}_T &= 4g_T\epsilon_T \quad , \end{aligned}$$

The bounds resulting from the phenomenological extractions of the tensor charge [65, 67, 69] on the observability of new physics are still huge compared to that the lattice calculations can nowadays achieve. An analysis of the precise projection of these bounds, together with the expected precision of future hadronic structure dedicated experiments is ongoing [78].

## 8. Conclusions

Hadronic physics is the perfect framework to study the intersection between perturbative and non-perturbative QCD. The properties of hadrons reflect the features of the strong interactions and transition from chiral symmetry to confinement.

Non-perturbative QCD provides inputs for perturbative calculation, mainly through *initial conditions* for the QCD evolution equations. As described in these proceedings, Parton Distribution Functions in QCD have a scale dependence, through the RGE. Non-perturbative predictions often come from evaluation in models for the proton structure. This will, in turn, strongly affect the PDF fits: it is called *procedural bias* in Ref. [6] and is a problem known in the hadronic community for years, *e.g.* [8].

Besides the uncertainty related to the *hadronic scale* and the usual statistical errors, we have mentionned the poor knowledge on PDFs at large values of Bjorken-$x$. The uncertainty on the region $x \to 1$ [44] will affect the exclusion limit for heavy boson like $Z', W'$.

Finally we have mentionned hadronic matrix elements related to New Physics observables.

The understanding of parton distributions at low energy have important repercussions for the high energy physics. This is why this plethora of distributions and new information about the hadron structure require to be handled carefully via complementary theoretical approaches. Therefore, in that sense, QCD must be considered as a whole, from the infrared to the ultraviolet region.

[11] The analysis is finalized and is submitted to the CLAS collaboration internal committee.

**Acknowledgments**
I am grateful to Ted C. Rogers for priceless discussions about the impact of non-perturbative physics on perturbative QCD. I also thank Simonetta Liuti and Vicente Vento, for their advices and support ; Martín González Alonso for the explanations about his recent works ; Alexei Prokudin and Stefano Melis for sharing the Torino13 transversity. This work was funded by the Belgian Fund F.R.S.-FNRS via the contract of Chargée de recherches.

## References

[1] Moch S, Vermaseren J and Vogt A 2004 *Nucl.Phys.* **B688** 101–134 (*Preprint* hep-ph/0403192)
[2] Vogt A, Moch S and Vermaseren J 2004 *Nucl.Phys.* **B691** 129–181 (*Preprint* hep-ph/0404111)
[3] Vermaseren J, Vogt A and Moch S 2005 *Nucl.Phys.* **B724** 3–182 (*Preprint* hep-ph/0504242)
[4] Forte S and Watt G 2013 *Ann.Rev.Nucl.Part.Sci.* **63** 291–328 (*Preprint* 1301.6754)
[5] The durham hepdata project http://hepdata.cedar.ac.uk/pdfs/ .
[6] Jimenez-Delgado P 2012 *Phys.Lett.* **B714** 301–305 (*Preprint* 1206.4262)
[7] Gluck M, Jimenez-Delgado P and Reya E 2008 *Eur.Phys.J.* **C53** 355–366 (*Preprint* 0709.0614)
[8] Traini M, Mair A, Zambarda A and Vento V 1997 *Nucl.Phys.* **A614** 472–500
[9] Stratmann M 1993 *Z. Phys.* **C60** 763–772
[10] Parisi G and Petronzio R 1976 *Phys.Lett.* **B62** 331
[11] Cornwall J M 1982 *Phys.Rev.* **D26** 1453
[12] Fischer C S and Alkofer R 2003 *Phys.Rev.* **D67** 094020 (*Preprint* hep-ph/0301094)
[13] Aguilar A, Binosi D, Papavassiliou J and Rodriguez-Quintero J 2009 *Phys.Rev.* **D80** 085018 (*Preprint* 0906.2633)
[14] Shirkov D and Solovtsov I 1997 *Phys.Rev.Lett.* **79** 1209–1212 (*Preprint* hep-ph/9704333)
[15] Mattingly A and Stevenson P M 1992 *Phys.Rev.Lett.* **69** 1320–1323 (*Preprint* hep-ph/9207228)
[16] Bernard C W 1982 *Phys.Lett.* **B108** 431
[17] Parisi G and Petronzio R 1980 *Phys.Lett.* **B94** 51
[18] Aguilar A C and Papavassiliou J 2008 *Eur.Phys.J.* **A35** 189–205 (*Preprint* 0708.4320)
[19] Aguilar A C and Papavassiliou J 2006 *JHEP* **0612** 012 (*Preprint* hep-ph/0610040)
[20] Binosi D and Papavassiliou J 2009 *Phys.Rept.* **479** 1–152 (*Preprint* 0909.2536)
[21] Courtoy A, Scopetta S and Vento V 2011 *Eur.Phys.J.* **A47** 49 (*Preprint* 1102.1599)
[22] Jaffe R and Ross G G 1980 *Phys.Lett.* **B93** 313
[23] Scopetta S, Vento V and Traini M 1998 *Phys.Lett.* **B421** 64–70 (*Preprint* hep-ph/9708262)
[24] Scopetta S, Vento V and Traini M 1998 *Phys.Lett.* **B442** 28–37 (*Preprint* hep-ph/9804302)
[25] Bloom E D and Gilman F J 1970 *Phys.Rev.Lett.* **25** 1140
[26] Bloom E D and Gilman F J 1971 *Phys.Rev.* **D4** 2901
[27] Melnitchouk W, Ent R and Keppel C 2005 *Phys.Rept.* **406** 127–301 (*Preprint* hep-ph/0501217)
[28] Liang Y *et al.* (Jefferson Lab Hall C E94-110 Collaboration) 2004 (*Preprint* nucl-ex/0410027)
[29] Liuti S 2011 *Int.J.Mod.Phys.Conf.Ser.* **04** 190–199 (*Preprint* 1108.3266)
[30] Bianchi N, Fantoni A and Liuti S 2004 *Phys.Rev.* **D69** 014505 (*Preprint* hep-ph/0308057)
[31] Liuti S, Ent R, Keppel C and Niculescu I 2002 *Phys.Rev.Lett.* **89** 162001 (*Preprint* hep-ph/0111063)
[32] Courtoy A and Liuti S 2013 *Phys.Lett.* **B726** 320–325 (*Preprint* 1302.4439)
[33] Monaghan P, Accardi A, Christy M, Keppel C, Melnitchouk W *et al.* 2012 (*Preprint* 1209.4542)
[34] Whitlow L, Riordan E, Dasu S, Rock S and Bodek A 1992 *Phys.Lett.* **B282** 475–482
[35] Martin A, Stirling W, Thorne R and Watt G 2009 *Eur.Phys.J.* **C63** 189–285 (*Preprint* 0901.0002)
[36] Amati D, Bassetto A, Ciafaloni M, Marchesini G and Veneziano G 1980 *Nucl.Phys.* **B173** 429
[37] Roberts R 1999 *Eur.Phys.J.* **C10** 697–702 (*Preprint* hep-ph/9904317)
[38] Deur A, Burkert V, Chen J P and Korsch W 2007 *Phys.Lett.* **B650** 244–248 (*Preprint* hep-ph/0509113)
[39] Deur A, Burkert V, Chen J and Korsch W 2008 *Phys.Lett.* **B665** 349–351 (*Preprint* 0803.4119)
[40] Deur A 2013 *Private Communication*
[41] Boucaud P, Leroy J, Micheli J, Pene O and Roiesnel C 1998 *JHEP* **9810** 017 (*Preprint* hep-ph/9810322)
[42] Courtoy A 2013 (*Preprint* 1311.7017)
[43] Owens J, Accardi A and Melnitchouk W 2013 *Phys.Rev.* **D87** 094012 (*Preprint* 1212.1702)
[44] Brady L, Accardi A, Melnitchouk W and Owens J 2012 *JHEP* **1206** 019 (*Preprint* 1110.5398)
[45] 2012 Search for high-mass dilepton resonances in 6.1/fb of pp collisions at sqrt(s)= 8 TeV with the ATLAS experiment Tech. Rep. ATLAS-CONF-2012-129 CERN Geneva
[46] Bhattacharya T, Cirigliano V, Cohen S D, Filipuzzi A, Gonzalez-Alonso M *et al.* 2012 *Phys.Rev.* **D85** 054512 (*Preprint* 1110.6448)
[47] Weinberg S 1958 *Phys.Rev.* **112** 1375–1379

[48] Ellis J R, Olive K A and Savage C 2008 *Phys.Rev.* **D77** 065026 (*Preprint* 0801.3656)
[49] Jaffe R L and Ji X D 1992 *Nucl. Phys.* **B375** 527–560
[50] Gockeler M *et al.* (QCDSF) 2007 *Phys. Rev. Lett.* **98** 222001 (*Preprint* hep-lat/0612032)
[51] Green J, Engelhardt M, Krieg S, Negele J, Pochinsky A *et al.* 2012 (*Preprint* 1209.1687)
[52] Cloet I C, Bentz W and Thomas A W 2008 *Phys. Lett.* **B659** 214–220 (*Preprint* 0708.3246)
[53] Wakamatsu M 2007 *Phys. Lett.* **B653** 398–403 (*Preprint* 0705.2917)
[54] He H and Ji X D 1995 *Phys. Rev.* **D52** 2960–2963 (*Preprint* hep-ph/9412235)
[55] Pasquini B, Pincetti M and Boffi S 2007 *Phys. Rev.* **D76** 034020 (*Preprint* hep-ph/0612094)
[56] Gamberg L P and Goldstein G R 2001 *Phys. Rev. Lett.* **87** 242001 (*Preprint* hep-ph/0107176)
[57] Barone V, Drago A and Ratcliffe P G 2002 *Phys. Rept.* **359** 1–168 (*Preprint* hep-ph/0104283)
[58] Ralston J P and Soper D E 1979 *Nucl. Phys.* **B152** 109
[59] Soffer J 1995 *Phys. Rev. Lett.* **74** 1292–1294 (*Preprint* hep-ph/9409254)
[60] Bourrely C, Soffer J and Teryaev O V 1998 *Phys. Lett.* **B420** 375–381 (*Preprint* hep-ph/9710224)
[61] Vogelsang W 1998 *Phys. Rev.* **D57** 1886–1894 (*Preprint* hep-ph/9706511)
[62] Hayashigaki A, Kanazawa Y and Koike Y 1997 *Phys. Rev.* **D56** 7350–7360 (*Preprint* hep-ph/9707208)
[63] Kumano S and Miyama M 1997 *Phys. Rev.* **D56** 2504–2508 (*Preprint* hep-ph/9706420)
[64] Anselmino M, Boglione M, D'Alesio U, Kotzinian A, Murgia F, Prokudin A and Melis S 2009 *Nucl. Phys. Proc. Suppl.* **191** 98–107 (*Preprint* 0812.4366)
[65] Anselmino M, Boglione M, D'Alesio U, Melis S, Murgia F *et al.* 2013 *Phys.Rev.* **D87** 094019 (*Preprint* 1303.3822)
[66] Bacchetta A, Courtoy A and Radici M 2011 *Phys.Rev.Lett.* **107** 012001 (*Preprint* 1104.3855)
[67] Bacchetta A, Courtoy A and Radici M 2013 *JHEP* **1303** 119 (*Preprint* 1212.3568)
[68] Courtoy A, Bacchetta A, Radici M and Bianconi A 2012 *Phys.Rev.* **D85** 114023 (*Preprint* 1202.0323)
[69] Goldstein G R, Hernandez J O G and Liuti S 2014 (*Preprint* 1401.0438)
[70] Spokespersons: Avakian H, Anefalos Pereira S, Courtoy A, Radici M and Griffioen K 2012 *A 12 GeV Research Proposal to Jefferson Lab (PAC 39)* **PR-12-12-009**
[71] Zhang J and Courtoy A 2013 *A Letter of Intent to Jefferson Lab (PAC 40)* **LOI12-13-002**
[72] Gockeler M *et al.* (QCDSF Collaboration, UKQCD Collaboration) 2005 *Phys.Lett.* **B627** 113–123 (*Preprint* hep-lat/0507001)
[73] Hecht M B, Roberts C D and Schmidt S M 2001 *Phys.Rev.* **C64** 025204 (*Preprint* nucl-th/0101058)
[74] Roberts C Private communication
[75] EfremovV A and Schweitzer P 2003 *JHEP* **0308** 006 (*Preprint* hep-ph/0212044)
[76] Efremov A V, Goeke K and Schweitzer P 2003 *Phys. Rev.* **D67** 114014 (*Preprint* hep-ph/0208124)
[77] Courtoy A, Mirazita M and Pisano S 2014 (*Preprint* in preparation)
[78] Courtoy A, González-Alonso M and Liuti S 2014 (*Preprint* in preparation)

Journal of Physics: Conference Series **651** (2015) 012008 doi:10.1088/1742-6596/651/1/012008

# Canonical quantization of spin 1 *matter* fields. Preliminary results.

**Rodolfo Ferro-Hernández & Mauro Napsuciale**
Departamento de Física, Universidad de Guanajuato,León, Guanajuato, México
E-mail: `ferrohr2009@licifug.ugto.mx` , `mauro@fisica.ugto.mx`

**Abstract.** In this work we present preliminary results on the canonical quantization of spin 1 *matter* fields, i.e. massive fields transforming in the $(1,0)\oplus(0,1)$ representation of the Homogeneous Lorentz Group. The classical field theory is based on the projection onto subspaces of well defined parity in this representation.

## 1. Introduction

The Standard Model (SM) is constructed using only a few of the representations of the Homogeneous Lorentz Group (HLG): the (0,0) for the Higgs, $\left(\frac{1}{2},0\right)$and $\left(0,\frac{1}{2}\right)$ for quarks and leptons and $\left(\frac{1}{2},\frac{1}{2}\right)$ for the gauge bosons. The possible inclusion of higher spin matter fields in extensions of the SM requires to have consistent quantum field theory for $j>1$ which is a long standing problem. Recently, the covariant structure of operators acting on the $(j,0)\oplus(0,j)$ representation was elucidated based on the covariant properties of parity [1] and a proposal for the description of fields transforming in this representation was put forth based on the projection onto subspaces of well defined parity [2]. Here we present preliminary results on the canonical quantization of the simplest case, $j=1$.

## 2. Parity and covariant basis for $(1,0)\oplus(0,1)$

The Homogeneous Lorentz Group (HLG) generated by rotations and boosts $\{\mathbf{J},\mathbf{K}\}$is isomorphic to the group $SU(2)_A\otimes SU(2)_B$ generated by

$$\mathbf{A}=\tfrac{1}{2}(\mathbf{J}+i\mathbf{K}),\quad \mathbf{B}=\tfrac{1}{2}(\mathbf{J}-i\mathbf{K})\ . \tag{1}$$

The quantum states with well defined transformation properties under the HLG can be labeled as $|a,m_a;b,m_b\rangle$ corresponding to the eigenvalues of $\{\mathbf{A}^2,A_3,\mathbf{B}^2,B_3\}$ and the irreducible representations are labeled by the corresponding two $SU(2)$ good quantum numbers $(a,b)$.
Under parity $\mathbf{J}\to\mathbf{J}$, $\mathbf{K}\to$-$\mathbf{K}$ and $\mathbf{A}\to\mathbf{B}$, $\mathbf{B}\to\mathbf{A}$. Parity maps $(a,b)\to(b,a)$, i.e. for $a\neq b$, the simple representations $(a,b)$ are not invariant subspaces under this transformation. The construction of theories invariant under parity requires to use fields transforming in the $(a,b)\oplus(b,a)$. For the special case $b=0$, we get $\mathbf{A}=\mathbf{J}$, i.e. the representation is $(j,0)\oplus(0,j)$. In this case, using the basis $|j,m\rangle_R$ and $|j,m\rangle_L$ for the "right" $(j,0)$ and "left" $(0,j)$ representations, the generators of the HLG are given by

$$M_{ij}=\epsilon_{ijk}\begin{pmatrix}\tau_k & 0\\ 0 & \tau_k\end{pmatrix}\quad M^{0i}=i\begin{pmatrix}\tau_i & 0\\ 0 & -\tau_i\end{pmatrix} \tag{2}$$

where $\tau_i$ are the conventional rotation matrices for spin $j$. In general for the $(a,b)$ representations we can define the chirality operator

$$\chi = \frac{i}{4a\left(a+1\right) - 4b\left(b+1\right)} M_{\mu\nu}\widetilde{M}^{\mu\nu}. \tag{3}$$

Since $\chi$ is proportional to a Casimir operator of the HLG, we have the commutation rule

$$[\chi, M^{\mu\nu}] = 0. \tag{4}$$

In the rest frame, the parity and chirality operators for the specific $(j,0)\oplus(0,j)$ representations are given by

$$\chi = \begin{pmatrix} \mathbf{1}_{2j+1} & 0 \\ 0 & -\mathbf{1}_{2j+1} \end{pmatrix}, \qquad \Pi = \begin{pmatrix} 0 & \mathbf{1}_{2j+1} \\ \mathbf{1}_{2j+1} & 0 \end{pmatrix}. \tag{5}$$

Using the explicit form of the boost generators in Eq. (2), $K^i = \tau^i$ and the Jordan algebra satisfied by these matrices it is possible to explicitly construct the boost operator. Using it on the rest frame states $u\left(0,\lambda\right)$ we obtain the states of arbitrary momentum $u\left(\mathbf{p},\lambda\right)$.

It was shown in [1] that the set $\left\{1, \chi, S^{\mu\nu}, \chi S^{\mu\nu}, M^{\mu\nu}, C^{\mu\nu\alpha\beta}\right\}$ form a covariant basis for operators acting on the $(1,0)\oplus(0,1)$ representation where

$$S^{\mu\nu} = \Pi\left[\eta^{\mu\nu} - i\left(\eta^{0\mu}M^{0\nu} + \eta^{0\nu}M^{0\mu}\right) - \left\{M^{0\mu}, M^{0\nu}\right\}\right], \tag{6}$$

is a symmetric traceless tensor $\left(S^{\mu}{}_{\mu} = 0\right)$. The $C$ tensor is defined as

$$C^{\mu\nu\alpha\beta} = 4\left\{M^{\mu\nu}, M^{\alpha\beta}\right\} + 2\left\{M^{\mu\alpha}, M^{\nu\beta}\right\} - 2\left\{M^{\mu\beta}, M^{\nu\alpha}\right\} - 8\left(\eta^{\mu\alpha}\eta^{\nu\beta} - \eta^{\mu\beta}\eta^{\nu\alpha}\right). \tag{7}$$

It has the symmetries of the Weyl tensor

$$C^{\mu\nu\alpha\beta} = -C^{\nu\mu\alpha\beta} = -C^{\mu\nu\beta\alpha}, \qquad C^{\mu\nu\alpha\beta} = C^{\alpha\beta\mu\nu}. \tag{8}$$

The contraction of any index pair vanishes and it satisfies the Bianchi identity

$$C^{\mu\nu\alpha\beta} + C^{\mu\alpha\beta\nu} + C^{\mu\beta\nu\alpha} = 0. \tag{9}$$

The $S$ tensor satisfy the following commutation relations

$$\begin{aligned} i\left[S^{\mu\nu}, S^{\alpha\beta}\right] &= \eta^{\mu\alpha}M^{\nu\beta} + \eta^{\nu\alpha}M^{\mu\beta} + \eta^{\mu\beta}M^{\nu\alpha} + \eta^{\nu\beta}M^{\mu\alpha}, \\ \left\{S^{\mu\nu}, S^{\alpha\beta}\right\} &= \frac{4}{3}\left(\eta^{\mu\alpha}\eta^{\nu\beta} + \eta^{\nu\alpha}\eta^{\mu\beta} - \frac{1}{2}\eta^{\mu\nu}\eta^{\alpha\beta}\right) - \frac{1}{6}\left(C^{\mu\alpha\nu\beta} + C^{\mu\beta\nu\alpha}\right), \end{aligned} \tag{10}$$

and anticommutes with the chirality operator

$$\{\chi, S^{\mu\nu}\} = 0. \tag{11}$$

States of well defined parity satisfy the corresponding projection condition in the rest frame

$$\frac{1}{2}\left(1+\Pi\right)u\left(0,\lambda\right) = u\left(0,\lambda\right). \tag{12}$$

Boosting this equation we get

$$(S^{\mu\nu}p_\mu p_\nu - m^2)u\left(\mathbf{p},\lambda\right) = 0. \tag{13}$$

Journal of Physics: Conference Series **651** (2015) 012008 doi:10.1088/1742-6596/651/1/012008

Defining the corresponding wave function as usual

$$\psi(x) = u(p, \lambda)e^{-ip\cdot x} \tag{14}$$

we get

$$(S^{\mu\nu}\partial_\mu\partial_\nu + m^2)\psi(x) = 0. \tag{15}$$

This equation was considered by Weinberg and Joos long ago [3]. It can be derived from the following Lagrangian

$$\mathcal{L} = \partial_\mu\bar{\psi}S^{\mu\nu}\partial_\nu\psi - m^2\bar{\psi}\psi \tag{16}$$

where $\bar{\psi} = \psi^\dagger\Pi$. This Lagrangian has an obvious global $U(1)$ symmetry whose conserved current is

$$J^\alpha = i\; q[(\partial_\mu\bar{\psi})S^{\mu\alpha}\psi - \bar{\psi}S^{\alpha\nu}\partial_\nu\psi]. \tag{17}$$

The free Lagrangian has a chiral decomposition

$$\mathcal{L} = \partial_\mu\bar{\psi}_R S^{\mu\nu}\partial_\nu\psi_R + \partial_\mu\bar{\psi}_L S^{\mu\nu}\partial_\nu\psi_L - m^2\left(\bar{\psi}_R\psi_L + \bar{\psi}_L\psi_R\right) \tag{18}$$

where $\psi_R = \frac{1}{2}\left(1+\chi\right)\psi$ and $\psi_L = \frac{1}{2}\left(1-\chi\right)\psi$, such that in the massless case it is symmetric under independent rotations of the left and right components, i.e. , it is chiral symmetric.

## 3. Canonical quantization

First we expand the corresponding fields

$$\Psi(x) = \sum_{p,r}\frac{1}{\sqrt{2Vp_0}}\left[c_r(\mathbf{p})u_r(\mathbf{p})e^{-ipx} + d_r^+(\mathbf{p})u_r^c(\mathbf{p})e^{ipx}\right], \tag{19}$$

$$\bar{\Psi}(x) = \sum_{p,r}\frac{1}{\sqrt{2Vp_0}}\left[c_r^+(\mathbf{p})\bar{u_r}(\mathbf{p})e^{ipx} + d_r(\mathbf{p})\bar{u}_r^c(\mathbf{p})e^{-ipx}\right]. \tag{20}$$

Here, $r$ stands for the polarization and we use the shorthand notation $u_r(\mathbf{p}) = u(\mathbf{p}, r)$. The charged conjugated spinor is given by [2]

$$u^c(p, r) = \Gamma u^*(p, r) \tag{21}$$

with the matrix

$$\Gamma = \begin{pmatrix} 0 & U \\ U^{-1} & 0 \end{pmatrix} \tag{22}$$

where $U = e^{-i\pi J_2}$. The coefficients $c_r\left(\mathbf{p}\right), d_r\left(\mathbf{p}\right), c_r^+\left(\mathbf{p}\right), d_r^+\left(\mathbf{p}\right)$ satisfy the usual commutation relations. The spinors $u, u^c$ satisfy

$$\left(S\left(p\right) - m^2\right)u_r(\mathbf{p}) = 0, \qquad \left(S\left(p\right) - m^2\right)u_r^c(\mathbf{p}) = 0, \tag{23}$$

where $S\left(p\right) \equiv S^{\mu\alpha}p_\mu p_\alpha$. Notice that both particle and antiparticle satisfy the same free equation, this occurs because charge conjugation and parity commute [2]. In spite of this, it can be shown that the interacting equations satisfied by the field and its charge conjugated differ in the sign of the coupling.

The momentum densities of the field are

$$\bar{\Pi}_d = \frac{\partial\mathcal{L}}{\partial\bar{\Psi}_{d,0}} = S_{da}^{0\mu}\left(\partial_\mu\Psi\right)_a, \tag{24}$$

$$\Pi_d = \frac{\partial \mathcal{L}}{\partial \Psi_{d,0}} = \left(\partial_\mu \bar{\Psi}\right)_a S^{\mu 0}_{ad}. \tag{25}$$

The commutators of these momenta with their respective fields are given by

$$[\Pi_d, \Psi_b] = \left(\partial_\mu \bar{\Psi}\right)_a S^{\mu 0}_{ad} \Psi_b - \Psi_b \left(\partial_\mu \bar{\Psi}\right)_a S^{\mu 0}_{ad}, \tag{26}$$

and

$$\left[\bar{\Pi}_d, \bar{\Psi}_b\right] = S^{0\mu}_{da} \left(\partial_\mu \Psi\right)_a \bar{\Psi}_b - \bar{\Psi}_b S^{0\mu}_{da} \left(\partial_\mu \Psi\right)_a. \tag{27}$$

In the following we give some details of the calculation of Eq. (26). The calculation for the other commutator is quite similar. After substituting the Fourier expansion for the field and momentum density we get

$$[\Pi_d(x_1), \Psi_b(x_2)] = -i \sum_{\mathbf{p},r} \frac{p_\mu}{2V p_0} \left[\bar{u}_{ra}(\mathbf{p}) u_{rb}(\mathbf{p}) S^{\mu 0}_{ad} e^{ip(x_1 - x_2)} + \bar{u}^c_{ra}(\mathbf{p}) u^c_{rb}(\mathbf{p}) S^{\mu 0}_{ad} e^{ip(x_2 - x_1)}\right]. \tag{28}$$

The states of well defined parity satisfy

$$\sum_r u_{ra}(\mathbf{p}) \bar{u}_{rb}(\mathbf{p}) = \left(\frac{S(p) + m^2}{2m^2}\right)_{ab}. \tag{29}$$

Since the charge conjugated states have the same parity, they satisfy the same relation

$$\sum_r u^c_{ra}(\mathbf{p}) \bar{u}^c_{rb}(\mathbf{p}) = \left(\frac{S(p) + m^2}{2m^2}\right)_{ab}. \tag{30}$$

Substituting this expression in Eq. (28) and evaluating at equal times $x_1^0 = x_2^0 = 0$ we obtain

$$[\Pi_d(\mathbf{x}_1), \Psi_b(\mathbf{x}_2)]_{x^0_{1,2}=0} = -i \sum_{\mathbf{p}} \frac{p_\mu}{2V p_0} \left(\frac{S(p) + m^2}{2m^2}\right)_{ba} \left[S^{\mu 0}_{ad} e^{ip_i(x_1^i - x_2^i)} + S^{\mu 0}_{ad} e^{-ip_i(x_1^i - x_2^i)}\right] \tag{31}$$

changing $p_i \to -p_i$ in the second term after some algebra we get

$$\begin{aligned} [\Pi_d(\mathbf{x}_1), \Psi_b(\mathbf{x}_2)]_{x^0_{1,2}=0} &= -i \sum_{\mathbf{p}} e^{ip_i(x_1^i - x_2^i)} \frac{1}{2V p_0} [p_0 \left[\left(\frac{S^{ij} p_i p_j + S^{00} p_0 p_0 + m^2}{m^2}\right)_{ba} S^{00}_{ad}\right] \\ &+ p_i \left(\frac{2S^{0j} p_0 p_j}{m^2}\right)_{ba} S^{i0}_{ad}]. \end{aligned} \tag{32}$$

In order to reduce this equation we use Eq. (10) to show that

$$S^2(p) = p^4. \tag{33}$$

Now, we rewrite the term $S^2(p)$ as

$$\begin{aligned} S^2(p) &= S^{ij} S^{lm} p_l p_m p_i p_j + 2\left\{S^{ij}, S^{0l}\right\} p_l p_0 p_i p_j + \left\{S^{ij}, S^{00}\right\} p_0^2 p_i p_j \\ &+ 4S^{0i} S^{0l} p_0^2 p_l p_i + 2\left\{S^{0i}, S^{00}\right\} p_0^3 p_i + p_0^4. \end{aligned} \tag{34}$$

Using Eq. (33) and comparing terms with equal powers of $p_0$ we obtain

$$\left\{S^{ij}, S^{00}\right\} p_i p_j + 4S^{0i} S^{0l} p_l p_i = -2\mathbf{p}^2. \tag{35}$$

This result can be ever reduced by taking advantage of Eq. (10) with $\mu = i, \nu = j, \alpha = \beta = 0$ which yields $[S^{ij}, S^{00}] = 0$. The final result is

$$\left[\Pi_d(\mathbf{x}_1), \Psi_b(\mathbf{x}_2)\right]_{x^0_{1,2}=0} = -i\delta(\mathbf{x}_1 - \mathbf{x}_2)\frac{(\delta_{bd} + S^{00}_{bd})}{2} = -i\delta(\mathbf{x}_1 - \mathbf{x}_2)\Lambda^+_{bd}(0). \quad (36)$$

where $\Lambda^+(0)$ is the positive parity projector in the rest frame. This is an unusual but not an unexpected result since our particle and antiparticle live in the same parity subspace of $(1,0) \oplus (0,1)$ and the projector $\Lambda^+(0)$ is like the identity in this subspace.

A similar calculation for the remaining commutator yields

$$\left[\bar{\Pi}_d(\mathbf{x}), \bar{\Psi}_b(\mathbf{y})\right]_{x^0_{1,2}=0} = -i\delta(\mathbf{x} - \mathbf{y})\frac{(\delta_{db} + S^{00}_{db})}{2} = -i\delta(\mathbf{x} - \mathbf{y})\Lambda^+_{db}(0). \quad (37)$$

The stress tensor is given by

$$T^{\mu\nu} = \partial^\nu\bar{\Psi}S^{\mu\alpha}\partial_\alpha\Psi + \partial_\alpha\bar{\Psi}S^{\alpha\nu}\partial^\mu\Psi - \eta^{\mu\nu}\mathcal{L}. \quad (38)$$

Integrating $T^{00}$ we get the total energy as

$$H = \int N\left\{T^{00}\right\}d^3x = \int N\left\{\partial_0\bar{\Psi}S^{00}\partial_0\Psi - \partial_i\bar{\Psi}S^{ij}\partial_j\Psi + m^2\bar{\Psi}\Psi\right\}d^3x. \quad (39)$$

Substituting the Fourier expansion in this equation some factors like

$$\bar{u}_r(\mathbf{p})\left(S^{0\alpha}p_0p_\alpha\right)u_s(\mathbf{p}) \quad (40)$$

appear. By taking advantage of the property:

$$\bar{u}_r(\mathbf{p}) = \bar{u}_r(\mathbf{p})\frac{S^{\mu\nu}p_\mu p_\nu}{m^2}, \quad u_s(\mathbf{p}) = \frac{S^{\mu\nu}p_\mu p_\nu}{m^2}u_s(\mathbf{p}) \quad (41)$$

these terms can be reduced. First we insert Eq.(41) in Eq.(40), then we commute the $S$'s and use again Eq.(41) to obtain

$$\bar{u}_r(\mathbf{p})\left(S^{0\alpha}p_0p_\alpha\right)u_s(\mathbf{p}) = \frac{1}{2m^2}\bar{u}_r(\mathbf{p})\left\{S^{\mu\nu}, S^{0\alpha}\right\}p_0p_\alpha u_s(\mathbf{p})p_\mu p_\nu. \quad (42)$$

Using Eq.(10) the result for the commutators inside Eq.(42) is

$$\left\{S^{\mu\nu}, S^{0\alpha}\right\} = \frac{4}{3}\left(\eta^{\mu0}\eta^{\nu\alpha} + \eta^{\nu0}\eta^{\mu\alpha} - \frac{1}{2}\eta^{\mu\nu}\eta^{0\alpha}\right) - \frac{1}{6}\left(C^{\mu0\nu\alpha} + C^{\mu\alpha\nu0}\right). \quad (43)$$

Using this result we get

$$\bar{u}_r(\mathbf{p})\left(S^{0\alpha}p_0p_\alpha\right)u_s(\mathbf{p}) = \frac{1}{2m^2}\bar{u}_r(\mathbf{p})\left(2m^2\left(p^0\right)^2 - \frac{1}{6}\left(C^{\mu0\nu\alpha} + C^{\mu\alpha\nu0}\right)p_\mu p_\nu p_0p_\alpha\right)u_s(\mathbf{p}), \quad (44)$$

and using now symmetry properties in Eq.(8) we obtain

$$\bar{u}_r(\mathbf{p})\left(S^{0\alpha}p_0p_\alpha\right)u_s(\mathbf{p}) = \left(p^0\right)^2\bar{u}_r(\mathbf{p})u_s(\mathbf{p}) = \left(p^0\right)^2\delta_{rs}. \quad (45)$$

Finally, inserting this result in Eq. (39) it is straightforward to show that

$$H = \sum_r\int d^3p[c^+_r(\mathbf{p})c_r(\mathbf{p}) + d^+_r(\mathbf{p})d_r(\mathbf{p})]p_0. \quad (46)$$

Journal of Physics: Conference Series **651** (2015) 012008 doi:10.1088/1742-6596/651/1/012008

Notice that the algebraic properties of covariant basis of operators are crucial in order to get this result.

A similar calculation for $T^{oi}$ yield the momentum of the field as

$$P_i = \sum_r \int d^3\mathbf{p} p_i \left[c_r^+(\mathbf{p})c_r(\mathbf{p}) + d_r^+(\mathbf{p})d_r(\mathbf{p})\right]. \tag{47}$$

Also, the total charge associated to the current in Eq. (17) is

$$Q = \int d^3xN\left\{J^0\right\} = q\sum_r \int d^3\mathbf{p}\left(d_r^+(\mathbf{p})d_r(\mathbf{p}) - c_r^+(\mathbf{p})c_r(\mathbf{p})\right). \tag{48}$$

## 4. Propagator

The Feynman propagator is defined as

$$i\Gamma_F(x-y)_{ab} \equiv \langle 0|T\left\{\phi_a(x)\bar{\phi}_b(y)\right\}|0\rangle \tag{49}$$

Using the expansion of the fields and $c_r(\mathbf{p})|0\rangle = d_r(\mathbf{p})|0\rangle = 0$, we get

$$i\Gamma_F(x-y)_{ab} = \begin{cases} \sum\limits_{\mathbf{p},r} \frac{1}{2Vp_0} u_{ra}(\mathbf{p})\bar{u}_{rb}(\mathbf{p})e^{-ip(x-y)} & x_0 > y_0 \\ \sum\limits_{\mathbf{p},r} \frac{1}{2Vp_0} u^c_{ra}(\mathbf{p})\bar{u}^c_{rb}(\mathbf{p})e^{ip(x-y)} & y_0 > x_0 \end{cases} \tag{50}$$

Using Eq.(29) and Eq.(30) in the propagator we obtain

$$i\Gamma_F(x-y)_{ab} = \begin{cases} \sum\limits_{\mathbf{p}} \frac{1}{2Vp_0}\left(\frac{S(p)+m^2}{2m^2}\right)_{ab} e^{-ip(x-y)} & x_0 > y_0 \\ \sum\limits_{\mathbf{p}} \frac{1}{2Vp_0}\left(\frac{S(p)+m^2}{2m^2}\right)_{ab} e^{ip(x-y)} & y_0 > x_0. \end{cases} \tag{51}$$

A four-dimensional integral representation can be obtained following the conventional $i\epsilon$ prescription to handle the poles in the energy plane. We obtain

$$i\Gamma_F(x-y) = \frac{i}{(2\pi)^4}\int d^4p \frac{\left(S(p)+m^2\right)e^{-ip(x-y)}}{2m^2\left(p^2-m^2+i\varepsilon\right)}. \tag{52}$$

However, using Eq.(33) it can be shown that this is not the appropriate integral representation for the two-point Green function for our theory since

$$\begin{aligned} \left(S\left(i\partial_x\right)-m^2\right)\Gamma_F(x) &= \frac{1}{(2\pi)^4}\int \frac{\left(S^2(p)-m^4\right)e^{-ip(x)}d^4p}{2m^2\left(p^2-m^2+i\varepsilon\right)} \\ &= \frac{1}{(2\pi)^4}\int \frac{\left(p^2+m^2\right)e^{-ip(x)}d^4p}{2m^2} \neq \delta(x). \end{aligned} \tag{53}$$

Clearly, due to the algebraic properties of the operator $S(p)$, an appropriate integral representation must contain the $p^4 - m^4$ denominator but in such a way that the tachyonic pole at $p^2 = -m^2$ do not contribute to the Green function. This can be formally done by an appropriate choice of the contour integral. Choosing the following contour

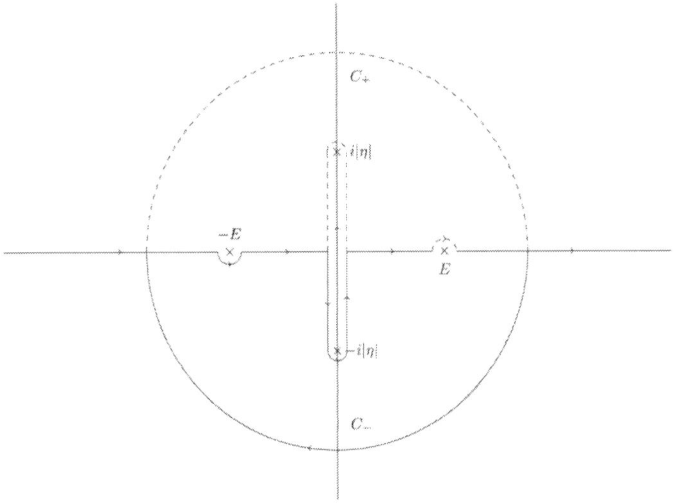


we get the integral representation

$$\Gamma_F(x-y) = \frac{1}{(2\pi)^4} \int_C \frac{\left(S\left(p\right) + m^2\right) e^{-ip(x-y)} d^4p}{\left(p^4 - m^4\right)}. \tag{54}$$

We can trace this problem to the fact that Eq.(12) is a projector only on the mass-shell. For arbitrary values of $p^2$ the covariant parity projectors is

$$P\left(p\right) = \frac{1}{2}\left(1 + \frac{S\left(p\right)}{p^2}\right), \tag{55}$$

thus a possibility is to use the mass-projector combined with the parity projector, i.e. to consider

$$\frac{p^2}{m^2}\frac{1}{2}\left(1 + \frac{S\left(p\right)}{p^2}\right)\psi\left(\mathbf{p},\lambda\right) = \psi\left(\mathbf{p},\lambda\right). \tag{56}$$

This amounts to change

$$S^{\mu\nu} \to \Sigma^{\mu\nu} = \frac{1}{2}\left(S^{\mu\nu} + \eta^{\mu\nu}\right) \tag{57}$$

in our starting equation of motion. In this case the calculations go along the ones outlined here and we obtain a two-point Green function with the conventional $p^2 - m^2$ denominator. This theory is actually equivalent to the formalism used in R$\chi$PT for the effective description of $j = 1$ hadrons [4]. The eventual disadvantage of this formulation is that in this case we loose the chiral decomposition in Eq.(18) and we do not get a chiral theory in the massless limit.

## 5. Conclusions

We present preliminary results on the canonical quantization of spin 1 fields with well definite (positive) parity transforming in the $(1,0) \oplus (0,1)$ representation of the HLG. Conventional results are obtained for the energy, momentum and vector charge. The algebraic structure of the covariant basis of operators acting on this space is crucial in getting these results. In the simplest formulation we get a chiral structure in the mass-less limit and an unconventional propagator whose properties are worth to explore further. A conventional propagator is obtained for a formulation using the true (off-mass-shell) projector onto positive parity subspace combined with a mass-projector. In this case, however we loose chiral symmetry in the massless case.

Journal of Physics: Conference Series **651** (2015) 012008 doi:10.1088/1742-6596/651/1/012008

## References

[1] S. Gómez-Ávila and M. Napsuciale, Phys. Rev. D **88**, 096012 (2013) [arXiv:1307.4711 [hep-ph]].
[2] M. Napsuciale and S. Gómez-Ávila, arXiv:1309.0968 [hep-ph].
[3] H. Joos, Fortsch.Phys., 10, 65-146, 1962; S. Weinberg, Phys.Rev., 133, B1318-B1332, 1964.
[4] G. Ecker, J. Gasser, A. Pich and E. de Rafael, Nucl. Phys. B **321**, 311 (1989).

## Acknowledgements

Work supported by CONACyT-México under grant 156618. R. F. H. acknowledges support from CONACyT with a SNI III grant for young researchers. We thank Selim Gómez-Ávila for discussions on some aspects of this work.

# Yukawa couplings in string theory: the case for F-theory GUT's

**Anamaría Font**

Departamento de Física, Centro de Física Teórica y Computacional, Facultad de Ciencias, Universidad Central de Venezuela, A.P. 20513, Caracas 1020-A, Venezuela

E-mail: afont@fisica.ciens.ucv.ve

**Abstract.** We study the pattern of ukawa couplings in local F-theory SU(5) GUT's. Couplings for the third family of quarks and leptons appear at the perturbative level, but to reproduce the observed couplings for the lighter families requires non-perturbative dynamics. We show that corrections due to instanton effects do lead to a ukawa matrix with a hierarchical structure. Our results apply to both down-like $\mathbf{10} \times \bar{\mathbf{5}} \times \bar{\mathbf{5}}$ and up- like $\mathbf{10} \times \mathbf{10} \times \mathbf{5}$ couplings. The models include magnetic fluxes needed for a chiral spectrum and for symmetry breaking down to the Standard Model. We compute the holomorphic couplings via residues and then obtain the physical couplings taking into account the normalization of wavefunction pro les. Combining non-perturbative corrections and magnetic fluxes allows to t the measured masses and hierarchies of the third and second generations in the Standard Model.

## 1. Introduction

An outstanding open question in the Standard odel (S ) is how to explain the structure of fermion masses and mixings. If the S arises as a low-energy limit of a fundamental string theory [1], it should be possible to understand the observed pattern in terms of basic parameters characterizing the string vacuum. In particular the values of Yukawa couplings in the string context are determined by the geometric properties of extra compactified dimensions. In principle an explicit computation of Yukawa coupling constants seems then rather complicated since we would need a complete knowledge of the underlying compact space. owever, due to the localization properties of branes, certain quantities such as Yukawa couplings, do not depend on the full geometry of the compactification space but rather on local data around the region in which the S fields are localized. This key feature allows to follow a bottom-up road to reproduce the S in the framework of string theory [2].

The bottom-up idea can be implemented in type IIB compactifications with D3 and/or D7-branes. In these local constructions gauge coupling unification emerges naturally because all gauge interactions originate from the same region of the internal space and all S gauge couplings typically depend on the same closed string modulus. oreover, the chiral spectrum of the resulting local G T models, such as S (5) G T's, can be systematically realized with D-branes. owever, in type IIB S (5) G T's the up-like coupling $\mathbf{10} \times \mathbf{10} \times \mathbf{5}$, denoted $Y_U$, is forbidden at perturbative level by charge conservation [1]. This particular Yukawa coupling could be induced by D-brane instantons [3], but the large experimental value for the top Yukawa rather suggests to consider a set-up, such as F-theory [4], in which both up-like $Y_U$, as well as down-like Yukawas $\mathbf{10} \times \bar{\mathbf{5}} \times \bar{\mathbf{5}}$, denoted $Y_D$, can appear on equal footing.

F-theory provides a promising starting point to construct realistic models and a comprehensive formalism to build F-theory G T's has been developed in the last years [5 8] (see [9 13] for recent reviews). F-theory and type IIB models share properties such as localization of gauge and matter fields, gauge coupling unification, and moduli stabilization. oreover, in F-theory, unlike in type IIB, exceptional gauge groups and matter representations such as spinorials of SO(10) are allowed, thereby enlarging the possibilities for spectra and interactions. In local F-theory S (5) G T's the gauge degrees of freedom live in a 4-cycle $S_{\rm GUT}$ of the internal space. The chiral matter fields, transforming as **10**, $\bar{\mathbf{5}}$, or **5**, are supported at certain 2-cycles of $S_{\rm GUT}$. At these curves the gauge symmetry is enhanced, for instance to S (6) or SO(10), and at points where the curves intersect there is further enhancing, for instance to SO(12) or $\rm E_6$. The triple intersection of quarks/leptons/ iggs curves, gives rise to Yukawa couplings that are given by the overlap integral of the internal wavefunctions for the chiral matter fields. Since this integral is dominated by the wavefunction profiles around the intersection point $p$, only data in a small region around $p$ is necessary to determine Yukawa couplings in local F-theory models. In fact, calculation of holomorphic couplings can be achieved via computation of residues at $p$ [14].

One general result established earlier is that fermion mass hierarchies appear automatically by restricting the number of intersection points. Indeed, as shown in [14] (see also [15 17]) the matrix of down-like Yukawa couplings $Y_D$ will have rank one if there is a single intersection point $p_{\rm down}$, with a similar statement for up-like Yukawas. This ensures a avor structure in which one family is much heavier, while the remaining two acquire smaller masses through non-perturbative effects. Indeed, such masses can be induced by D3-brane instantons or a gaugino condensate supported on a different 4-cycle of the compactification, as proposed in [18].[1]

A systematic analysis of this scenario was initiated in [23] where a toy example with (1) gauge symmetry was considered. It was continued in [24] where a more realistic local F-theory model with S (5) symmetry enhanced to SO(12) at the intersection point was explored in depth. In this model D-quark, or lepton, Yukawa couplings $Y_D$ are allowed. It was found that non-perturbative effects distort the wavefunction profiles near the intersection point in such a way that a hierarchy of fermion mass eigenvalues of the form $(1, \epsilon, \epsilon^2)$ is generated. ere $\epsilon$ is a small parameter that measures the size of the non-perturbative effects. The hierarchy is already manifest at the level of holomorphic Yukawa couplings which are independent of the magnetic uxes in local F-theory models. In contrast, the physical Yukawa couplings do depend on uxes, and in particular on the hypercharge ux needed to trigger further symmetry breaking down to S (3) × S (2) × $(1)_Y$. This hypercharge dependence then allows to explain the difference between D-quark and lepton Yukawas at the unification scale.

-quark $Y_U$ couplings require an S (5) model with enhancing to $\rm E_6$ at the intersection point $p_{up}$. This case is more subtle because a non-trivial local geometry is necessary to reproduce the observed pattern. Indeed, if the fields in the coupling come from three distinct matter curves there can be no diagonal -quark couplings. This problem could be avoided with a self-intersecting matter curve associated to the **10**, as suggested in [6, 7, 15]. owever, as noticed in [26], if the two branches of the **10** wavefunction are independent the Yukawa matrix would be rank two with no automatic hierarchy. In order to obtain the desired rank one structure one must either introduce 7-brane monodromy [25] or describe the matter curves via non-Abelian 7-brane profiles [26, 27] called T-branes. In [28] the T-brane approach was taken to compute up-like Yukawa couplings in presence of the kind of non-perturbative effects described in [18]. A hierarchical structure $(1, \epsilon, \epsilon^2)$ occurs again at the level of holomorphic Yukawa couplings and translates into the eigenvalues of the physical Yukawa matrix. The local model parameters can be chosen to attain a $\mathcal{O}(1)$ Yukawa for the top quark, thus ustifying the initial motivation that led to consider F-theory instead of type IIB S (5) models.

[1] For different mechanisms to generate fermion mass hierarchies in F-theory see e.g. [1 22].

This paper is based on the work in [23, 24, 28]. We will report the main results, but will skip many details that can be found in the original articles. The organization is as follows. In section 2 we review the construction of local F-theory models. The S (5) models are presented in section 3, emphasizing the geometrical and field theoretical data behind the Yukawa couplings. In section 4 we first describe the non-perturbative effects that can induce corrections to the Yukawa couplings. The holomorphic couplings are then computed by means of a residue formula. Section 5 is devoted to discussing the physical couplings. Some concluding remarks are collected in section 6.

## 2. Review of local F-theory models

In this section we brie y outline the standard approach to F-theory model building [5 8] (see [9 13] for recent reviews). We first describe the geometry behind the matter content and their interactions, and then discuss the corresponding effective theory.

In F-theory G T's the gauge theory arises from 7-branes wrapping a compact four-dimensional surface $S_{\rm GUT}$ in the six-dimensional base $B$ of an elliptically-fibered eight-dimensional Calabi-Yau. The gauge group $G_{\rm GUT}$ depends on the singularity type of the elliptic fiber. Together with the stack of 7-branes on $S_{\rm GUT}$, a semi-realistic F-theory model contains extra 7-branes wrapping other surfaces $S_i$, which intersect $S_{\rm GUT}$ on certain curves $_i$. On these curves the singularity type of the elliptic fiber is enhanced to a higher rank group $G_{\Sigma_i} \supset G_{\rm GUT}$. As expected, there will be additional degrees of freedom due to open strings stretching between the intersecting 7-branes. ore precisely, there will be chiral matter multiplets charged under $G_{\rm GUT}$, localized at the so-called matter curves $_i$. The charged representations can be found decomposing the ad oint of the enhanced group $G_{\Sigma_i}$ under $G_{\rm GUT}$. For example, in the case $G_{\rm GUT}$ = S (5), one can have curves $_{\bf 10}$ with enhancement to SO(10) and matter in the **10** or $\overline{\bf 10}$ representations, or curves $_{\bf 5}$ with enhancement to S (6) and matter in the **5** or $\bar{\bf 5}$. Finally, when two or more matter curves meet at a point $p$ there will be further enhancement to a group $G_p$, which signals the appearance of interactions among the matter fields localized on the curves meeting at $p$. The couplings allowed at $p$ depend on the the enhanced group $G_p$ which contains $G_{\rm GUT}$ and each of the $G_{\Sigma_i}$ involved. For $G_{\rm GUT}$ = S (5), down-like $\mathbf{10} \times \bar{\mathbf{5}} \times \bar{\mathbf{5}}$ Yukawa couplings arise at points of SO(12) enhancement, whereas for up-like $\mathbf{10} \times \mathbf{10} \times \mathbf{5}$ Yukawas enhancement to $E_6$ is expected.

The above geometric picture has a corresponding formulation in terms of the 8d supersymmetric action on the worldvolume of the 7-branes [5 8]. The observable 4d theory emerges upon compactification of the 8d theory on the surface $S$ wrapped by the 7-branes. The complex coordinates of $S$ are denoted $x$ and $y$. The 8d bosonic fields are the gauge field $A$ and the (2,0) form $\Phi$, namely $\Phi = \Phi_{xy} dx \wedge dy$. $\Phi$ is associated to motion transverse to the 7-branes. Both $A$ and $\Phi$ transform in the ad oint of a non-Abelian gauge group $G$ that contains $G_{\rm GUT}$ and $G_{\Sigma_i}$, which for the purposes of analyzing Yukawa couplings at $p$ can be taken to be $G_p$. The 4d bosonic fields are the zero modes of $A$ and $\Phi$ and the 4d action is obtained by dimensional reduction. In particular, the Yukawa couplings among 4d matter fields can be computed from the superpotential

$$W = m_*^4 \int_S {\rm tr}\,(F \wedge \Phi) \tag{1}$$

where $m_*$ is the F-theory characteristic scale, and $F = dA - iA \wedge A$. The dynamics also depends on the D-term

$$D = \int_S \omega \wedge F + \frac{1}{2}[\Phi, \bar{\Phi}] \tag{2}$$

where $\omega$ is the fundamental form of $S$. The equations of motion that follow from the

Journal of Physics: Conference Series **651** (2015) 012009 doi:10.1088/1742-6596/651/1/012009

superpotential and the D-term turn out to be

$$\partial_A \Phi = \partial\Phi - i[A_{0,1}, \Phi] = 0 \tag{3a}$$

$$F_{0,2} = 0 \tag{3b}$$

for the F-term equations and

$$\omega \wedge F + \frac{1}{2}[\Phi, \Phi] = 0 \tag{4}$$

for the D-term equation. The equations of motion do apply to the background and the uctuations.

To derive the equations of motion for the bosonic uctuations we define

$$\Phi_{xy} = \langle\Phi_{xy}\rangle + \varphi_{xy} \quad ; \quad A_{\bar{m}} = \langle A_{\bar{m}}\rangle + a_{\bar{m}} \tag{5}$$

and then expand (3) and (4) to first order in the uctuations $(\varphi, a)$. In this way we find

$$\partial_{\langle A\rangle}\varphi + i[\langle\Phi\rangle, a] = 0 \tag{6a}$$

$$\partial_{\langle A\rangle} a = 0 \tag{6b}$$

$$\omega \wedge \partial_{\langle A\rangle} a - \frac{1}{2}[\langle\Phi\rangle, \varphi] = 0 \tag{6c}$$

where $a = a_{\bar{x}}d\bar{x} + a_{\bar{y}}d\bar{y}$, and $\varphi = \varphi_{xy}dx \wedge dy$. In the local analysis we can take the Kähler form to be

$$\omega = \frac{i}{2}(dx \wedge d\bar{x} + dy \wedge d\bar{y}) \tag{7}$$

A crucial result of the theory is that the equations (6) admit zero mode solutions that are localized on the matter curves.

The possible matter curves are in turn determined by the background of $\Phi$ which in absence of uxes depends holomorphically on the complex coordinates of $S$, as implied by (3a). A non-trivial vev $\langle\Phi\rangle$ such that its rank changes at curves $_i$ implies that instead of a single $S$ there are intersecting surfaces $S_{\text{GUT}}$ and $S_i$. At a generic point on $S$ $\langle\Phi\rangle$ breaks $G_p$ to $G_{\text{GUT}}$ times some additional factors, typically (1)'s, due to the 7-branes wrapping the $S_i$. At $_i$ there are extra commuting generators whose associated uctuations give rise to matter localized on $_i$, as shown solving the equations of motion. At the point $p$ where the matter curves intersect there is further enhancing to $G_p$.

In a local model we can also include a background $\langle A\rangle$, or equivalently a worldvolume ux $\langle F\rangle$. A ux can produce a chiral spectrum by choosing it so that at a given matter curve only zero modes of a definite 4d chirality are fully localized. Besides, turning extra ux along the hypercharge generator breaks $G_{\text{GUT}}$ to the S group.

It is instructive to look at a simple example to illustrate the formalism we have been describing. To this end we consider the toy model with $G_p =$ $(3)^3$ discussed in detail in [23] in which

$$\langle\Phi_{xy}\rangle = \frac{1}{3}m^2\,\text{diag}(-2x + y, x + y, x - 2y) \tag{8}$$

where $m$ is a mass parameter. We see that this vev breaks (3) to $(1)^3$ at generic points in $S$. It is also clear that the group is enhanced to S (2) × $(1)^2$ at the curves $_a = \{x = 0\}$, $_b = \{y = 0\}$, and $_c = \{x = y\}$. For instance, the generators $E_{a^\pm}$ given by

$$E_{a^+} = \begin{pmatrix} 0 & 1 & 0 \\ 0 & 0 & 0 \\ 0 & 0 & 0 \end{pmatrix} \quad ; \quad E_{a^-} = E^\dagger_{a^+} \tag{9}$$

commute with $\langle\Phi\rangle$ when $x = 0$. To induce chirality we also include the ux

$$\langle F\rangle = -\frac{i}{3}M\,(dx\wedge d\bar{x} - dy\wedge d\bar{y})\ \mathrm{diag}(1,-2,1) \tag{10}$$

where $M$ is a constant parameter. To solve the equations of motion it is convenient to choose a holomorphic gauge [16], $\langle A_{0,1}\rangle = 0$, such that the ux corresponds to

$$\langle A_{1,0}\rangle^{\mathrm{hol}} = \frac{i}{3}M(\bar{x}dx - \bar{y}dy)\,\mathrm{diag}(1,-2,1) \tag{11}$$

The solutions in real gauge, in which $\langle A\rangle$ is real, are easily found by means of a gauge transformation. otice that the backgrounds (8) and (10) satisfy the equations of motion (3) and (4).

Given the backgrounds (8) and (10) it can be shown that the equations of motion (6) admit localized zero mode solutions [23]. For instance, the zero modes localized on $_a = \{x = 0\}$ have $a_{\bar{y}} = 0$, $a_{\bar{x}} \propto \varphi_{xy}$, and $\varphi_{xy} = \chi_{a^\pm}E_{a^\pm}$, where in real gauge the scalar wavefunction $\chi_{a^\pm}$ becomes

$$\chi_{a^\pm} = e^{-\sqrt{\left(\frac{M}{2}\right)^2+m^4}\,|x|^2}\, e^{\mp\frac{M}{2}|y|^2} f_{a^\pm}(y) \tag{12}$$

with $f_{a^\pm}$ an arbitrary holomorphic function of the intersecting coordinate $y$. Therefore, the zero modes are indeed localized at $x = 0$, whereas convergence in the $y$ direction depends on the ux. To select a definite chirality we can choose $M > 0$ so that only the zero modes in sector $a^+$ are fully normalizable.

Summarizing, in the language of the 8d field theory, to specify an F-theory local model we need to choose vevs $\langle\Phi\rangle$ and $\langle A\rangle$ such that there are localized zero modes on the matter curves $_i$ determined by $\langle\Phi\rangle$. Concerning Yukawa couplings, the tree-level superpotential (1) includes the trilinear term

$$W_{Yuk} = -im_*^4\int_S \mathrm{tr}\,(A\wedge A\wedge\Phi) \tag{13}$$

that leads to 4d couplings among the zero modes of $A$ and $\Phi$. otice that the couplings are given by an overlap integral of the zero mode wavefunctions. Since the zero modes are localized, the integrals can be restricted to a region of $S$ near the point $p$ where the matter curves intersect. In fact, as found in [14] and explained in section 4.2, the couplings can be computed by residues evaluated at $p$. We will verify that the resulting Yukawa matrix has rank one, thus implying that two families of quarks and leptons would be massless. To evade this rank one problem we will introduce non-perturbative corrections to the superpotential that in turn contribute to the couplings.

In order to obtain a hierarchical pattern of fermion masses we follow the proposal of [15] and consider a setup in which all up-like Yukawas are generated from a single Yukawa point $p_{\mathrm{up}}$, and all down-like Yukawas from $p_{\mathrm{down}}$. Thus, to compute down-like or up-like Yukawa couplings we only need to describe the F-theory G T model in the vicinity of a single point. It would be interesting to engineer a model in which both $p_{\mathrm{up}}$ and $p_{\mathrm{down}}$ are present but we have only studied each case separately. In section 3 we will describe two local F-theory models with $G_{\mathrm{GUT}} =$ S (5) but with different $G_p$. In one model $G_p = \mathrm{SO}(12)$ and down-like couplings $\mathbf{10}\times\bar{\mathbf{5}}\times\bar{\mathbf{5}}$ are allowed [24]. In the other model $G_p = \mathrm{E}_6$ and up-like couplings $\mathbf{10}\times\mathbf{10}\times\mathbf{5}$ are generated [28].

## 3. Local F-theory SU(5) GUT's

In this section we present the F-theory local models with $G_p = \mathrm{SO}(12)$ and $G_p = \mathrm{E}_6$ which contain down-like and up-like Yukawa couplings [24, 28]. In each case we first define the

Journal of Physics: Conference Series **651** (2015) 012009 doi:10.1088/1742-6596/651/1/012009

background $\langle\Phi\rangle$ that determines the matter curves and break $G_p$ to S  (5). We then specify the  ux $\langle F\rangle$ responsible for 4d chirality and breaking to S  (3) × S  (2) ×  $(1)_Y$.

Since $\langle\Phi\rangle$ and $\langle A\rangle$ transform in the ad oint of $G_p$ they can be written in terms of generators that can be decomposed as $\{H_i, E_\rho\}$, where the $H_i$, $i = 1, \cdots, 6$, belong to the Cartan subalgebra and the $E_\rho$ are step generators. Recall that

$$[H_i, E_\rho] = \rho_i E_\rho \tag{14}$$

where $\rho_i$ is the $i$-th component of the root $\rho$. The 60 non-trivial roots of SO(12) read

$$\underline{(\pm1, \pm1, 0, 0, 0, 0)} \tag{15}$$

 ere and below underlying means all possible permutations of the vector entries. On the other hand, the 72 non-zero roots of $E_6$ are given by

$$(0, \underline{\pm1, \pm1, 0, 0, 0}) \tag{16}$$

together with

$$\frac{1}{2}(\pm\sqrt{3}, \pm1, \pm1, \pm1, \pm1, \pm1) \tag{17}$$

with even number of +'s.

### *3.1. The SO(12) model*

At the intersection point of matter curves the group is $G_p$ = SO(12) but away from this point the symmetry is broken to S  (5) ×  $(1)^2$ by the vev $\langle\Phi\rangle$.  oreover, this vev must be such that at some curves there is an enhancement to either SO(10) ×  (1) or S  (6) ×  (1). In this way, we can identify $G_S$ = S  (5) as the G  T gauge group and the enhancement curves as matter curves where chiral matter wavefunctions are localized.

To obtain the above symmetry breaking pattern the vev of the transverse position field $\Phi = \Phi_{xy} dx \wedge dy$ is chosen to be

$$\langle\Phi_{xy}\rangle = m^2(xQ_x + yQ_y) \tag{18}$$

where $m$ is a mass parameter, and $(x, y)$ are the complex coordinates of the 4-cycle $S$. The charge operators $Q_x$ and $Q_y$ are the following combinations of generators of elements of the $SO(12)$ Cartan subalgebra

$$Q_x = -H_1 \quad ; \quad Q_y = \frac{1}{2}(H_1 + H_2 + H_3 + H_4 + H_5 + H_6) \tag{19}$$

The implications of (18) can be understood from the commutator

$$[\langle\Phi_{xy}\rangle, E_\rho] = m^2 q_\Phi(\rho) E_\rho \tag{20}$$

with $q_\Phi$ a holomorphic function of $x$ and $y$. The subgroup of SO(12) not broken by the presence of this vev corresponds to those generators that commute with $\langle\Phi\rangle$ at any point in $S$. This set includes the Cartan subalgebra of SO(12) and those step generators $E_\rho$ such that $q_\Phi(\rho) = 0$ for all $(x, y)$. It is easy to see that such unbroken roots are given by

$$(0, \underline{1, -1, 0, 0, 0}) \tag{21}$$

together with the Cartan generators. Therefore, from the symmetry group SO(12) only the subgroup S  (5) ×  $(1)^2$ remains as a gauge symmetry, and we can identify $G_S$ = S  (5) as our G  T gauge group.

The remaining generators of SO(12), that have $q_\Phi \neq 0$ for generic $(x, y)$, precisely describe the pattern of matter curves and the charged matter localized on them. The broken roots and their charges $q_\Phi$ are displayed in table 1. From this table we see that there are three complex curves within $S$ where the bulk symmetry S (5) × (1)$^2$ is enhanced, in the sense that there $q_\Phi = 0$ for an additional set of roots. Concretely, for $x = 0$ there are 10 additional roots that together with those in (21) complete the S (6) root system. We have labeled such matter curve as ${}_a$, so that in the language of the previous section we would have that $G_{\Sigma_a}$ = S (6) × (1). These extra set of roots whose $q_\Phi$ vanishes at ${}_a$ can be split into subsets that have different $q_\Phi$ away from ${}_a$. It is easy to convince oneself that each of these subsectors must fall into complete weight representations of S (5), which in turn correspond to the matter localized at the curve. In the case of ${}_a$, there are two sectors $a^+$ and $a^-$ that correspond to the representations $\bar{\mathbf{5}}$ and **5** of S (5), respectively, as shown in table 1. Similarly to ${}_a$, at the curve ${}_b = \{y = 0\}$ there are 20 extra unbroken roots and $G_{\Sigma_b}$ = SO(10) × (1), giving rise to the representations **10** and $\overline{\mathbf{10}}$. The third matter curve is given by ${}_c = \{x = y\}$, where there is also an enhancement to S (6) × (1). Finally, notice that the three curves meet at $p_{\text{down}} = \{x = y = 0\}$.

**Table 1.** SO(12) broken generators

| $\rho$ | root | $q_\Phi$ | $SU(5)$ rep. |
|---|---|---|---|
| $a^+$ | $(1, \underline{-1, 0, 0, 0, 0})$ | $-x$ | $\bar{\mathbf{5}}$ |
| $a^-$ | $(-1, \underline{1, 0, 0, 0, 0})$ | $x$ | $\mathbf{5}$ |
| $b^+$ | $(0, \underline{1, 1, 0, 0, 0})$ | $y$ | $\mathbf{10}$ |
| $b^-$ | $(0, \underline{-1, -1, 0, 0, 0})$ | $-y$ | $\overline{\mathbf{10}}$ |
| $c^+$ | $(-1, \underline{-1, 0, 0, 0, 0})$ | $x - y$ | $\bar{\mathbf{5}}$ |
| $c^-$ | $(1, \underline{1, 0, 0, 0, 0})$ | $-(x - y)$ | $\mathbf{5}$ |

Let us now discuss the uxes needed to achieve 4d chirality and trigger symmetry breaking to the Standard odel group. It is convenient to assemble the ux in various steps. We first add a ux $\langle F_1 \rangle$ to obtain a chiral spectrum on the curves ${}_a$ and ${}_b$, selecting the sectors $a^+$ and $b^+$, as opposed to $a^-$ and $b^-$. Then we add a piece $\langle F_2 \rangle$ such that the matter curve ${}_c$ also has a chiral spectrum including a iggs. either of these two uxes breaks S (5) so that we must also include a ux $\langle F_Y \rangle$ along the hypercharge generator. This last ux further enforces doublet-triplet splitting in the iggs sector.

To proceed we then consider the ux

$$\langle F_1 \rangle = -iM\,(dx \wedge d\bar{x} - dy \wedge d\bar{y})\, Q_F \tag{22}$$

where

$$Q_F = \frac{1}{2}(H_1 - H_2 - H_3 - H_4 - H_5 - H_6) \tag{23}$$

To simplify we take the parameter $M$ to be constant. Since the unbroken roots in (21) are neutral under $\langle F_1 \rangle$ this ux does not break S (5). On the contrary, the roots in sectors $a$ and $b$ are charged. Thus, if the ux permeates the curves ${}_a$ and ${}_b$, namely if the integral of (22) over each of these curves does not vanish, they will each support zero modes of definite chirality.

We will assume that this occurs so that $\langle F_1 \rangle$ leads to a a net chiral spectrum of three $\bar{\mathbf{5}}$'s in the curve ${}_a$ and three $\mathbf{10}$'s in the curve ${}_b$. In fact, in [24] it is shown that only zero modes in the sectors $a^+$ and $b^+$ are fully localized as long as $M > 0$, in agreement with the general arguments in [29].

The spectrum in the $c$ sector is not affected by $\langle F_1 \rangle$ because the associated roots are neutral under (23). ow, to generate a down-like Yukawa coupling at the intersection of three matter curves we do need one iggs $\bar{\mathbf{5}}$ in ${}_c$, but no $\mathbf{5}$ to avoid unwanted $\bar{\mathbf{5}}\mathbf{5}$ mass terms. These conditions can be met by adding the extra ux

$$\langle F_2 \rangle = i\,(dx \wedge d\bar{y} + dy \wedge d\bar{x})\,(N_a Q_x + N_b Q_y) \tag{24}$$

where $Q_x$ and $Q_y$ are given in (19). The matter on ${}_c$ is charged under this ux and the required iggs $\bar{\mathbf{5}}$ appears in the $c^+$ sector provided that $N_a > N_b$, as shown in appendix B of [24]. Although the matter on the curves ${}_a$ and ${}_b$ is also charged under (24) it can be proved that the number of (local) families in these curves is independent of the ux $\langle F_2 \rangle$ [24].

nlike (22) and (24) the last piece of worldvolume ux to be added does break S (5) down to the SS gauge group. As usual, such ux should be turned along the hypercharge generator $Q_Y$, and it can generically be taken to be

$$\langle F_Y \rangle = i\left[(dx \wedge d\bar{y} + dy \wedge d\bar{x})\,\tilde{N}_Y + (dy \wedge d\bar{y} - dx \wedge d\bar{x})\,N_Y\right] Q_Y \tag{25}$$

where

$$Q_Y = \frac{1}{3}\,(H_2 + H_3 + H_4) - \frac{1}{2}\,(H_5 + H_6) \tag{26}$$

The component proportional to $N_Y$ could in principle upset the net chirality in sectors $a^+$ and $b^+$ but this is prevented choosing $-1 < N_Y/M < 3$ [24]. A final constraint comes from the condition of doublet-triplet splitting in the $c^+$ sector. It can be verified that the triplets are vector-like, thus generically massive, if $\tilde{N}_Y = 3(N_a - N_b)$ [24].

In summary, the total worldvolume ux on the local SO(12) model is given by

$$\langle F \rangle = i(dy \wedge d\bar{y} - dx \wedge d\bar{x})Q_P + i(dx \wedge d\bar{y} + dy \wedge d\bar{x})Q_S \tag{27}$$

where we have defined

$$Q_P = MQ_F + N_Y Q_Y \tag{28}$$

$$Q_S = N_a Q_x + N_b Q_y + \tilde{N}_Y Q_Y \tag{29}$$

The vev of the corresponding vector potential $A$ can be written in the holomorphic gauge, defined in [16], as

$$\langle A_{1,0} \rangle^{\text{hol}} = i\,(\bar{x}Q_P - \bar{y}Q_S)\,dx - i\,(\bar{y}Q_P + \bar{x}Q_S)\,dy \tag{30}$$

This is the quantity that enters into the D-term equation of motion for the zero mode wavefunctions at the curves ${}_a$, ${}_b$ and ${}_c$. Finally, notice that compared to table 1 the matter in the sectors $a$, $b$ and $c$ splits into S (3) × S (2) × (1)$_Y$ representations once the total ux is taken into account. The resulting content of charged particles is displayed in table 2 of [24].

otice that the backgrounds for $\langle \Phi \rangle$ and $\langle F \rangle$ fulfill the equations of motion. Since both vevs are Abelian the F-terms (3) ust require $\bar{\partial}\langle \Phi \rangle = 0$ and indeed (18) depends holomorphically on the coordinates. Concerning the ux, observe that (27) has $\langle F_{0,2} \rangle = 0$. oreover, the total ux is primitive, i.e. $\omega \wedge \langle F \rangle = 0$, so that the D-term (4) is satisfied because $[\langle \Phi \rangle, \langle \bar{\Phi} \rangle] = 0$.

To end this section we wish to stress that provided the backgrounds (18) and (27), and the constraints on the ux parameters discussed above, it can be shown that the equations of motion (6) admit zero mode solutions of definite chirality localized on the matter curves. The uctuations are of the form $\varphi = \varphi_\rho E_\rho$, $a = a_\rho E_\rho$, where $E_\rho$ is the generator corresponding to a root $\rho$ in a given sector. We refer the reader to appendix D of [24] for the explicit computation of the $(\varphi_\rho, a_\rho)$ and their normalizations.

### *3.2. The* E$_6$ *model*

We will follow the approach of [27] and realize the breaking pattern by means of a non-Abelian background for $\langle\Phi\rangle$ . In this way the correct hierarchy of up-quark masses can be obtained as argued in [27], and verified in the next sections.

Let us consider the T-brane background

$$\langle\Phi_{xy}\rangle = m(E^+ + m\,xE^-) + \mu^2(x-y)Q \tag{31}$$

where $m$ and $\mu$ are mass parameters. The charge $Q$ is given by the combination of Cartan generators

$$Q = \frac{1}{2}\left(\frac{5}{\sqrt{3}}H_1 - H_2 - H_3 - H_4 - H_5 - H_6\right) \tag{32}$$

The main new ingredients are the non-commuting generators $E^\pm$ whose corresponding roots, also denoted $E^\pm$, are defined as

$$E^\pm = \pm\frac{1}{2}(\sqrt{3},1,1,1,1,1) \tag{33}$$

These generators satisfy the relation $[E^+,E^-]=P$, where

$$P = \frac{1}{2}(\sqrt{3}H_1 + H_2 + H_3 + H_4 + H_5 + H_6) \tag{34}$$

ore precisely, the triplet $\{E^+,E^-,P\}$ spans the $\mathfrak{su}(2)$ factor of a $\mathfrak{su}(5)\oplus\mathfrak{su}(2)\oplus\mathfrak{u}(1)$ maximal Lie subalgebra of $\mathfrak{e}_6$, under which the $E_6$ ad oint decomposes as

$$\mathbf{78} \to (\mathbf{24},\mathbf{1})_0 \oplus (\mathbf{1},\mathbf{3})_0 \oplus (\mathbf{1},\mathbf{1})_0 \oplus (\mathbf{10},\mathbf{2})_{-1} \oplus (\overline{\mathbf{10}},\mathbf{2})_1 \oplus (\mathbf{5},\mathbf{1})_2 \oplus (\overline{\mathbf{5}},\mathbf{1})_{-2} \tag{35}$$

where the subindex is the $Q$ charge. This decomposition shows that there is a pair of **10**'s transforming as a doublet of the S (2) generated by $\{E^+,E^-,P\}$. In particular, if we define

$$\begin{array}{ll} E_{\mathbf{10}^+} = (0,\underline{1,1,0,0,0}) & E_{\mathbf{10}^-} = \frac{1}{2}(-\sqrt{3},\underline{1,1,-1,-1,-1}) \\ E_{\overline{\mathbf{10}}^+} = -(0,\underline{1,1,0,0,0}) & E_{\overline{\mathbf{10}}^-} = -\frac{1}{2}(-\sqrt{3},\underline{1,1,-1,-1,-1}) \end{array} \tag{36}$$

we have the relations

$$[E^\pm, E_{\mathbf{10}^\mp}] = E_{\mathbf{10}^\pm}, \qquad [E^\pm, E_{\mathbf{10}^\pm}] = 0, \qquad [P, E_{\mathbf{10}^\pm}] = \pm E_{\mathbf{10}^\pm} \tag{37}$$

We now analyze the resulting gauge symmetry group and the structure of matter curves due to (31) by looking at the commutant of $\langle\Phi\rangle$ as a function of the coordinates $(x,y)$. The gauge group is the commutant at generic points while changes in its rank signal the matter curves [27]. In this case one can easily check that the set of roots of the subalgebra $\mathfrak{su}(5)\oplus\mathfrak{u}(1)\subset\mathfrak{e}_6$ do commute at generic points in $S$ and so we can identify the G T gauge group with S (5).

Regarding the matter curves, we find that at $_{\mathbf{5}} = \{x-y=0\}$ there are additional roots commuting with $\langle\Phi\rangle$ given by

$$E_{\mathbf{5}} = \frac{1}{2}(\sqrt{3},\underline{1,-1,-1,-1,-1}) \qquad ; \qquad E_{\overline{\mathbf{5}}} = \frac{1}{2}(-\sqrt{3},\underline{-1,1,1,1,1}) \tag{38}$$

which correspond respectively to $(\mathbf{5},\mathbf{1})_2$ and $(\overline{\mathbf{5}},\mathbf{1})_{-2}$ in the decomposition (35). ence, at $_{\mathbf{5}}$ the symmetry group enhances to S (6) × (1). On the other hand, the action of $\langle\Phi\rangle$ on the sector $(\mathbf{10},\mathbf{2})_{-1}$ is given by

$$[\langle\Phi\rangle, R_+E_{\mathbf{10}^+} + R_-E_{\mathbf{10}^-}] = \begin{pmatrix} -\mu^2(x-y) & m \\ m^2x & -\mu^2(x-y) \end{pmatrix}\begin{pmatrix} R_+E_{\mathbf{10}^+} \\ R_-E_{\mathbf{10}^-} \end{pmatrix} \tag{39}$$

where $R_\pm$ are functions on $S$. For the con ugate sector $(\overline{\mathbf{10}}, \mathbf{2})_1$ there is an analogous result. At $\ _{\mathbf{10}} = \{\mu^4(x-y)^2 = m^3 x\}$ the matrix (39) has vanishing determinant. ence, there are additional roots commuting with $\langle\Phi\rangle$, and therefore a ump in the rank of the symmetry group. Therefore $\ _{\mathbf{10}}$ is identified with the **10** curve of this T-brane background. Observe that the curves $\ _{\mathbf{5}}$ and $\ _{\mathbf{10}}$ meet at the Yukawa point $p_{\rm up} = \{x = y = 0\}$.[2]

The non-Abelian T-brane background is such that $[\langle\Phi\rangle, \langle\Phi\rangle^\dagger] \neq 0$. In consequence, a non-primitive ux $\langle F_{\rm np}\rangle$, i.e. $\omega \wedge \langle F_{\rm np}\rangle \neq 0$, is required to fulfill the D-term equation (4). Proceeding as in [27] we deduce that the F-terms (3) are still verified by

$$\langle\Phi_{xy}\rangle = m(e^f E^+ + mxe^{-f}E^-) + \mu^2(bx-y)Q, \quad \langle A_{0,1}\rangle = -\frac{i}{2}\bar\partial f P, \quad \langle F_{1,1}\rangle = -i\partial\bar\partial f P \quad (40)$$

It then transpires that to satisfy the D-term the function $f$ must satisfy a non-linear differential equation of the Painlevé III type. This complicates the task of solving for the zero mode wavefunctions, but as shown in [28] approximate solutions can still be found.

Besides the non-primitive ux in (40), we can still add contributions to the background worldvolume ux $\langle F\rangle$ provided they do not spoil the F-term and D-term conditions. The simplest possibility is to introduce primitive $(1,1)$ uxes in the Cartan of $E_6$. Such uxes are generically present and are important for the phenomenology of the model. Indeed they can be chosen to generate 4d chirality for the S (5) spectrum, and to further break S (5) to S (3) × S (2) × (1)$_Y$. Let us then consider consider the worldvolume ux

$$\langle F_Q\rangle = i\left[-M(dy\wedge d\bar y - dx\wedge d\bar x) + N(dx\wedge d\bar y + dy\wedge d\bar x)\right]Q \quad (41)$$

where $Q$ is given in (32), and $M$ and $N$ are ux densities near the Yukawa point that we approximate by constants. It is easy to check that the equations of motion are still satisfied for any value of $M$, $N$, which can be considered as real parameters of the model. oreover, $\langle F_Q\rangle$ induces 4d chirality in the matter curves because, unlike the background in (40), it does distinguish the modes of opposite chirality **5**, $\bar{\mathbf{5}}$ and **10**, $\overline{\mathbf{10}}$. Thus, it can select normalizable modes of one chirality or the other according to the sign of $M$ and $N$. A more detailed discussion of the local chirality index can be found in appendix B of [24].

We also switch on a ux along the hypercharge generator. Concretely

$$\langle F_Y\rangle = i\Big[N_Y(dy\wedge d\bar y - dx\wedge d\bar x) + N_Y(dx\wedge d\bar y + dy\wedge d\bar x)\Big]Q_Y \quad (42)$$

where $N_Y$, $N_Y$ are local ux densities and

$$Q_Y = \frac{1}{3}(H_2 + H_3 + H_4) - \frac{1}{2}(H_5 + H_6) \quad (43)$$

is the hypercharge generator. As in the SO(12) model the hypercharge ux splits the particles within the same S (5) multiplet according to their hypercharge.

Summarizing, the total ux of this model is given by

$$\langle F\rangle = \langle F_{\rm p}\rangle + \langle F_{\rm np}\rangle \quad (44)$$

ere $\langle F_{\rm np}\rangle$ is the non-primitive ux provided in (40). The primitive piece $\langle F_{\rm p}\rangle$ is the sum of (41) and (42) that can be conveniently cast as

$$\langle F_{\rm p}\rangle = iQ_R(dy\wedge d\bar y - dx\wedge d\bar x) + iQ_S(dx\wedge d\bar y + dy\wedge d\bar x) \quad (45)$$

[2] Note that for this local model $\langle\Phi_{xy}\rangle \neq 0$ at $p_{\rm up}$, and so the symmetry group is no longer $E_6$ at the ukawa point. As discussed in [2 ] this is a general feature of T-brane con gurations. By abuse of terminology, we will still refer to this point as the $E_6$ point of the local model.

where we have defined

$$Q_R = -MQ + N_Y Q_Y, \qquad Q_S = NQ + N_Y Q_Y \tag{46}$$

Given the corresponding roots it is straightforward to compute the charges of the fields localized on the matter curves (c.f. table 1 in [28]). The ux parameters are constrained by the conditions of chirality and doublet-triplet splitting. In [28] it was found that chirality in the **10** sector requires $-3/2 < N_Y/M < 6$. In the **5** sector doublet-triplet splitting imposes $N_Y = -6N$, whereas chirality demands $N > 0$.

The equations of motion with given backgrounds (40) and (45) have been studied in [28] and shown to have localized solutions on the matter curves $_{\mathbf{5}}$ and $_{\mathbf{10}}$. In the **5** sector, which is neutral under S (2), $\langle\Phi\rangle$ is effectively Abelian and the zero modes that are found are analogous to those in the $c$ sector in the SO(12) model. The **10** sector is more complicated because the zero modes belong in a S (2) doublet. ence, these modes feel the non-Abelian piece of $\langle\Phi\rangle$ and are charged under the non-primitive ux that depends on the Painlevé transcendent $f$. evertheless, in [28] we were able to find approximate solutions in the limit $m \gg \mu$. With the zero modes at hand one can then compute the normalizations that appear in the physical Yukawa couplings.

## 4. Non-perturbative effects and Yukawa couplings

In this section we will first review how the non-perturbative effects proposed in [18] can be incorporated in local F-theory models. We will then study corrections to the Yukawa couplings due to the non-perturbative effects. Following [14, 27] we will obtain the holomorphic couplings by means of a residue formula.

### *4.1. Non-perturbative effects*

Besides the surfaces $S_{\rm GUT}$ and $S_i$ that enter in the local G T model, a global F-theory compactification will contain other divisors of the six-dimensional base $B$ that are also wrapped by branes. Typical examples are hidden sector 7-branes that develop a gaugino condensate, or Euclidean 3-branes. In both cases the additional branes generically wrap a surface, denoted $S_{\rm np}$, defined by a function $h$, namely $S_{\rm np} = \{h = 0\}$. As noticed in [18] the extra branes can source non-perturbative effects that induce corrections to the tree-level superpotential (1). Generalizing the analysis in [18] the non-perturbative superpotential was found to be [24]

$$W_{\rm np} = m_*^4 \left[ \frac{\epsilon}{2} \sum_{n\in\mathbb{N}} \int_S \theta_n \, {\rm STr} \left( \Phi_{xy}^n F \wedge F \right) \right] \tag{47}$$

where $\theta_n$ is proportional to the $n^{\rm th}$ derivative of $h$ normal to $S = \{z = 0\}$ and $\epsilon$ is a small parameter that measures the strength of the non-perturbative effect. ore precisely,

$$\epsilon = \mathcal{A} e^{-T_{\rm np}} h_0^{N_{\rm D3}} \tag{48}$$

where $T_{np}$ is the complexified Kähler modulus of $S_{\rm np}$, $h_0 = \int_S h$, $\mathcal{A}$ depends on the bulk moduli of the base $B$, and $N_{\rm D3} = (8\pi^2)^{-1} \int_S {\rm tr}(F \wedge F) \in \mathbb{N}$. We refer to Appendix C of [24] for more details about these quantities, the derivation of (47), and the proof that these non-perturbative effects do not modify the 7-brane D-term.

Observe that (47) is written in terms of the G T 7-brane fields $\Phi$ and $A$, ust like the tree-level superpotential (1). ence, we can add up both expressions and apply the ultra-local approach to compute 7-brane zero mode wavefunctions near a Yukawa point. This program was first carried out in a toy model with $G_p =$ (3) but considering only $\theta_1 \neq 0$. The reason for such simplification was the assumption made in [18] that the two 4-cycles $S$ and $S_{\rm np}$ do not intersect, so that $h|_S \equiv h_0$ and $\theta_0 = 0$, while other $\theta_n$ are more suppressed. ow, for $G_p = \mathrm{SO}(12)$ or $G_p = \mathrm{E}_6$ these assumptions have to be revisited because the corrections to Yukawa couplings due to the term proportional to $\theta_1$ in $W_{\rm np}$ are proportional to the symmetric trace of three group generators which vanishes identically for SO(12) and $\mathrm{E}_6$.

In [24] it was argued that $\theta_0 = (4\pi^2 m_*)^{-1}[\log h/h_0]_{z=0}$ can be non-trivial and it was included together with $\theta_2$ in the computation of down-type Yukawa point $p_{\rm down}$. In fact, it was found that ust $\theta_0$ was enough to generate a hierarchical rank 3 matrix of Yukawas. In [28] it was also shown that the least suppressed parameter $\theta_0$ solves the rank 1 problem for up-type Yukawa couplings. ence, in order to simplify the discussion we only consider the first term of $W_{\rm np}$ and work with the total superpotential

$$W_{\rm total} = m_*^4 \int_S \mathrm{tr}\,(F \wedge \Phi) + \epsilon\, \frac{\theta_0}{2} \mathrm{Tr}\,(F \wedge F) \tag{49}$$

In the following we will base our local analysis on the superpotential (49) and the D-term (2) which does not receive non-perturbative corrections.

The specific form of $\theta_0$ depends on $S_{\rm np}$ and thus on the global completion of the local model but we can still assume it to be a general holomorphic function of $(x, y)$. ear the intersection point, that can be chosen to be $x = y = 0$, we can make the linear approximation

$$\theta_0 \;=\; i(\theta_{00} + \theta_x x + \theta_y y) \tag{50}$$

where $\theta_{00}$, $\theta_x$ and $\theta_y$ are constant parameters. otice that for $\theta_0$ non-constant the second term in (49) is not a total derivative, and we can expect a non-trivial effect on the 7-brane fields. In fact, the non-perturbative correction changes the F-term equations (3) to

$$\partial_A \Phi + \epsilon\, \partial\theta_0 \wedge F = 0 \tag{51a}$$
$$F_{0,2} = 0 \tag{51b}$$

The $\theta_0$ dependent contribution already indicates that the backgrounds $\langle\Phi\rangle$ and $\langle F\rangle$ must be shifted from their original values in order to satisfy the new F-terms. Clearly, the equations of motion for the uctuations will also be modified as discussed below.

To solve the equations we perform a perturbative expansion in the small parameter $\epsilon$ of the form

$$\Phi \;=\; \Phi^{(0)} + \epsilon\, \Phi^{(1)} + \cdots \tag{52}$$

and similarly for the gauge field. To begin let us solve the F-terms at the level of the background. To zeroth order in $\epsilon$ we use a holomorphic gauge $\langle A_{0,1}\rangle^{(0)} = 0$. The F-terms are then satisfied as long as $\bar\partial\langle\Phi\rangle^{(0)} = 0$, so that $\langle\Phi\rangle^{(0)} = 0$ is a holomorphic function of the coordinates. To first order in $\epsilon$ the F-terms still allow to maintain a holomorphic gauge, namely $\langle A_{0,1}\rangle^{(1)} = 0$. owever, $\langle\Phi\rangle$ must be shifted according to

$$\langle\Phi\rangle \;=\; \langle\Phi\rangle^{(0)} + \epsilon\, \partial\theta_0 \wedge \langle A_{1,0}\rangle^{(0)} + \mathcal{O}(\epsilon^2) \tag{53}$$

Recall that in general $\langle A_{1,0}\rangle^{(0)} \neq 0$, for instance in the SO(12) model it is given in (30).

Let us now derive the equations of motion for the bosonic uctuations $(\varphi, a)$ defined by (5). In turn $(\varphi, a)$ have a perturbative expansion in $\epsilon$ of the form (52). From the F-terms, and taking into account the shift in the background (53), we find the zero mode equations

$$\bar\partial a = 0 + \mathcal{O}(\epsilon^2) \tag{54a}$$
$$\bar\partial\varphi - i[a, \langle\Phi\rangle^{(0)}] = -\epsilon\, \partial\theta_0 \wedge \partial a + \mathcal{O}(\epsilon^2) \tag{54b}$$

ere we have disregarded the uctuations $a^\dagger$ since they do not contribute to the Yukawa couplings as shown in appendix D of [24]. The D-term does not receive non-perturbative corrections, meaning that the uctuations still satisfy the equation (6c). owever, this D-term equation will acquire a $\theta_0$ dependence once the shift in the background (53) is taken into account.

To order $\epsilon$ the zero mode equations (54) have the general solution

$$a = \partial\xi \tag{55a}$$

$$\varphi = h - i[\langle\Phi\rangle^{(0)}, \xi] - \epsilon\partial\theta_0 \wedge \partial\xi \tag{55b}$$

where $\xi$ is a 0-form and $h$ is a $(2,0)$-form such that $\bar{\partial}h = 0$. Both $\xi$ and $h$ transform in the ad oint representation of $G_p$. As in [14] $h$ can be fixed up to normalization by exploiting the invariance of the tree-level superpotential (1) under the gauge transformations

$$\begin{array}{rcl} a & \to & a + \bar{\partial}_{\langle A\rangle}\chi \\ \varphi & \to & \varphi - i\,[\langle\Phi\rangle, \chi\,] \end{array} \tag{56}$$

Concretely, to fix $h$ we can perform transformations with $\chi$ holomorphic.

The function $\xi$ will enter prominently in the computation of the Yukawa couplings. It can be found in terms of $\varphi$ as follows. First we introduce the holomorphic matrix $\Psi$ defined by

$$[\langle\Phi\rangle^{(0)}, \xi] = \Psi\xi\, dx \wedge dy \tag{57}$$

Then we rewrite (55b) as

$$\Psi\xi\, dx \wedge dy \;=\; i(\varphi - h + \epsilon\partial\theta_0 \wedge \partial\xi) \tag{58}$$

This last equation can be solved perturbatively to obtain

$$\xi = \xi^{(0)} + i\epsilon\,\Psi^{-1}\left(\partial_x\theta_0\partial_y\xi^{(0)} - \partial_y\theta_0\partial_x\xi^{(0)}\right) + \mathcal{O}(\epsilon^2) \tag{59a}$$

$$\xi^{(0)} = i\Psi^{-1}\left(\varphi^{(0)}_{xy} - h_{xy}\right) \tag{59b}$$

The function $\xi$ is required to be regular at the matter curves.

Substituting $\xi$ in (59) in $a = \bar{\partial}\xi$ we can find $a$ in terms of $\varphi$ and then plug in the D-term equation (6c) to obtain the solution for $\varphi$. In this way the non-perturbative zero modes are derived. We could then insert these zero modes in the superpotential to deduce the Yukawa couplings. owever, we will see that the couplings can be more e ciently determined by a residue formula which is independent of the exact form of the corrected zero modes. We only need to know that the corrected zero modes are localized and this is indeed the case as shown in [24] for the SO(12) model and in [28] for the $E_6$ model. Although the zero modes receive corrections at order $\epsilon$ it can be proved that the contributions to the norms appear only at $\mathcal{O}(\epsilon^2)$ and would only affect the physical Yukawa couplings a this order. This means that to compute the couplings at $\mathcal{O}(\epsilon)$ it su ces to use the normalization of the tree-level zero modes.

### *4.2. Holomorphic Yukawa couplings*

The Yukawa couplings are found by expanding the superpotential to cubic order in uctuations, substituting the zero mode solutions, and finally performing the integration. In principle there are contributions due to the tree-level superpotential plus the non-perturbative term. In the case of the total superpotential (49), with a non-perturbative correction depending only on $\theta_0$, it can be shown that expanding to cubic order gives

$$Y \;=\; -im_*^4 \int_S \mathrm{Tr}\,(\varphi \wedge a \wedge a) \tag{60}$$

Journal of Physics: Conference Series **651** (2015) 012009 doi:10.1088/1742-6596/651/1/012009

Although there is no additional piece proportional to $\epsilon$, the couplings will depend on $\theta_0$ through the corrected zero modes.

To perform the integration we first use the solution to the F-terms (55) to arrive at

$$Y = -\frac{im_*^4}{3}\int_S \mathrm{Tr}\left\{h \wedge a \wedge a - \partial(\varphi \wedge [a,\xi]) - \epsilon\partial(\theta_0\partial(a \wedge a\,\xi))\right\} \tag{61}$$

The last two terms are total derivatives that vanish upon integration because they involve localized fields $(\varphi, a)$. sing (54a) once more then yields

$$Y = -\frac{im_*^4}{3}\int_S \mathrm{Tr}\left(h \wedge \partial\xi \wedge \partial\xi\right) \tag{62}$$

Since $h$ is holomorphic the integrand is still a total derivative. Thus, upon integration, $Y$ can be converted into a surface integral around the intersection point. As in [14, 27] the localized modes $\varphi_{xy}$ that appear in (59) do not contribute, and we deduce that

$$Y = m_*^4 f_{abc}\int_{\mathcal{R}} (\eta^a \eta^b h^c_{xy})\, dx \wedge dy \tag{63}$$

where $f_{abc}$ are structure constants of the symmetry group $G_p$ at the Yukawa point $p$, $\mathcal{R}$ is diffeomorphic to the product of two circles surrounding $p$, and $\eta$ are the auxiliary holomorphic functions

$$\eta = -i\Psi^{-1}h_{xy} + \epsilon\Psi^{-1}\left(\partial_x\theta_0\partial_y(\Psi^{-1}h_{xy}) - \partial_y\theta_0\partial_x(\Psi^{-1}h_{xy})\right) + \mathcal{O}(\epsilon^2) \tag{64}$$

related to $\xi$ by removing the dependence on $\varphi_{xy}$. Finally, we can express (63) as a residue formula evaluated at the Yukawa point $p$

$$Y = m_*^4\pi^2 f_{abc}\,\mathrm{Res}_p(\eta^a\eta^b h^c) \tag{65}$$

where for simplicity we have dropped the subindices to $h_{xy}$. In the following we will apply this residue formula to the SO(12) and E$_6$ local models constructed in section 3.

It is important to realize that the residue formula was obtained inserting only the solution of the F-terms in the superpotential as advocated in [14]. In this way only the holomorphic piece of the Yukawa couplings is computed. To evaluate the physical Yukawa couplings we also need to solve the D-term equations to find the zero mode wavefunctions and impose that they give canonically normalized kinetic terms. This last condition will fix the arbitrary constants that appear in the holomorphic couplings.

*4.2.1. Holomorphic couplings in the* SO(12) *model* To determine the couplings we first have to find the auxiliary functions $\eta$ defined in (64). In the SO(12) model the matrix $\Psi$ is proportional to the identity. To see this we write $\xi = \xi_\rho E_\rho$ where $\rho$ is a root in one of the $a^+$,$b^+$, or $c^+$ sector. From (20) and (57) it follows that we can replace $\Psi$ by $m^2 q_\Phi(\rho)$, where the $q_\Phi(\rho)$ are given in table 1. ence, in the SO(12) case (64) reduces to

$$\eta_\rho = -\frac{i}{m^2 q_\Phi(\rho)}h_\rho + \frac{i\epsilon}{m^4 q_\Phi(\rho)}\left[\theta_x\partial_y\left(\frac{h_\rho}{q_\Phi(\rho)}\right) - \theta_y\partial_x\left(\frac{h_\rho}{q_\Phi(\rho)}\right)\right] + \mathcal{O}(\epsilon^2) \tag{66}$$

where we have already used (50) for $\theta_0$.

We also need to know the so called holomorphic representatives $h_\rho$. Consider for example the $a^+$ sector in which $q_\Phi(\rho) = -x$. In principle $h_{a^+}$ is an arbitrary function of $(x, y)$ but by selecting $\chi$ appropriately in (56) we can gauge away all $x$ dependence. ence, $h_{a^+}$ is a holomorphic function of $y$ and following standard practice [15] we can choose a basis $\{1, y, y^2\}$ to represent 3 families. The $b^+$ and $c^+$ sectors are analogous. Concretely, $h_{b^+}$ is a holomorphic function of $x$ and we pick a basis $\{1, x, x^2\}$, whereas $h_{c^+}$ depends on $(x+y)$ and we take it to be a constant since in this sector there is only one zero mode corresponding to the iggs. Altogether we have

$$h_{a^+} = \gamma_a^i y^{3-i} \quad ; \quad h_{b^+} = \gamma_b^j x^{3-j} \quad ; \quad h_{c^+} = \gamma_c \tag{67}$$

where $i, j, = 1, 2, 3$.

At this stage the normalization constants in (67) are arbitrary, but they are actually related to the normalization of the zero mode wavefunctions through (59b). For example,

$$\xi_{a^+}^{i\,(0)} = -\frac{i}{m^2 x}\left(\varphi_{a^+}^{i\,(0)} - \gamma_a^i y^{3-i}\right) \tag{68}$$

Regularity at the curve $_a = \{x = 0\}$ where the modes are localized requires that

$$\varphi_{a^+}^{i\,(0)} \to \gamma_a^i y^{3-i} \tag{69}$$

as $x \to 0$.

Inserting (67) in (66) yields

$$\eta_{a^+}^i = \frac{i\gamma_a^i}{m^2 x}\left\{y^{3-i} + \frac{\epsilon}{m^2 x}\left[(3-i)\theta_x y^{2-i} + \frac{\theta_y}{x} y^{3-i}\right]\right\} + \mathcal{O}(\epsilon^2) \tag{70a}$$

$$\eta_{b^+}^j = -\frac{i\gamma_b^j}{m^2 y}\left\{x^{3-j} + \frac{\epsilon}{m^2 y}\left[(3-j)\theta_y x^{2-j} + \frac{\theta_x}{y} x^{3-j}\right]\right\} + \mathcal{O}(\epsilon^2) \tag{70b}$$

We can now plug these functions, together with $h_{c^+} = \gamma_c$, in the formula (65) and then extract the residue at $(x, y) = (0, 0)$ to evaluate the couplings to $\mathcal{O}(\epsilon^2)$. The resulting couplings are collected in the Yukawa matrix $Y_{a^+b^+c^+}^{ij}$

$$Y_{a^+b^+c^+} = \frac{\pi^2\gamma_c}{\rho_m^2} f_{a^+b^+c^+} \begin{pmatrix} 0 & 0 & \epsilon\theta_x m_*^2 \gamma_a^1 \gamma_b^3 \\ 0 & \epsilon(\theta_x + \theta_y) m_*^2 \gamma_a^2 \gamma_b^2 & 0 \\ \epsilon\theta_y m_*^2 \gamma_a^3 \gamma_b^1 & 0 & \gamma_a^3 \gamma_b^3 \end{pmatrix} + \mathcal{O}(\epsilon^2) \tag{71}$$

where $\rho_m = m^2/m_*^2$.

We clearly observe that at tree-level, $\epsilon = 0$, the Yukawa matrix has rank one. Including non-perturbative effects so that $\epsilon \neq 0$ increases the rank to three. In fact, the eigenvalues have a hierarchical structure $(\mathcal{O}(1), \mathcal{O}(\epsilon), \mathcal{O}(\epsilon^2))$.

*4.2.2. Holomorphic couplings in the* $\mathrm{E}_6$ *model* To find the auxiliary functions it is convenient to treat the **5** and **10** sectors separately because the matrix $\Psi$ has a different form on each. In the **5** sector, neutral under S (2), the action of $\langle\Phi_{xy}\rangle^{(0)}$ in (31) is such that effectively $\Psi = 2\mu^2(x-y)$. ence

$$i\eta_{\mathbf{5}}/\gamma_{\mathbf{5}} = \frac{1}{2\mu^2(x-y)} - \epsilon\frac{\theta_x + \theta_y}{4\mu^4(x-y)^3} + \mathcal{O}(\epsilon^2) \tag{72}$$

ere we have used that $h_{\mathbf{5}} = \gamma_{\mathbf{5}}$ which can be attained by a gauge transformation of type (56) that allows to gauge away any dependence on $(x-y)$. Thus, $h_{\mathbf{5}}$ can be taken to be an arbitrary holomorphic function of the orthogonal coordinate $(x+y)$ and since there is only one iggs zero mode we choose it to be a constant, according to the usual practice [15].

Journal of Physics: Conference Series **651** (2015) 012009 doi:10.1088/1742-6596/651/1/012009

In the **10** sector the curve is given by $_{\mathbf{10}} = \{\mu^4(x-y)^2 = m^3 x\}$ and the localized modes live in the root subspace spanned by (36). The uxes can be chosen so that the chiral zero modes belong in the subspace spanned by $E_{\mathbf{10}^+}$ and $E_{\mathbf{10}^-}$ [28]. We will assume that there are exactly three chiral families associated with three matter **10**-plets in our S (5) G T model. ow $\Psi$ can be represented by a $2\times 2$ matrix. From (31) we read

$$\Psi\begin{pmatrix} E_{\mathbf{10}^+} \\ E_{\mathbf{10}^-}\end{pmatrix} = \begin{pmatrix} -\mu^2(x-y) & m \\ m^2 x & -\mu^2(x-y)\end{pmatrix}\begin{pmatrix} E_{\mathbf{10}^+} \\ E_{\mathbf{10}^-}\end{pmatrix} \tag{73}$$

since $Q=-1$ acting on the doublet.

We also need to specify $h_{\mathbf{10}}$ which is now an S (2) doublet of arbitrary holomorphic functions. By performing a gauge transformation (56) it can be brought in the form

$$h^i_{\mathbf{10}} = \begin{pmatrix} 0 \\ m_*^{3-i}\gamma^i_{\mathbf{10}}(x-y)^{3-i}\end{pmatrix} \tag{74}$$

with $i=1,2,3$. The reason is that the gauge parameter $\chi$, now a doublet, can be chosen so that the upper component of $h_{\mathbf{10}}$ is gauged away while the lower component is made to depend only on $(x-y)$. As usual we take a base of monomials to realize 3 families. Substituting (74) in (64) then yields

$$\begin{aligned} i\eta^i_{\mathbf{10}}/\gamma^i_{\mathbf{10}} &= -\left[\frac{m_*^{3-i}(x-y)^{3-i}}{\mu^4(x-y)^2-m^3x}\right]\begin{pmatrix} m \\ \mu^2(x-y)\end{pmatrix} + \mathcal{O}(\epsilon^2) \\ &+ \epsilon\frac{2\mu^4(\theta_x+\theta_y)(x-y)+m^3\theta_y}{(\mu^4(x-y)^2-m^3x)^3}m_*^{3-i}(x-y)^{3-i}\begin{pmatrix} 2m\mu^2(x-y) \\ (m^3x+\mu^4(x-y)^2)\end{pmatrix} \\ &+ \epsilon\frac{(\theta_x+\theta_y)}{(\mu^4(x-y)^2-m^3x)^2}m_*^{3-i}(x-y)^{2-i}\begin{pmatrix} 2m\mu^2(x-y)(6-i) \\ m^3x(3-i)+(4-i)\mu^4(x-y)^2\end{pmatrix}\end{aligned} \tag{75}$$

The $\mathcal{O}(\epsilon)$ correction to $\eta^i_{\mathbf{10}}$ is seemingly complicated but there will be simplifications when inserting in the residue formula.

We can now apply the explicit expressions for $(h_{\mathbf{5}},\eta_{\mathbf{5}})$ and $(h_{\mathbf{10}},\eta_{\mathbf{10}})$ to the residue formula (65) for the Yukawa couplings. The calculation is made easier from the fact that the structure constants of $E_6$ satisfy

$$\mathrm{Tr}([E_{\mathbf{5}\,i}, E_{\mathbf{10}\,jk}^{\;M}]E_{\mathbf{10}\,lm}^{\;N}) = \epsilon_{ijklm}\epsilon^{MN} \tag{76}$$

where $i,j,k,l,m$ are S (5) indices and $M,N=\pm$ are S (2) indices. Then, the non-trivial contributions to the $\mathbf{10}\times\mathbf{10}\times\mathbf{5}$ Yukawa will be of the form

$$Y = m_*^4\pi^2\mathrm{Res}_{(0,0)}\left(\epsilon_{MN}\eta_{\mathbf{5}}\eta^M_{\mathbf{10}}h^N_{\mathbf{10}}\right) = m_*\pi^2\mathrm{Res}_{(0,0)}\left(\eta_{\mathbf{5}}\eta^+_{\mathbf{10}}h^-_{\mathbf{10}}\right) \tag{77}$$

where the contractions of the S (5) indices have been left implicit. In the first equality we have used that any other contribution will contain a term of the form $\epsilon_{MN}\eta^M_{\mathbf{10}}\eta^N_{\mathbf{10}}$ and thus it will vanish identically, and in the second equality we have used that in our solution (74) $h^+_{\mathbf{10}}=0$. ence, even if (76) has a complicated expression only the terms proportional to $E_{\mathbf{10}^+}$ will be relevant when computing up-like Yukawa couplings.

Let us proceed by computing (77) explicitly. At zeroth order in $\epsilon$ we have a contribution

$$\begin{aligned} Y^{ij}_{\text{tree}} &= m_*^4\pi^2\gamma_{\mathbf{5}}\gamma^i_{\mathbf{10}}\gamma^j_{\mathbf{10}}\,\mathrm{Res}_{(0,0)}\left[\frac{m(m_*(x-y))^{6-i-j}}{2\mu^2(x-y)(\mu^4(x-y)^2-m^3x)}\right] \\ &= -\frac{m_*^4\pi^2}{2m^2\mu^2}\gamma_{\mathbf{5}}\gamma^i_{\mathbf{10}}\gamma^j_{\mathbf{10}}\,\delta_{i3}\delta_{j3}\end{aligned} \tag{78}$$

ence at this level only $Y^{33}$ is non-zero. At order $\mathcal{O}(\epsilon)$ we find terms of the form

$$Y^{ij}_{\rm np} = \epsilon \frac{m_*^6 \pi^2}{4m^2\mu^4}(\theta_y + \theta_x)\gamma_{\bf 5}\gamma^i_{\bf 10}\gamma^j_{\bf 10}\,\delta_{(i+j)4} \tag{79}$$

from the $\mathcal{O}(\epsilon)$ correction to $\eta_{\bf 5}$. In fact, one can check that the $\mathcal{O}(\epsilon)$ correction to $\eta_{\bf 10}$ does not contribute to (77). In the end we are left with the following $\mathbf{10}\times\mathbf{10}\times\mathbf{5}$ Yukawa couplings

$$Y^{ij} = \frac{\pi^2\gamma_{\bf 5}}{4\rho_\mu\rho_m}\begin{pmatrix} 0 & 0 & \epsilon(\theta_x+\theta_y)\rho_\mu^{-1}\gamma^1_{\bf 10}\gamma^3_{\bf 10} \\ 0 & \epsilon(\theta_x+\theta_y)\rho_\mu^{-1}\gamma^2_{\bf 10}\gamma^2_{\bf 10} & 0 \\ \epsilon(\theta_x+\theta_y)\rho_\mu^{-1}\gamma^1_{\bf 10}\gamma^3_{\bf 10} & 0 & -2\gamma^3_{\bf 10}\gamma^3_{\bf 10}\end{pmatrix} + \mathcal{O}(\epsilon^2) \tag{80}$$

where $\rho_\mu = \mu^2/m_*^2$, and $\rho_m = m^2/m_*^2$.

As anticipated, the Yukawa matrix has rank one in the absence of non-perturbative effects, but when taking them into account increases its rank to three. oreover, the eigenvalues of this matrix display a hierarchical pattern $(\mathcal{O}(1), \mathcal{O}(\epsilon), \mathcal{O}(\epsilon^2))$.

## 5. Physical couplings and mass hierarchies

The physical Yukawa couplings are computed via the overlap integral of zero mode wavefunctions which are normalized such that the kinetic terms have canonical form. After considerable work it can be shown that this calculation gives couplings that precisely match the holomorphic couplings up to normalization [24, 28]. We have also explained that the arbitrary $\gamma$ constants introduced in the computation of the holomorphic couplings, e.g. $\gamma^i_a$ in (67), are in turn related to the normalization of the zero modes, e.g. according to (69). We then conclude that the matrices (71) and (80) give the physical couplings provided the $\gamma$ constants are fixed by requiring that the zero modes have canonical norms. Roughly speaking we demand

$$|\gamma^i_a|^{-2} \sim \int_S |\varphi^i_{a^+}|^2 \tag{81}$$

oreover, it can also be argued that to obtain the physical Yukawa couplings at order $\epsilon$ order it is enough to use the norms of the tree-level zero modes [24, 28].

A very important fact is that the zero mode norms depend on the uxes and the associated charges. For example, in the SO(12) model [24]

$$|\gamma^i_a|^2 \sim \left(\frac{M + q_Y N_Y}{m_*^2}\right)^{4-i} \tag{82}$$

where $q_Y$ is the hypercharge. ence, as already remarked in [23], the physical couplings depend on the uxes, and the corresponding charges, through the norms. In particular the couplings $\mathbf{10}\times\bar{\mathbf{5}}\times\bar{\mathbf{5}}$ depend on hypercharge thereby explaining the difference between D-quark and lepton couplings. Explicit formulas for the normalization constants are given in [24] for the SO(12) and in [28] for the $E_6$ model. The full expressions are necessary to evaluate for instance the third generation couplings. owever, mass ratios such as $m_s/m_b$ only involve quotients such as $\gamma^2_a/\gamma^3_a$ which are rather simple.

The Yukawa couplings also depend on the mass parameters $m$ and $\mu$ that enter in the vev $\langle\Phi\rangle$, and on the F-theory scale $m_*$ related to the string scale $m_{st} = 2\pi\alpha'$ according to $m_{st}^4 = g_s(2\pi)^3 m_*^4$, where $g_s$ is the string coupling [24]. To analyze the physical couplings it is convenient to write the quantities $\rho_m$ and $\rho_\mu$ as

$$\rho_m = \left(\frac{m}{m_*}\right)^2 = (2\pi)^{3/2} g_s^{1/2}\sigma_m \quad ; \quad \rho_\mu = \left(\frac{\mu}{m_*}\right)^2 = (2\pi)^{3/2} g_s^{1/2}\sigma_\mu \tag{83}$$

ere $\sigma_m = (m/m_{st})^2$ and $\sigma_\mu = (\mu/m_{st})^2$ are the 7-brane intersection slopes which are assumed to be small. The value of $g_s$ is constrained by gauge coupling unification $\alpha_G = 2\pi^2 g_s/(m_{st}^4 V_S)$, where $V_S$ is the volume of the 7-brane surface $S$ (see e.g. [1]). Besides, ux quantization requires $M/\sqrt{V_S} \sim 2\pi$. Setting $\alpha_G \simeq 1/24$ leads to the estimate for the uxes

$$\frac{M}{m_{st}^2} = \left(\frac{2\alpha_G}{g_s}\right)^{1/2} \simeq \frac{0.29}{g_s^{1/2}} \tag{84}$$

aving diluted uxes imposes $M < m_{st}^2$. We then conclude that $\sqrt{g_s}$ cannot be arbitrarily small. The conditions of small intersection slope and uxes are needed to ustify the effective description of the 7-brane theory.

It is interesting to contrast the theoretical results against experimental values. Our expressions for Yukawa couplings apply at the unification-string scale, presumably of order $10^{16}$ GeV, so that in order to compare with experimental fermion masses it is necessary to run the data up to the unification scale. Our reference will be the updated two-loop analysis for this running within the SS performed in [30]. Below we will brie y discuss how the unification scale data can be reproduced in our models. ore details can be found in [24, 28].

### 5.1. *Physical couplings in the SO(12) model*

The third generation Yukawa can be read from the 33 element in (71). sing (83) and the exact form of the normalization factors yields that $Y_{b/\tau}$ is proportional to $\sigma_m\sqrt{g_s}$ with a small coe cient $\mathcal{O}(1)$. Taking a typical value $\sigma_m\sqrt{g_s} \sim 0.1$ indicates that the third generation Yukawa couplings are large, of the same order of the Yukawa coupling of the top quark. This points to a large value of $\tan\beta \simeq 20 - 50$ in the SS scheme, according to the data in [30].

We can also consider the ratio $Y_\tau/Y_b$ to see how these couplings differ at the unification scale. Taking into account the dependence on hypercharge of the normalization constants, it can be shown that there is a range of parameters for which $Y_\tau/Y_b \sim 1.37$, in good agreement with the data.

Identifying the first and second eigenvalues with the third and second generations gives to leading order in $\epsilon$ at the unification scale

$$\frac{m_s}{m_b} = \frac{M}{m^2}\left(1 - \frac{N_Y}{3M}\right)^{1/2}\left(1 + \frac{N_Y}{6M}\right)^{1/2} \epsilon\,(\theta_x + \theta_y) \tag{85a}$$

$$\frac{m_\mu}{m_\tau} = \frac{M}{m^2}\left(1 + \frac{N_Y}{2M}\right)^{1/2}\left(1 + \frac{N_Y}{M}\right)^{1/2} \epsilon\,(\theta_x + \theta_y) \tag{85b}$$

The ratio of the above quantities is independent of the non-perturbative effects and depends on uxes only through $N_Y/M$. Experimentally $\frac{m_\mu/m_\tau}{m_s/m_b} = 3.3 \pm 1$ [30], which can be fitted taking e.g. $N_Y/M \sim 1.8$. Substituting this value in (85b), and using $m_\mu/m_\tau = 5.4 \pm 0.6 \times 10^{-2}$, yields the estimate

$$\frac{M}{m^2}\,\epsilon\,(\theta_x + \theta_y) \simeq (2.3 \pm 0.2) \times 10^{-2} \tag{86}$$

ote that this value is quite small, consistent with a non-perturbative origin.

### 5.2. *Physical couplings in the* $\mathrm{E}_6$ *model*

The top Yukawa is given in (78). The ux and mass parameters can be chosen so that $Y_t \sim 0.5$ to match the experimental data [30].

Identifying the first and second eigenvalues with the third and second generations of -quarks gives

$$\frac{m_c}{m_t} = \frac{M}{2\mu^2}\left(1+\frac{2N_Y}{3M}\right)^{1/2}\left(1-\frac{N_Y}{6M}\right)^{1/2}\epsilon\,(\theta_x+\theta_y) \tag{87}$$

It is possible to ad ust the uxes to reproduce the hierarchy between the charm and the top quark with a small non-perturbative parameter $\epsilon$. For example, we may take $N_Y/M \sim 1.8$ as in the SO(12) model for the down-like Yukawa point $p_{\rm down}$, since, if the two Yukawa points $p_{\rm up}$ and $p_{\rm down}$ are not far away the ux densities should be alike. Substituting this value in (87) and requiring a realistic mass ratio, say $m_c/m_t \sim 2.5 \times 10^{-3}$, then gives

$$\frac{M}{\mu^2}\epsilon\,(\theta_x+\theta_y) \simeq 4\times 10^{-3} \tag{88}$$

which is similar to the estimate (86) in the SO(12) model.

It is encouraging that the data can be fitted with similar parameters as in the SO(12) model. owever, a more systematic analysis would entail embedding both Yukawa points $p_{\rm up}$ and $p_{\rm down}$ in the same local model, possibly in a region of $E_7$ or $E_8$ enhancement.

## 6. Final remarks

In this paper we have studied the structure of Yukawa couplings in F-theory S (5) models, taking into account the contribution of non-perturbative effects. We have considered explicit local models in order to compute down-like and up-like couplings. In one model there is an Abelian background that gives rise to three intersecting curves supporting localized matter transforming as **10**, $\bar{\mathbf{5}}$, and $\bar{\mathbf{5}}$. At the point where the curves intersect there is enhancing to SO(12) and down-like couplings $\mathbf{10}\times\bar{\mathbf{5}}\times\bar{\mathbf{5}}$ emerge. In the second model there is a non-Abelian background such that up-like $\mathbf{10}\times\mathbf{10}\times\mathbf{5}$ couplings are generated by the intersection of **10** and **5** matter curves at a point where S (5) is enhanced to $E_6$. The models include worldvolume uxes needed to guarantee the existence of localized chiral modes, as well as doublet-triplet splitting in the iggs sector.

From the general results of [14, 27] one expects to obtain a rank one matrix of Yukawa couplings when the couplings arise in a configuration with matter curves intersecting at one point. We have verified that this is indeed the case in our models. To overcome this situation we have then added non-perturbative effects sourced by gaugino condensation or D3-brane instantons. As argued in [18] these effects can be written in terms of the fields in the effective 8d theory and can thus be incorporated in the formalism that allows to compute Yukawa couplings. We have found that in both models the non-perturbative effects do lead to Yukawa matrices of rank three. Therefore, the non-perturbative effects induce non-trivial Yukawas for the two lighter families of quarks and leptons. oreover, the corrected couplings naturally reproduce a hierarchical mass structure among the three families.

At our level of approximation we are only able to predict the ratios of second to third generation couplings. Ad usting the free parameters in the models allows to match these ratios with the experimental values. We also find that the third generation Yukawa couplings are of the right order of magnitude.

Our results can be extended in various directions. For instance, we could examine the contribution of additional terms in the expansion (47) underlying the non-perturbative effects. From the results of [24] terms proportional to $\theta_n$ with $n>0$ should not alter the hierarchical structure that we have obtained with only $\theta_0$, but they could be relevant in the computation of the Yukawa couplings of the lightest generation. To this aim, we would have to take our approximations to order $\epsilon^2$ in the non-perturbative parameter $\epsilon$. Besides, improving the level of approximation would be required to determine the CK matrix for these models. Actually, because the CK matrix involves considering both Yukawa points $p_{\rm up}$ and $p_{\rm down}$ simultaneously, we could consider implementing our approach in local models of $E_7$ or $E_8$ enhancement [27, 31].

## Acknowledgments

I am deeply indebted to Luis Aparicio, Luis Ibánez, Fernando archesano, Diego Regalado, and Gianluca Zoccarato for a very enlightening collaboration on the sub ect of this paper. I thank Saúl Ramos-Sánchez for the invitation to the Workshop. Financial support from the organizers is gratefully acknowledged.

## References

[1] For an overview of string phenomenology see: Ibáñez L E and Uranga A M 2012 *String Theory and Particle Physics. An Introduction to String Phenomenology* (Cambridge: Cambridge University Press)
[2] G. Aldazábal, L. E. Ibáñez, F. Quevedo and A. M. Uranga, *"D-branes at singularities: A Bottom up approach to the string embedding of the standard model,"* HEP **0008**, 002 (2000) [ar iv:hep-th/000506 ].
[3] R. Blumenhagen, M. Cvetič, D. Lüst, R. Richter and T. Weigand, *"Non-perturbative Yukawa Couplings from String Instantons,"* Phys. Rev. Lett. **100**, 061602 (2008) [ar iv:0 0 .18 1 [hep-th]].
[4] C. Vafa, *"Evidence for F theory,"* Nucl. Phys. B **469**, 403 (1 6) [hep-th/ 602022].
[5] R. Donagi, M. Wi nholt, *"Model Building with F-Theory,"* [ar iv:0802.2 6 [hep-th]].
[6] C. Beasley, . . Heckman and C. Vafa, *"GUTs and Exceptional Branes in F-theory - I,"* HEP **0901** (200 ) 058 [ar iv:0802.33 1 [hep-th]].
[ ] C. Beasley, . . Heckman and C. Vafa, *"GUTs and Exceptional Branes in F-theory - II: Experimental Predictions,"* HEP **0901** (200 ) 05 [ar iv:0806.0102 [hep-th]].
[8] R. Donagi and M. Wi nholt, *"Breaking GUT Groups in F-Theory,"* [ ar iv:0808.2223 [hep-th]].
[ ] . . Heckman, *"Particle Physics Implications of F-theory,"* ar iv:1001.05 [hep-th].
[10] T. Weigand, *"Lectures on F-theory compactifications and model building,"* Class. Quant. Grav. **27**, 214004 (2010) [ar iv:100 .34 [hep-th]].
[11] G. . Leontaris, *"Aspects of F-Theory GUTs,"* PoS CORFU **2011** (2011) 0 5 [ar iv:1203.62 [hep-th]].
[12] L. E. Ibáñez, *"From Strings to the LHC: Les Houches Lectures on String Phenomenology,"* ar iv:1204.52 6 [hep-th].
[13] A. Maharana and E. Palti, *"Models of Particle Physics from Type IIB String Theory and F-theory: A Review,"* Int. . Mod. Phys. A **28**, 1330005 (2013) [ar iv:1212.0555 [hep-th]].
[14] S. Cecotti, M. C. N. Cheng, . . Heckman and C. Vafa, *"Yukawa Couplings in F-theory and Non-Commutative Geometry,"* [ar iv:0 10.04 [hep-th]].
[15] . . Heckman and C. Vafa, *"Flavor Hierarchy From F-theory,"* Nucl. Phys. B **837** (2010) 13 [ar iv:0811.241 [hep-th]].
[16] A. Font and L. E. Ibáñez, *"Matter wave functions and Yukawa couplings in F-theory Grand Unification,"* HEP **0909**, 036 (200 ) [ar iv:0 0 .48 5 [hep-th]].
[1 ] . P. Conlon and E. Palti, *"Aspects of Flavour and Supersymmetry in F-theory GUTs,"* HEP **1001**, 02 (2010) [ar iv:0 10.2413 [hep-th]].
[18] F. Marchesano and L. Martucci, *"Non-perturbative effects on seven-brane Yukawa couplings,"* Phys. Rev. Lett. **104**, 231601 (2010) [ar iv:0 10.54 6 [hep-th]].
[1 ] A. Font and L. E. Ibanez, *"Yukawa Structure from U(1) Fluxes in F-theory Grand Unification,"* HEP **0902**, 016 (200 ) [ar iv:0811.215 [hep-th]].
[20] E. Dudas and E. Palti, *"Froggatt-Nielsen models from E(8) in F-theory GUTs,"* HEP **1001**, 12 (2010) [ar iv:0 12.0853 [hep-th]].
[21] S. rippendorf, M. . Dolan, A. Maharana and F. Quevedo, *"D-branes at Toric Singularities: Model Building, Yukawa Couplings and Flavour Physics,"* HEP **1006**, 0 2 (2010) [ar iv:1002.1 0 [hep-th]].
[22] G. . Leontaris and G. G. Ross, *"Yukawa couplings and fermion mass structure in F-theory GUTs,"* HEP **1102**, 108 (2011) [ar iv:100 .6000 [hep-th]].
[23] L. Aparicio, A. Font, L. E. Ibáñez, F. Marchesano, *"Flux and Instanton Effects in Local F-theory Models and Hierarchical Fermion Masses,"* HEP **1108** (2011) 152 [ar iv:1104.260 [hep-th]].
[24] A. Font, L. E. Ibáñez, F. Marchesano and D. Regalado, *"Non-perturbative effects and Yukawa hierarchies in F-theory SU(5) Unification,"* HEP **1303**, 140 (2013) [ar iv:1211.652 [hep-th]].
[25] H. Hayashi, T. awano, R. Tatar and T. Watari, *"Codimension-3 Singularities and Yukawa Couplings in F-theory,"* Nucl. Phys. B **823**, 4 (200 ) [ar iv:0 01.4 41 [hep-th]].

Journal of Physics: Conference Series **651** (2015) 012009 doi:10.1088/1742-6596/651/1/012009

[26] H. Hayashi, T. awano, . Tsuchiya and T. Watari, *"Flavor Structure in F-theory Compactifications,"* HEP **1008**, 036 (2010) [ar iv:0 10.2 62 [hep-th]].
[2 ] S. Cecotti, C. Cordova, . . Heckman and C. Vafa, *"T-Branes and monodromy,"* HEP **1107**, 030 (2011) [ar iv:1010.5 80 [hep-th]].
[28] A. Font, F. Marchesano, D. Regalado and G. occarato, *Up-type quark masses in SU(5) F-theory models,* HEP **11**, 125 (2013). [ar iv:130 .808 [hep-th]].
[2 ] E. Palti, *"Wavefunctions and the Point of $E_8$ in F-theory,"* HEP **1207**, 065 (2012) [ar iv:1203.44 0 [hep-th]].
[30] G. Ross and M. Serna, *"Unification and fermion mass structure,"* Phys. Lett. B **664** (2008) [ar iv:0 04.1248 [hep-ph]].
[31] C. -C. Chiou, A. E. Faraggi, R. Tatar and W. Walters, *"T-branes and Yukawa Couplings,"* HEP **1105**, 023 (2011) [ar iv:1101.2455 [hep-th]].

# New insight into the Berezinskii-Kosterlitz-Thouless phase transition

**Urs Gerber** [1], **Wolfgang Bietenholz** [1] **and Fernando G Rejón-Barrera** [2]

[1] Instituto de Ciencias Nucleares, Universidad Nacional Autónoma de México
A.P. 0-543, C.P. 04510 Distrito Federal, Mexico
[2] Institute for Theoretical Physics, University of Amsterdam
Science Park 04, Postbus 4485, 10 0 GL Amsterdam, The Netherlands

E-mail: gerber@correo.nucleares.unam.mx, wolbi@nucleares.unam.mx, f.rejon@student.uva.nl

**Abstract.** We investigate the 2d model by using the constraint angle action, which belongs to the class of topological lattice actions. These actions violate important features usually demanded for a lattice action, such as the correct classical continuum limit and the applicability of perturbation theory. Nevertheless, they still lead to the same universal quantum continuum limit and show excellent scaling behavior. By using the constraint angle action we gain new insight into the Berezinskii- osterlitz-Thouless phase transition of the 2d model. This phase transition is of special interest since it is one of the few examples of a phase transition beyond second order. It is of in nite order and therefore an essential phase transition. In particular, we observe an excellent scaling behavior of the helicity modulus, which characterizes this phase transition. We also observe that the mechanism of (un)binding vortex anti-vortex pairs follows the usual pattern, although free vortices do not require any energy in the formulation of the 2d model using the constraint angle action.

## 1. 2d XY model and the constraint angle action

The 2d XY model has fascinated people since the early seventies, in particular because of its description of systems in condensed matter physics, such as super uid helium films [1] and superconducting films [2]. Further applications include the Coulomb gas model [3], Josephson unction arrays [4,5] and nematic liquid crystals [6]. The 2d XY model is of particular conceptual interest because it involves an essential phase transition, i.e. a phase transition of infinite order. Berezinskii [7,8], as well as Kosterlitz and Thouless [1] realized that the key to the understanding of this phase transition are the vortices and the anti-vortices.

In the 2d XY model a classical spin $\vec{e}_x$ is attached to each site $x = (x_1, x_2)$ e.g. on an $L \times L$ square lattice

$$\vec{e}_x = \begin{pmatrix} \cos\phi_x \\ \sin\phi_x \end{pmatrix} \in \mathbb{R}^2 , \tag{1}$$

such that $|\vec{e}_x| = 1 , \ \forall x$. In Figure 1 we show an example of a spin configuration on a $4 \times 4$ square lattice with periodic boundary conditions. We also indicate the lattice spacing $a$. The standard lattice action reads

$$S[\vec{e}] = \sum_{\langle xy \rangle} s(\vec{e}_x, \vec{e}_y) , \quad s(\vec{e}_x, \vec{e}_y) = \beta\Big(1 - \vec{e}_x \cdot \vec{e}_y\Big) = \beta\Big(1 - \cos(\phi_x - \phi_y)\Big) , \tag{2}$$

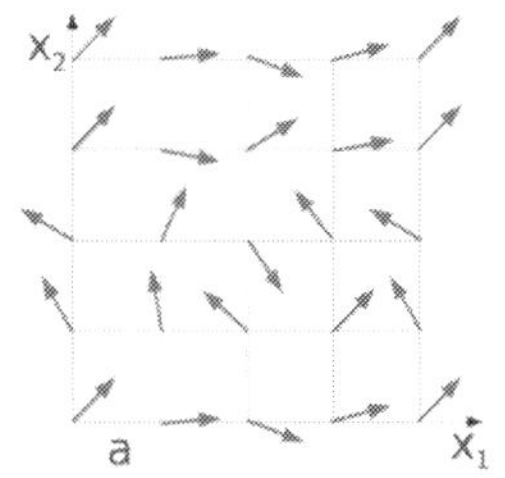


**Figure 1.** Example of a spin configuration on a $4 \times 4$ square lattice with lattice spacing $a$ and periodic boundary conditions.

with a positive inverse coupling constant $\beta > 0$. The sum runs over all nearest neighbor sites $x, y$. The standard lattice action is obtained by discretizing the derivatives of the continuum formulation of the 2d XY model. Due to the principle of universality, other lattice actions with additional spin couplings, such as the Villain action [3], lead to the same universal quantum continuum limit. The 2d XY model was also investigated by using the step action, which is not a continuous function in the angle $\phi_x$. Instead it is given by a step function

$$S[\vec{e}] = \sum_{\langle xy \rangle} s(\phi_x, \phi_y) , \quad s(\phi_x, \phi_y) = \begin{cases} -\beta & |\phi_x - \phi_y| < \pi/2 \\ \beta & \text{otherwise} \end{cases} . \tag{3}$$

It turns out that the step action indeed has the same universal quantum continuum limit as the standard lattice action [9, 10]. ere we study the 2d XY model by using the constraint angle action. This action belongs to the class of topological lattice actions. We define topological lattice actions as lattice actions that are invariant under (most) small deformations of a spin configuration [11]. In particular, for the constraint angle action the contribution of a nearest neighbor spin pair amounts to

$$s(\vec{e}_x, \vec{e}_y) = \begin{cases} 0 & \vec{e}_x \cdot \vec{e}_y > \cos\delta \\ +\infty & \text{otherwise} \end{cases} . \tag{4}$$

It does not have any couplings at all. earest neighbor spins are constrained to some relative maximum angle $\delta$. All configurations that violate this constraint for at least one spin pair $(\vec{e}_x, \vec{e}_y)$ have an infinite action (and are therefore excluded), while all other configurations have the same action $S[\vec{e}] = \sum_{\langle x,y \rangle} s(\vec{e}_x, \vec{e}_y) = 0$. For $\beta \to \infty$ the step action corresponds to the constraint angle action with $\delta = \pi/2$.

It should be pointed out that the idea of using a constraint angle action is not new. Patrascioiu and Seiler [12, 13] as well as Aizenman [14] have used an action with an angle constraint to simplify the proof for the existence of a massless phase in the 2d XY model. Furthermore, Patrascioiu and Seiler have also used a constraint angle action in their search for a massless phase in the 2d $O(3)$ model [15]. References [15, 16] presented numerical studies of the constraint angle action.

In the following sections we discuss our results and the new insight into the Berezinskii-Kosterlitz-Thouless phase transition. This proceeding contribution summarizes our publication [17].

## 2. Berezinskii-Kosterlitz-Thouless phase transition

The key to the understanding of the Berezinskii-Kosterlitz-Thouless (BKT) phase transition are the vortices and anti-vortices. On a square lattice, each plaquette has a winding number 0, +1 (vortex) or −1 (anti-vortex). This is illustrated in Figure 2.

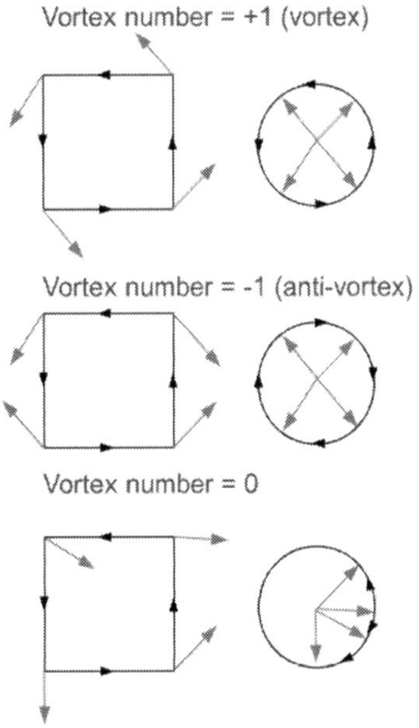


**Figure 2.** Vortices and anti-vortices: On a square lattice each plaquette has a winding number 0, +1 (vortex) or −1 (anti-vortex).

First an orientation on the plaquettes is chosen. ere we choose it counter-clockwise. The four spins on the plaquette are then mapped on the unit circle. The shortest arc between the neighboring spins are identified. If the circle is closed counter-clockwise we identify a vortex. On the other hand, if the circle is closed clockwise we deal with an anti-vortex. The last possibility is that no circle is covered. In this case we have neither a vortex nor an anti-vortex. Only these three cases are possible. In particular it is not possible that the circle is covered more than once.

In Figure 3 we show typical configurations of a $L \times L$ lattice with $L = 64$ by using the constraint angle action. The vortices are illustrated as black squares whereas the anti-vortices as white diamonds. On top left, for $\delta = 1.85$ we observe that most of the vortices and anti-vortices appear as tightly bound vortex anti-vortex pairs. In fact it is the density of free vortices (not bound in pairs) that is crucial for the BKT phase transition. There is clearly some ambiguity for an explicit definition of the free vortex density. A possibility is to count those vortices which are not accompanied by any anti-vortex (and vice versa) within some Euclidean distance $r$. In the next section we discuss the free vortex density quantitatively.

For $\delta < \delta_c$ the free vortex density is low and a long-range order emerges. The tightly bound vortex anti-vortex pairs appear neutral from a large-scale perspective. The correlations only decay with a power law, and the correlation length $\xi$ is infinite. Since $m = 1/\xi$, this phase is called massless phase. The qualitative situation for the standard lattice action at $\beta > \beta_c$ is the same.

Journal of Physics: Conference Series **651** (2015) 012010 doi:10.1088/1742-6596/651/1/012010

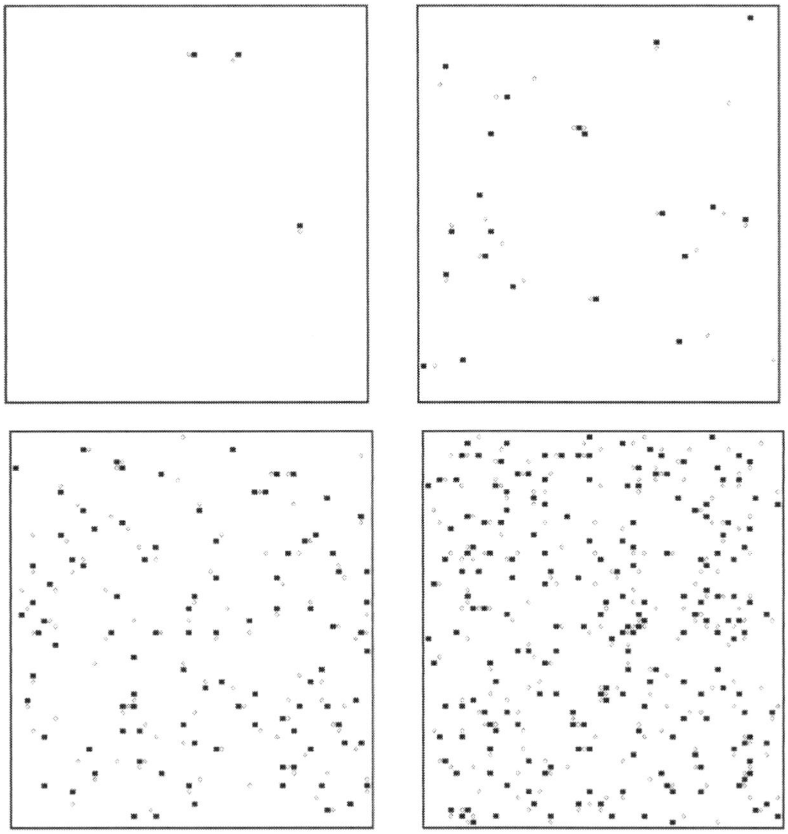

**Figure 3.** Typical configurations of a $L \times L$ lattice with $L = 64$ at $\delta = 1.85$, 2 (upper plots), 2.15 and 2.3 (lower plots). The vortices are illustrated as black squares and the anti-vortices as white diamonds.

If we reach the region where $\delta \gtrsim \delta_c$, the vortex anti-vortex pairs dissociate and the free vortex density umps up significantly. This destroys the long-range order, the correlation length $\xi$ becomes finite and we enter the massive phase. Again, for the standard lattice action the situation is qualitatively the same for $\beta \lesssim \beta_c$.

The infinite order phase transition is characterized by the exponential divergence of the correlation length $\xi$ at $\beta \lesssim \beta_c$

$$\xi \propto \exp\left(\frac{\text{const.}}{(\beta_c - \beta)^{\nu}}\right), \qquad \nu_c = 1/2 \,. \tag{5}$$

For the constraint angle action at $\delta \gtrsim \delta_c$ the divergence of the correlation length can be fitted well to [18]

$$\xi \propto \exp\left(\frac{\text{const.}}{\sqrt{\delta - \delta_c}}\right), \tag{6}$$

which results in a critical angle $\delta_c = 1.775(1)$ [17]. For the standard lattice action the BKT phase transition occurs at $\beta_c = 1.1199(1)$ [19].

## 3. Vortex density and vortex–anti-vortex pairs

In the previous section we discussed the vortices and the anti-vortices in a qualitative way. In this section we would like to discuss them quantitatively. We perform onte Carlo simulations of the 2d XY model in the formulation with the constraint angle action. To update the spins we use the very e cient Wolff cluster algorithm [20].

Since we use periodic boundary conditions, Stokes' Theorem implies that the total vorticity always vanishes

$$\sum_{\square} v_{\square} = 0 \; . \tag{7}$$

We measured the vortex density $\rho$, which is given by the number of plaquettes with $|v_{\square}| = 1$, divided by the volume. In Figure 4 we show the results on a $L \times L$ lattice with $L = 128$ using the constraint angle action. For $\delta > \pi/2$ vortices (and anti-vortices) are possible. We mentioned

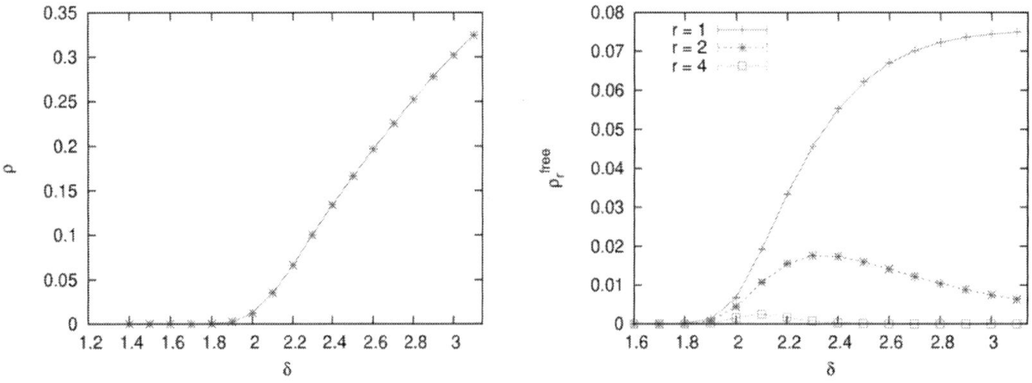


**Figure 4.** Left: Vortex density $\rho$, which is given by the number of plaquettes with $|v_{\square}| = 1$, divided by the volume. Right: Free vortex density, which can be defined as the density of vortices which are not accompanied by any anti-vortex (and vice versa) within some Euclidean distance $r$. Both plots are obtained on a $L \times L$ lattice with $L = 128$ using the constraint angle action.

before that it is the free vortex density (rather than the vortex density), which drives the BKT transition. In Figure 3 we indeed see qualitatively that for values $\delta \gtrsim \delta_c$ the tightly bound vortex anti-vortex pairs dissociate and the vortices and anti-vortices have no opposite neighbor. We show the free vortex densities for $\rho_r^{\rm free}$ for $r = 1$, $r = 2$ and $r = 4$ in Figure 4. For $r = 1$ the curve of the free vortex density increases monotonously for $\delta > \pi/2$. For $r = 2$ and $r = 4$ the curves decrease above some value of $\delta$. This can be explained by the fact that the system is filled up by many vortices and anti-vortices and therefore the probability to have an opposite partner within the radius $r$ rises again. For $\delta \gtrsim \delta_c$ we observe that all the curves grow rapidly which confirms the unbinding of the vortex anti-vortex pairs.

Furthermore, we studied the pair formation of the vortices and anti-vortices in the following way. Given a configuration, we first identify its $N$ vortices and $N$ anti-vortices, and search for the optimal pairing. This optimization minimizes the quantity

$$D^2 = \frac{1}{N} \sum_{i=1}^{N} d_{{\rm VA},\,i}^2 \; , \tag{8}$$

where $d_{{\rm VA},\,i}$ are the Euclidean distances that separate the vortex anti-vortex partners. The direct method of checking all possibilities is only applicable up to $N \approx 14$. We work again at lattice size $L = 128$ with constraint angles up to $\delta = 2.05$, where typically $N$ is close to 200. In order to still identify the optimal pairing (with high probability), we applied the technique of simulated annealing [21]. As a further reference quantity, we add the quantities

$$R^2_{\rm VV} = \frac{2}{N}\sum_{i=1}^{N/2} d^2_{{\rm VV},i} \quad \text{and} \quad R^2_{\rm AA} = \frac{2}{N}\sum_{i=1}^{N/2} d^2_{{\rm AA},i} \ , \tag{9}$$

where $d_{{\rm VV},i}$ ($d_{{\rm AA},i}$) are the distances between two vortices (two anti-vortices) We show the ratio between the quantities $D^2$, $R^2_{\rm VV}$ and $R^2_{\rm AA}$ for simulated configurations, and for random distributions, with $N$ vortices and $N$ anti-vortices. We observe a strong trend towards pair

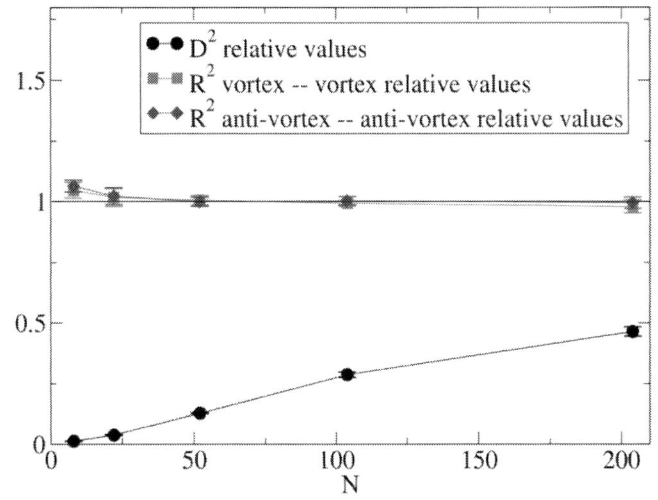


**Figure 5.** Ratio between the quantities $D^2$, $R^2_{\rm VV}$ and $R^2_{\rm AA}$ for simulated configurations, and for random distributions, with $N$ vortices and $N$ anti-vortices on a $L \times L$ lattice with $L = 128$ using the constraint angle action.

formation at small $N$. For increasing $N$ this trend fades away.

The established picture of the BKT transition also implies a sizable vorticity anti-correlation over short distances in the massless phase, in particular over distance 1. We consider the vorticity correlation function

$$C(r) = \langle\, v_{\square,(x_1,x_2)}\, v_{\square,(x_1+r,x_2)} \,\rangle \big|_{|v_{\square,(x_1,x_2)}|=1} \tag{10}$$

over distances $r = 1$, 2 and 3, at a set of constraint angles $\delta \geq \delta_c$. Indeed, we confirm a strong anti-correlation at distance 1 around $\delta_c$, which decreases rapidly as $\delta$ increases. The negative values for $C(r)$ can be explained since neighboring plaquettes contribute with opposite sign to the vorticity correlation function.

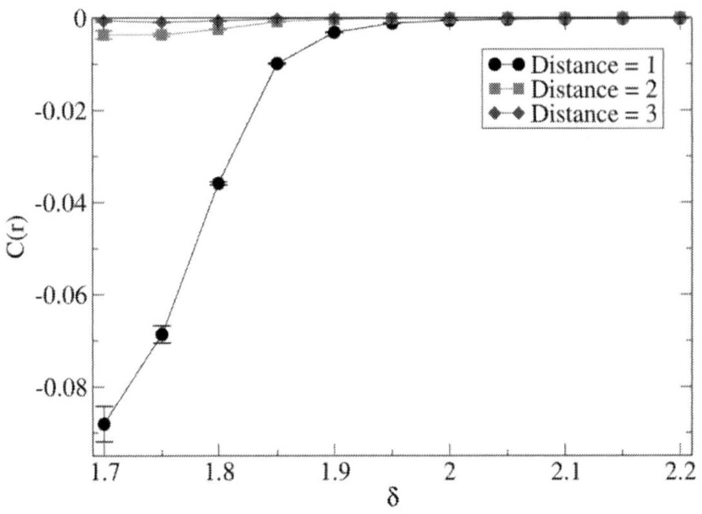


**Figure 6.** Vorticity correlation function $C(r)$ in eq. (10) over distances $r = 1, 2$ and 3 as a function of the constraint angle $\delta$. We observe a strong anti-correlation over the smallest distance $r = 1$ at $\delta \lesssim \delta_c$. This is consistent with the picture of the formation of vortex—anti-vortex pairs. As $\delta$ increases the pairs tend to dissociate.

## 4. Helicity modulus

The helicity modulus is a well-known quantity in condensed matter physics. It is proportional to the super uid density and in the literature sometime also called 'spin stiffness' or 'spin rigidity'. It is a measure for the sensitivity of a system to torsion, in particular to a variation of a twist in the boundary conditions. Furthermore, besides the critical exponents, this is often a useful indicator to characterize a universality class.

For a non-topological lattice action like the standard lattice action, the helicity modulus $\Upsilon$ can be defined as

$$\Upsilon = \frac{\partial^2}{\partial \alpha^2} F(\alpha)|_{\alpha=0} \, , \tag{11}$$

where $F = -\frac{1}{\beta} \ln Z$ is the free energy, and $Z$ the partition function. The boundary conditions are assumed to be periodic in one direction and twisted with the angle $\alpha$ in the other one. The helicity modulus corresponds to the curvature in the minimum at $\alpha = 0$.

Regarding topological lattice actions, and in particular for the constraint angle action this definition is problematic. The constraint angle action has no couplings and therefore the above definition is inappropriate. ence we consider a modification and define the dimensionless helicity modulus as

$$\Upsilon := \beta \Upsilon \, . \tag{12}$$

The critical value of the helicity modulus for the 2d XY model in a square volume was first predicted analytically in [22, 23] to be $\Upsilon_{\mathrm{c,\,theory}} = \frac{2}{\pi}$. Later it was noticed in [24] (see also [25]) that a tiny correction (below 0.2 per mille) due to winding configurations has to be taken into account,

$$\Upsilon_{\mathrm{c,\,theory}} = \frac{2}{\pi}\Big(1 - 16\pi e^{-4\pi}\Big) \simeq 0.636508 \, . \tag{13}$$

sing the standard lattice action, $\Upsilon$ at $\alpha = 0$ can be evaluated in a convenient way such that the generation of configurations can be restricted to periodic boundary conditions. In this way the most extensive study has been worked out by asenbusch in [25]. e performed simulations on lattices up to $L = 2048$, where he obtained at $\beta_c = 1.1199$ the value $\Upsilon_c = 0.67246(10)$. Compared to the analytic prediction this is 5.6 too large. (For convenience, we denote the dimensionless helicity modulus at the critical parameter by $\Upsilon_c$, even in finite volume.) In order to obtain an infinite volume extrapolation, he fitted his results for various sizes $L$ to the form

$$\Upsilon_c(L) = \Upsilon_{c,\,\text{theory}} + \frac{c_1}{\ln L + c_2}\,, \tag{14}$$

with free parameters $c_1$, $c_2$, which worked decently. Theoretical predictions in the spin wave limit [25] and based on the renormalization group [26] argue for the universality of $c_1$ and predict the value $c_1 \simeq 0.3189$. owever, it should be mentioned that in this calculation the parameter $\beta$ is inverted. Since the constraint angle action does not involve any $\beta$ parameter, it is not obvious if the universality arguments for $c_1$ still apply in this case.

For the constraint angle action the determination of $\Upsilon$ at $\alpha = 0$ fails. A small change in $\alpha$ does (in general) not affect $F$ at all (in a finite volume). On the first sight this seems to be problematic. owever, $\Upsilon$ can also be evaluated from a histogram for the $\alpha$ values, which describes their probability $p(\alpha)$. In this case $\alpha$ has to be treated as a dynamical variable in the simulation. Its probability density is related to $\Upsilon$ as [10]

$$\Upsilon = -\frac{\partial^2}{\partial \alpha^2} \ln p(\alpha)|_{\alpha=0}\,. \tag{15}$$

In practice, the idea is to determine the curvature in the maximum of $\ln p(\alpha)$ from a histogram up to moderate $|\alpha|$.

For the step action the convenient way of evaluating $\Upsilon$ at $\alpha = 0$ is not applicable either. In this case the corresponding histograms for $p(\alpha)$ have been studied in [10]. The BKT transition is observed around $\beta_c \approx 1.2 \dots 1.3$ [9, 10]. sing twisted boundary conditions, Olsson and olme measured $\Upsilon_c = 0.663(6)$, at $\beta_c$ on a $L \times L$ lattice with $L = 256$ lattice. Compared to the standard lattice action this is closer to the BKT value, but still 4.2 too large.

For the constraint angle action we performed onte Carlo simulations by using the etropolis algorithm to update the spins and the angle $\alpha$ at the twisted boundary. We also formulated a cluster algorithm [17]. owever it turned out that it is not consistent without performing additional accept/re ect decisions.

sing the constraint angle action we measured $\Upsilon_c$ in various volumes, focusing on the critical constraint angle $\delta_c = 1.775(1)$. We simulated the model on $L \times L$ lattices in the range $L = 8 \dots 256$ with dynamical boundary conditions. Our results are illustrated in Figure 7, and compared with those for the standard lattice action [25] and for the step action [10].

The deviation of $\Upsilon_c$ from the theoretical value $\Upsilon_{c,\,\text{theory}}$ is less than 2.4 for all sizes $L$ that we considered. Starting from $L \geq 64$ our results confirm the prediction within the errors. In view of earlier attempts to measure $\Upsilon_c$ with other lattice actions, this observation is highly remarkable.

Let us now have a look at the so-called helicity gap. For an infinite volume, the BKT theory predicts a discontinuity of the helicity modulus. As soon as the coupling exceeds its critical value, $\Upsilon$ drops to 0. In finite volume the function $\Upsilon(\beta)$ is continuous, but for increasing size $L$ the ump to 0 is approximated better and better. The situation is illustrated qualitatively for the standard lattice action in Figure 8.

For the standard lattice action [27] and for the step action [10] the observations are compatible with this property. We expect the same behavior for the constraint angle action, where $\Upsilon(\delta)$ should ump to 0 in an infinite volume when $\delta$ exceeds $\delta_c$. As a test, we measured $\Upsilon(\delta)$ in various

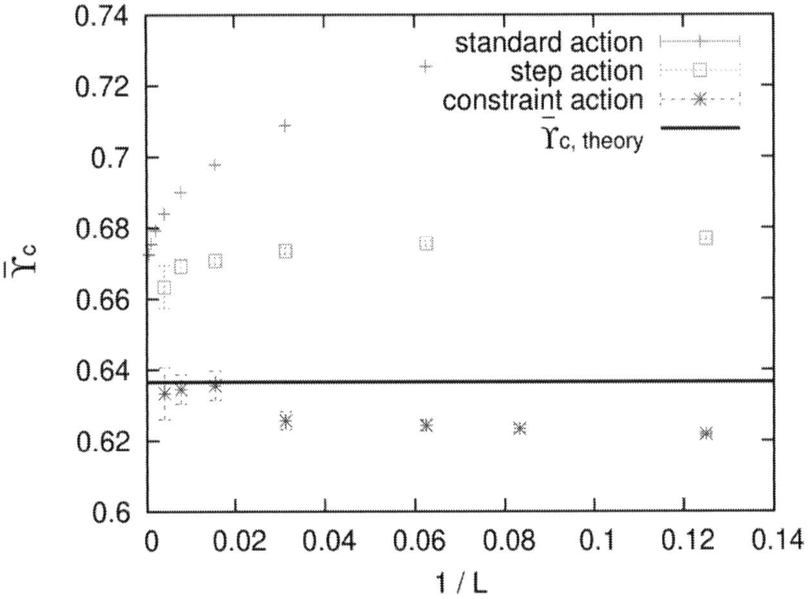


**Figure 7.** umerical results for $\bar{\Upsilon}_c$ on $L \times L$ lattices at the critical parameter, for the standard lattice action (data from [25]), the step action (data from [10]) and for the constraint angle action [17]. For the standard lattice action and the step action the results differ significantly from the theoretical BKT value $\bar{\Upsilon}_{c,\text{theory}}$ in eq. (13), which could only be attained with extended extrapolations. Only for the constraint action the results agree with the BKT prediction for $L \geq 64$, and the deviation is ust 1.9 (2.3 ) even at $L = 16$ ($L = 8$).

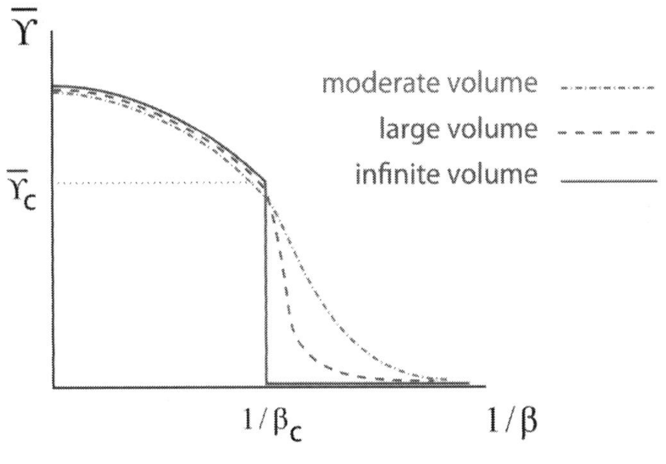


**Figure 8.** Qualitative picture of the expected coupling dependence of the helicity modulus in different volumes. In finite volume the function $\bar{\Upsilon}(\beta)$ is continuous, but for increasing size $L$ the ump to 0 is approximated better and better.

volumes. We show the results in Figure 9. Indeed, the data for the constraint angle action are also compatible with the expected qualitative picture.

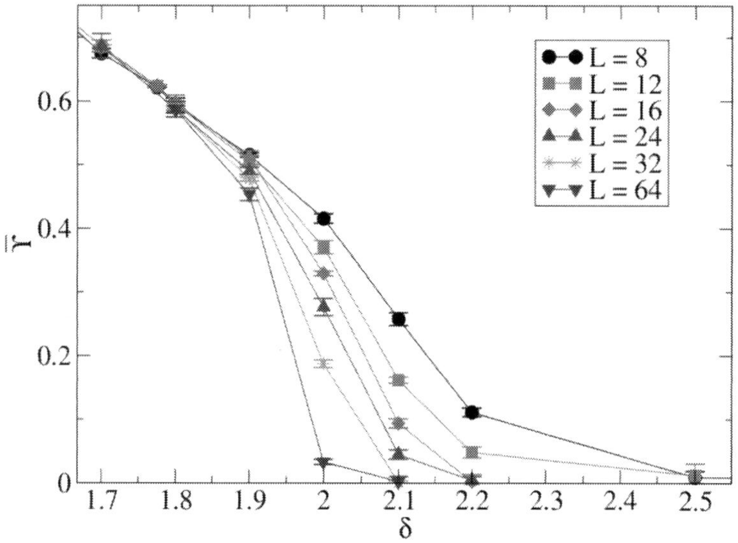


**Figure 9.** Results for $\Upsilon$ in six volumes, over a range of $\delta$ angles, which includes $\delta_c \simeq 1.775$. We observe the expected trend towards a ump down to 0 above $\delta_c$, in analogy to the schematic Figure 8.

## 5. Conclusions

sing the formulation of the constraint angle action, we have investigated the Berezinskii-Kosterlitz-Thouless phase transition of the 2d XY model. To this end, we performed onte Carlo simulations using dynamical boundary conditions. We confirmed the value of the dimensionless helicity modulus $\Upsilon_{c,\,\text{theory}}$, which was predicted at a BKT phase transition. sing the constraint angle action, the finite size effects are modest, in contrast to other lattice actions such as the standard lattice action and the step action. In particular, the value of $\Upsilon_c$ remains close to the BKT value down to volumes as small as $16 \times 16$, where the deviation is ust 1.9 . These facts eliminate any doubt that the constraint angle action belongs to the same universality class as the conventional lattice actions, which involve spin couplings (such as the standard lattice action).

Furthermore, we verified the picture of vortex anti-vortex pair (un)binding as the mechanism behind the BKT transition. Our results for the density of free vortices and anti-vortices (without an opposite partner up to some distance), for the vorticity correlation, as well as for the sum over pair separations squared, are all compatible with this picture.

It should be pointed out that the validity of this mechanism is highly non-trivial. Free vortices do not cost any energy (if the constraint allows them). Their suppression in the range $\pi/2 < \delta \lesssim \delta_c$ can only be explained by the combinatorial frequency of configurations carrying different vorticities. In this view, our results deviate from the established picture, since they demonstrate that a BKT transition can occur even without any Boltzmann factor suppression of free vortices.

## Acknowledgments
We thank ichael Bögli who has contributed to this work at an early stage. We further thank we-Jens Wiese for instructive discussions, and artin asenbusch, Silvano Romano, Erhard Seiler and lli Wolff for interesting remarks.

This work was supported by the exican *Consejo Nacional de Ciencia y Tecnología* (CO ACyT) through pro ect 155905/10 Física de Partículas por medio de Simulaciones uméricas and through the scholarship 312631 for graduate studies, as well as DGAPA- A . The simulations were performed on the cluster of the Instituto de Ciencias ucleares, A . We would like to thank Luciano Díaz and Enrique Palacios for technical support.

## References
[1] osterlitz M and Thouless D 1 3 *J. Phys.* C **6** 1181
[2] Beasley M R, Mooi E and Orlando T P 1 *Phys. Rev. Lett.* **42** 1165
[3] Fröhlich and Spencer T 1 81 *Commun. Math. Phys.* **81** 52
[4] Belousov A I and Lozovik u E 1 6 *Solid State Commun.* **100** 421
[5] Ro as C and osé V 1 6 *Phys. Rev.* B **54** 12361
[6] Pargellis A N, Green S and urke B 1 4 *Phys. Rev.* E **49** 4250
[ ] Berezinskii V L 1 1 *Sov. Phys. JETP* **32** 4 3
[8] Berezinskii V L 1 2 *Sov. Phys. JETP* **34** 610
[ ] enna R and Irving A C 1 *Nucl. Phys.* B **485** 583
[10] Olsson P and Holme P 2001 *Phys. Rev.* B **63** 05240
[11] Bietenholz W, Gerber U, Pepe M and Wiese U- 2010 *JHEP* **1012** 020
[12] Patrascioiu A and Seiler E 1 2 *Phys. Rev. Lett.* **68** 13 5
[13] Patrascioiu A and Seiler E 1 2 *J. Statist. Phys.* **69** 5 3
[14] Aizenman M 1 4 *J. Statist. Phys.* **77** 351
[15] Patrascioiu A and Seiler E 2002 *J. Statist. Phys.* **106** 811
[16] Hasenbusch M 1 6 *Phys. Rev.* D **53** 3445
[1 ] Bietenholz W, Gerber U and Re ón-Barrera F G 2013 *J. Stat. Mech.* P1200
[18] Bietenholz W, Bögli M, Niedermayer F, Pepe M, Re ón-Barrera F G and Wiese U- 2013 *JHEP* **1303** 141
[1 ] Hasenbusch M and Pinn 1 *J. Phys.* A **30** 63
[20] Wolff U 1 8 , *Phys. Rev. Lett.* **62** 361
[21] irkpatrick S, Gelatt C D and Vecchi M P *Science* **220** 6 1
[22] Nelson D R and osterlitz M 1 *Phys. Rev. Lett.* **39** 1201
[23] Minnhagen P and Warren G G 1 81 *Phys. Rev.* B **24** 2526
[24] Prokof'ev N V and Svistunov B V 2000 *Phys. Rev.* B **61** 11282
[25] Hasenbusch M 2005 *J. Phys.* A **38** 586
[26] Pelissetto A and Vicari E 2013 *Phys. Rev.* E **87** 032105
[2 ] Minnhagen P and im B 2003 *Phys. Rev.* B **67** 1 250

Journal of Physics: Conference Series **651** (2015) 012011 doi:10.1088/1742-6596/651/1/012011

# Proton structure functions at small $x$

**Martin Hentschinski**

Brookhaven National Laboratory, Physics Department, Upton, NY, 11973, U.S.A.

E-mail: hentsch@bnl.gov

**Abstract.** Proton structure functions are measured in electron-proton collision through inelastic scattering of virtual photons with virtuality $Q$ on protons; $x$ denotes the momentum fraction carried by the struck parton. Proton structure functions are currently described with excellent accuracy in terms of scale dependent parton distribution functions, defined in terms of collinear factorization and DGLAP evolution in Q. With decreasing x however, parton densities increase and are ultimately expected to saturate. In this regime DGLAP evolution will finally break down and non-linear evolution equations w.r.t x are expected to take over. In the first part of the talk we present recent result on an implementation of physical DGLAP evolution. Unlike the conventional description in terms of parton distribution functions, the former describes directly the Q dependence of the measured structure functions. It is therefore physical insensitive to factorization scheme and scale ambiguities. It therefore provides a more stringent test of DGLAP evolution and eases the manifestation of (non-linear) small x effects. It however requires a precise measurement of both structure functions $F_2$ and $F_L$, which will be only possible at future facilities, such as an Electron Ion Collider. In the second part we present a recent analysis of the small x region of the combined HERA data on the structure function $F_2$. We demonstrate that (linear) next-to-leading order BFKL evolution describes the effective Pomeron intercept, determined from the combined HERA data, once a resummation of collinear enhanced terms is included and the renormalization scale is fixed using the BLM optimal scale setting procedure. We also provide a detailed description of the $Q$ and $x$ dependence of the full structure functions $F_2$ in the small $x$ region, as measured at HERA. Predictions for the structure function $F_L$ are found to be in agreement with the existing HERA data.

## 1. Introduction

The description of the proton in terms its elementary constituents, quarks and gluons, remains one of the big unsovled problems of nuclear- and elementary particle physics. At the typical energy scale of the proton, which is of the order of $\Lambda_{\text{QCD}} \simeq 200$ MeV, Quantum Chromodynamics, the Quantum Field Theory description of strong interactions is strongly coupled, and quarks and gluons are subject to confinement. It is however possible to obtain very valuable information about the structure of the proton from collision processes of protons with leptonic projectiles, such as the electron. Due to the point-like structure of the electron and a very good theoretical understanding of electromagnetic interactions, the electron provides the perfect probe to explore the nucleon. To leading order in Quantum Electrodynamics (QED), scattering of the electron and the proton takes place through the exchange of a virtual photon with virtuality $q^2 = -Q^2$, see Fig. 1. If the photon virtuality is large, the proton is destroyed during the scattering and the process is generally referred to as Deep Inelastic Scattering (DIS). The cross-section for neutral-current DIS on unplolarized nucleons can be written in terms of

Journal of Physics: Conference Series **651** (2015) 012011 doi:10.1088/1742-6596/651/1/012011

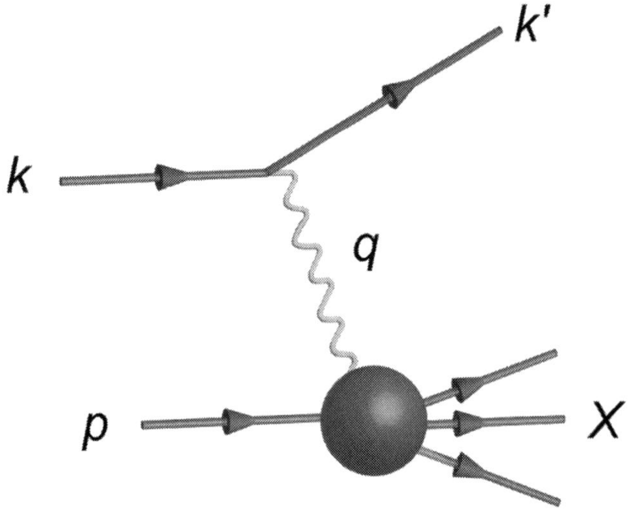


**Figure 1.** Schematic diagram of the Deep Inelastic Scattering (DIS) process. Measurement of the final state electron allows to determine inclusive nuclear structure functions in terms of the resolution scale $Q^2 = -q^2$ provided by the photon virtuality and Bjorken $x = Q^2/(2p \cdot q)$ (Figure taken from [1] ).

two Lorentz invariant structure functions $F_2$ and $F_L$ in the following way

$$\frac{d^2\sigma}{dx dQ^2} = \frac{4\pi\alpha^2_{\text{e.m.}}}{xQ^4}\left[\left(1 - y + \frac{y^2}{2}\right)F_2(x, Q^2) - \frac{y^2}{2}F_L(x, Q^2)\right] \tag{1}$$

Here $y = (q \cdot p)/(k \cdot p)$ denotes the inelasticity with $0 < y < 1$, see also Fig. 1. The structure functions themselves depend on only two Lorentz invariants, the photon virtuality $Q^2$ and Bjorken $x = Q^2/(2p \cdot q)$. Within the parton model, to be discussed below, $x$ denotes the momentum fraction of the parton hit by the virtual photon.

If the photon virtuality $Q^2$ is significantly larger than the non-perturbative energy scale of the proton, asymptotic freedom provides for such processes a weak strong coupling constant $\alpha_s(Q^2) \ll 1$ and a description within perturbative theory becomes possible. The conventional theoretical framework for such DIS processes is based on the collinear factorization theorem [2]. At leading order, the essential physics is captured by the parton model [3] of the proton. Within this model, the highly virtual photon interacts not with the entire proton with characteristic size $\sim 1/\Lambda_{\text{QCD}}$, but with a single, essentially point-like, parton, *i.e.*, a quark or gluon, with effective size $1/Q$. Interference effects with spectator quarks or gluons are on the other hand suppressed by powers of $Q^2$. To arrive at the complete cross-section, the "partonic" interaction of virtual photon and quark, needs to be convoluted with parton distribution functions (PDFs), $f_i(x, Q^2)$, $i = q, \bar{q}, g$, which encode the probability to find a parton with a certain proton moment fraction inside the proton. Higher order corrections to such partonic cross-sections, calculated within QCD perturbation theory, possess reveal then a new kind of singularity, apart from the conventional ultra-violet singularity, which is removed through renormalization of the QCD Lagrangian. This new singularity is of infra-red type and can be associated with configurations where an additionally emitted parton is collinear to the proton momentum. In physical terms, this initial state singularity reflects interference between the perturbative calculable partonic interactions at the hard scale $Q$, with spectator quarks and gluons, characterized by the hadronic scale $\Lambda_{\rm QCD}$. Collinear factorization provides then a systematic framework to remove such singularities from the perturbative hard cross-section, resulting into finite Wilson coefficients [4–7], absorbing them into parton distribution functions, which encode the long-distance, non-perturbative physics; for a recent review see, *e.g.*, [8].

To make the separation between long- and short-distance physics manifest, one needs to introduce some arbitrary factorization scale $\mu_f$, apart from the scale $\mu_r$ appearing in the renormalization of the strong coupling $\alpha_s$. The independence of physical observables such as $F_{2,L}$ on $\mu_f$ can be used to derive powerful renormalization group equations (RGEs) governing the scale dependence of PDFs in each order of perturbation theory, known as the Dokshitzer-Gribov-Lipatov-Altarelli-Parisi (DGLAP) evolution equation. The corresponding kernels are the anomalous dimensions or splitting functions associated with collinear two-parton configurations [9–11]. DGLAP evolution has been impressively confirmed by experiment, in particular through the very accurate DIS data on the structure function $F_2$ from the DESY-HERA experiment, see Fig. 2. In particular, parton distribution functions, fitted at some initial scale $Q_0 \simeq 1-2$ GeV

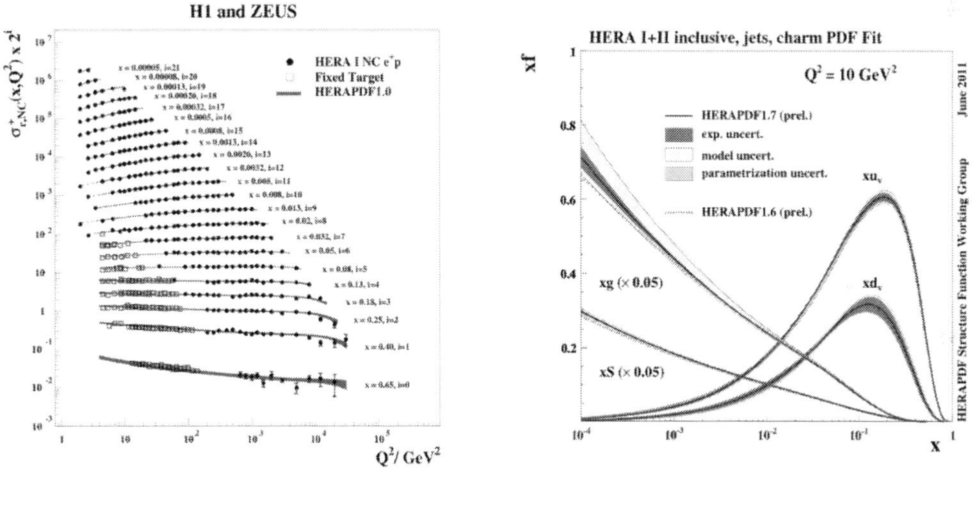


**Figure 2.** Left: Combined HERA data and parton distribution functions. Right: Proton parton distribution functions plotted as functions of Bjorken x. (Source [12])

and evolved with the DGLAP equations, have been shown to describe $F_2$ data over orders of magnitude, both in $x$ and $Q^2$ [13].

Despite of the impressive success of the DGLAP evolution equations, theoretical considerations suggests that at some point in phase space this description is supposed to break down. With decreasing $x$, logarithms $\ln 1/x$ increase and are capable of balancing the strong coupling, $\alpha_s \ln 1/x \sim 1$, leading to a break-down of the naïve perturbative expansion, see Fig. 3. The necessary resummation of enhanced terms $(\alpha_s \ln 1/x)^n$ is the achieved by the Balitsky-Fadin-Kuarev-Lipatov (BFKL) evolution equation [14]. One of the main predictions of BFKL evolution is a power-like rise of the gluon density. If continued to ultra-small $x$, the $1/Q^2$ expansion, on which collinear factorization is based, will eventually break down. A description

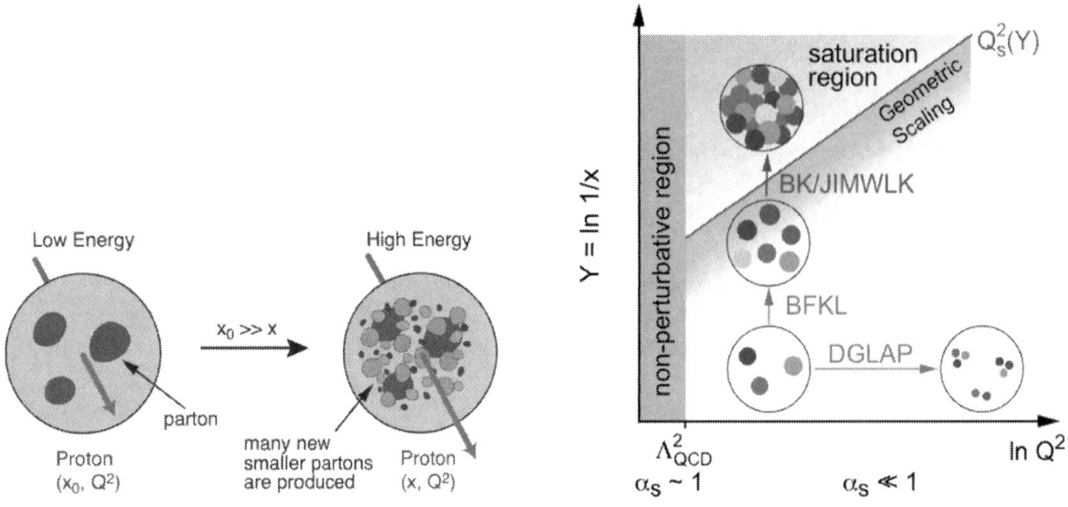


**Figure 3.** Left: The proton wave-function at small-x contains a large number of gluons as compared to the same wave-function at a larger x = x0. The figure is a projection on the plane transverse to the beam axis. Right: The map of high energy QCD in the (Q2 , Y = ln 1/x) plane. (Source [1])

of proton structure functions in this region of phase space is provided by the Balitsky-Kovchegov (BK) [15] and JIMWLK [16] evolution equations, which provide a non-linear extension of BFKL evolution, resumming corrections due to high gluon densities to all orders. While theoretical arguments suggests the relevance of such corrections already at current collider energies, current data provide no clear evidence for deviations from linear DGLAP evolution, which would provide a signature for the onset of a non-linear kinematic regime dominated by high, or saturated, gluon densities. Definite evidence for such a regime of QCD requires therefore new experiments such as a future high-luminosity electron-ion collider, *i.e..*, the EIC [1] and the LHeC [17] projects, whose physics case is currently studied. In particular these projects plan to measure both structure functions $F_2$ and $F_L$ and their scaling violations very precisely at small $x$ both in electron-proton and in electron-heavy ion collisions.

From the theory side this requires the development of suitable tools which allows to pin down possible deviations from DGLAP evolution. In particular it is necessary to reduce the large freedom in fitting initial conditions of parton distribution functions. With non-linear saturation effects most likely to manifest themselves at small values of $Q^2$, the large number of free parameters used for the description of initial conditions in PDF fits, does not allow to exclude the possibility that saturation effects, while present in reality, are currently hidden in the initial conditions of DGLAP evolution.

In the following we present two approaches which have the potential to restrict this large freedom at low scales. Section 2 is dedicated to the concept of physical evolution kernels, which allows to reduce the number of independent PDFs in DIS fits and eliminates scale- and scheme dependence in their definition. Section 3 contains results of a recently achieved BFKL fit of the combined HERA data. While both theoretically and experimentally less explored than collinear factorization, BFKL evolution has the potential to reveal the emergence of non-linear effects more easily than DGLAP evolution. Unlike DGLAP evolution, BFKL drives the system into the saturated regime, making a detection of high density effects more likely. For details we refer to [18–20]

Journal of Physics: Conference Series **651** (2015) 012011 doi:10.1088/1742-6596/651/1/012011

## 2. Physical evolution kernels for DIS observables

Since collinear factorization can be carried out in infinitely many different ways, one is left with an additional choice of the factorization scheme for which one usually adopts the $\overline{\text{MS}}$ prescription. Likewise, the RGE governing the running of $\alpha_s$ with $\mu_r$ can be deduced from taking the derivative of $F_{2,L}$ with respect to $\mu_r$. Upon properly combining PDFs and Wilson coefficients in the same factorization scheme, any residual dependence on $\mu_f$ is suppressed by an additional power of $\alpha_s$, i.e., is formally one order higher in the perturbative expansion but not necessarily numerically small.

Alternatively, it is possible to formulate QCD scale evolution equations directly for physical observables without resorting to auxiliary, convention-dependent quantities such as PDFs. This circumvents the introduction of a factorization scheme and $\mu_f$ and, hence, any dependence of the results on their actual choice. The concept of physical anomalous dimensions is not at all a new idea and has been proposed quite some time ago [4, 21, 22] but its practical aspects have never been studied in detail. The framework is suited best for theoretical analyses based on DIS data with the scale $\mu_r$ in the strong coupling being the only theoretical ambiguity. In addition, $F_{2,L}$ or their scaling violations can be parametrized much more economically than a full set of quark and gluon PDFs, which greatly simplifies any fitting procedure and phenomenological analysis. The determination of $\alpha_s$ from fits to DIS structure functions is the most obvious application, as theoretical scheme and scale uncertainties are reduced to a minimum.

Here we largely focus on the practical implementation of physical anomalous dimensions in analyses of DIS data up to next-to-leading order (NLO) accuracy. We shall study in detail potential differences with results obtained in the conventional framework based on scale-dependent quark and gluon densities, which could be caused by the way how the perturbative series is truncated at any given order.

### 2.1. Theoretical Framework

This gist of the factorization scheme-invariant framework amounts to combine any two DIS observables $\{F_A, F_B\}$ and determine their corresponding $2 \times 2$ matrix of physical anomalous dimensions instead of the scale-dependent quark singlet, $\Sigma \equiv \sum_q (q + \bar{q})$, and gluon distributions appearing in the standard, coupled singlet DGLAP evolution equations. Instead of using measurements of $F_2$ and $F_L$ (actually their flavor singlet parts), one can also utilize their variation with scale for any given value of $x$, i.e., $dF_{2,L}(x, Q^2)/d\ln Q^2$ as an observable. The required sets of physical anomalous dimensions for both $\{F_2, F_L\}$ and $\{F_2, dF_2/d\ln Q^2\}$ have been derived in [22] up to NLO accuracy. The additionally needed evolution equations for the non-singlet portions of the structure functions $F_{2,L}$ are simpler and not matrix valued. As we shall see below, the required physical anomalous dimensions comprise the inverse of coefficient and splitting functions and are most conveniently expressed in Mellin $n$ moment space. The Mellin transformation of a function $\phi$ given in Bjorken $x$ space, such as PDFs or splitting functions, is defined as

$$\phi(n) \equiv \int_0^1 dx\, x^{n-1} \phi(x)\,, \tag{2}$$

where $n$ is complex valued. As an added benefit, convolutions in $x$ space turn into ordinary products upon applying (2), which, in turn, allows for an analytic solution of QCD scale evolution equations for PDFs. The corresponding inverse Mellin transformation is straightforwardly performed numerically along a suitable contour in $n$ space, see, e.g., Ref. [23] for details. The necessary analytic continuations to non-integer $n$ moments are given in [24, 25], and an extensive list of Mellin transforms is tabulated in [26]. We will work in Mellin space throughout this review.

Assuming factorization, moments of DIS structure functions $F_I$ at a scale $Q$ can be expressed as

$$F_I(n,Q^2) = \sum_{k=q,\bar{q},g} e_k^2\, C_{I,k}\left(n, \alpha_s(\mu_r^2), \frac{Q^2}{\mu_f^2}, \frac{\mu_r^2}{\mu_f^2}\right) \cdot f_k\left(n, \alpha_s(\mu_r^2), \frac{\mu_f^2}{Q_0^2}, \frac{\mu_r^2}{\mu_f^2}\right) \tag{3}$$

where the sum runs over all contributing $n_f$ active quark flavors with electric charge squared $e_q^2$ and the gluon $g$, each represented by a PDF $f_k$. For $e_g^2$, the averaged quark charge factor $\bar{e}_q^2 = (1/n_f)\sum_q e_q^2$ has to be used. $\mu_r$ and $\mu_f$ specify renormalization and factorization scale, respectively. The scale $Q_0$ defines the starting scale for the PDF evolution, where a set of non-perturbative input distributions needs to be specified. For simplicity we identify in the following the renormalization scale with the factorization scale, i.e., $\mu_r = \mu_f \equiv \mu$. The coefficient functions $C_{I,k}$ are calculable in pQCD [4–7] and exhibit the following series in $a_s \equiv \alpha_s/4\pi$

$$C_{I,k}\left(n, \alpha_s(\mu^2), \frac{Q^2}{\mu^2}\right) = \sum_{m=m_0} a_s^m(\mu^2)\, C_{I,k}^{(m)}\left(n, \frac{Q^2}{\mu^2}\right), \tag{4}$$

where $m_0$ depends on the first non-vanishing order in $a_s$ in the expansion for the observable under consideration, e.g., $m_0 = 0$ for $F_2$ and $m_0 = 1$ for $F_L$.

Each PDF $f_k(n, \mu^2/Q_0^2)$ obeys the DGLAP evolution equation which reads

$$\frac{d}{d\ln\mu^2} f_k\left(n, \frac{\mu^2}{Q_0^2}\right) = \sum_{l=q,\bar{q},g} P_{kl}\left(n, \alpha_s(\mu^2)\right) f_l\left(n, \frac{\mu^2}{Q_0^2}\right) \tag{5}$$

where the $l \to k$ splitting functions have a similar expansion [9–11] as the coefficient functions in Eq. (4):

$$P_{kl}(n, \alpha_s(\mu^2)) = \sum_{m=0} a_s(\mu^2)^{1+m}\, P_{kl}^{(m)}(n)\,. \tag{6}$$

The $P_{kl}(n)$ relate to the corresponding anomalous dimensions through $\gamma_{kl}(n) = -2P_{kl}(n)$ in the normalization conventions we adopt, where we use the leading order (LO) and NLO expressions for $\gamma_{kl}(n)$ given in App. B of the first reference in [10]. We note that the same normalization is used in the publicly available PEGASUS evolution code [23]. In practice one distinguishes a $2\times 2$ matrix-valued DGLAP equation evolving the flavor singlet vector comprising $\Sigma(n, \mu^2/Q_0^2)$ and $g(n, \mu^2/Q_0^2)$ and a set of $n_f - 1$ RGEs for the relevant non-singlet quark flavor combinations.

The scale-dependent strong coupling itself obeys another RGE governed by the QCD beta function

$$\frac{da_s(\mu)}{d\ln\mu^2} = \beta(a_s) = -\sum_m a_s^{m+2}\beta_m \tag{7}$$

with $\beta_0 = 11 - 2n_f/3$ and $\beta_1 = 102 - 38n_f/3$ up to NLO accuracy. To compare below with the results for the physical anomalous dimensions in Ref. [22] we also introduce the evolution variable

$$t \equiv -\frac{2}{\beta_0}\ln\left(\frac{a_s(Q^2)}{a_s(Q_0^2)}\right). \tag{8}$$

Instead of studying $F_I(n,Q^2)$ in (3) in terms of scale-dependent PDFs, which are obtained from solving the singlet and non-singlet DGLAP equations (5) in a fit to data [13], one can also derive evolution equations directly in terms of the observables $F_I(n,Q^2)$. To this end, we consider a pair of DIS observables $F_A$ and $F_B$, to be specified below, whose scale dependence is governed by a coupled matrix-valued equation

$$\frac{d}{d\ln Q^2}\begin{pmatrix} F_A^{(S)}(n,Q^2) \\ F_B^{(S)}(n,Q^2)\end{pmatrix} = \begin{pmatrix} K_{AA} & K_{AB} \\ K_{BA} & K_{BB}\end{pmatrix}(n, a_s(Q^2)) \cdot \begin{pmatrix} F_A^{(S)}(n,Q^2) \\ F_B^{(S)}(n,Q^2)\end{pmatrix} \tag{9}$$

for the flavor singlet (S) parts of $F_{A,B}$ and a set of non-singlet (NS) equations

$$\frac{dF_{A,B}^{(NS)}(n,Q^2)}{d\ln Q^2} = K^{(NS)}(n, a_s(Q^2)) \cdot F_{A,B}^{(NS)}(n,Q^2) \tag{10}$$

for the remainders.

The required physical anomalous dimensions in Eqs. (9) and (10), obey a similar perturbative expansion in $a_s$ as in (6). The singlet kernels in (9) are constructed by substituting

$$\begin{pmatrix} F_A^{(S)}(n,Q^2) \\ F_B^{(S)}(n,Q^2) \end{pmatrix} = \bar{e}_f^2\, C\left(n, \alpha_s(\mu^2), \frac{Q^2}{\mu^2}\right) \cdot \begin{pmatrix} \Sigma(n, \mu^2/Q_0^2) \\ g(n, \mu^2/Q_0^2) \end{pmatrix} \tag{11}$$

into the left-hand side of Eq. (9) and taking the derivatives. Note that we have normalized the quark singlet part of $F_{A,B}$ with the same averaged charge factor $\bar{e}_f^2$ which appears in the gluonic sector. Upon making use of the RGEs for PDFs and the strong coupling in Eq. (5) and (7), respectively, one arrives at

$$\begin{aligned} K_{ij}(n, \alpha_s(Q^2)) = \Bigg[\Bigg(&\beta(a_s(Q^2))\frac{\partial C(n, \alpha_s(Q^2), 1)}{\partial a_s(Q^2)} \\ &+ C(n, \alpha_s(Q^2), 1) \cdot P(n, a_s(Q^2))\Bigg) \cdot C^{-1}(n, \alpha_s(Q^2), 1)\,,\Bigg]_{ij} \end{aligned} \tag{12}$$

where we have introduced $2 \times 2$ matrices

$$C = \begin{pmatrix} C_{A,q} & C_{A,g} \\ C_{B,q} & C_{B,g} \end{pmatrix}, \qquad P = \begin{pmatrix} P_{qq} & 2n_f P_{qg} \\ P_{gq} & P_{gg} \end{pmatrix}. \tag{13}$$

for the relevant singlet coefficient and splitting functions, respectively. An analogous, albeit much simpler expression holds for the NS kernel $K^{(NS)}$ in (10). As has been demonstrated in [22], the kernels (12) are independent of the chosen factorization scheme and scale but do depend on $\mu_r$ and the details of the renormalization procedure. We also note that the inverse $C^{-1}$ in (12), appearing upon re-expressing all PDFs by $F_{A,B}$, can be straightforwardly computed only in Mellin moment space.

### 2.2. *Example I: $F_2$ and $F_L$*

Let us first consider the evolution of the pair of observables $\{F_2, F_L\}$. A precise determination of $F_L$ in a broad kinematic regime is a key objective at both an EIC [1] and the LHeC [17]. Since the perturbative series for $F_L$ only starts at $\mathcal{O}(a_s)$, one wants to account for this offset by actually considering the evolution of either $\{F_2, F_L/(a_s C_{L,g}^{(1)})\}$ or $\{F_2, F_L/(a_s C_{L,q}^{(1)})\}$. Both sets of kernels $K_{AB}$ show a rather different behavior with $n$, as we shall illustrate below, but without having any impact on the convergence properties of the inverse Mellin transform needed to obtain $x$ dependent structure functions. The kernels $K_{AB}$ at LO and NLO accuracy for $\{F_A, F_B\} = \{F_2, F_L/(a_s C_{L,g}^{(1)})\}$ can be found in [22]. Note that evolution in [22] is expressed in terms of $t$. Using (8), $d/da_s = -2/(a_s\beta_0)d/dt$, and (7) to compute to extra terms proportional to $\beta(a_s)$, we fully agree with their results.

For $\{F_A, F_B\} = \{F_2, F_L/(a_s C_{L,q}^{(1)})\}$ one finds

$$K_{22}^{(0)} = P_{qq}^{(0)} - \frac{C_{L,q}^{(1)}}{C_{L,g}^{(1)}} P_{qg}^{(0)}, \qquad K_{2L}^{(0)} = \frac{C_{L,q}^{(1)}}{C_{L,g}^{(1)}} P_{qg}^{(0)}, \tag{14}$$

$$K_{L2}^{(0)} = \left( \frac{C_{L,g}^{(1)}}{C_{L,q}^{(1)}} - \frac{C_{L,q}^{(1)}}{C_{L,g}^{(1)}} \right) P_{qg}^{(0)} - P_{gg}^{(0)} + P_{qq}^{(0)}, \qquad K_{LL}^{(0)} = \frac{C_{L,q}^{(1)}}{C_{L,g}^{(1)}} P_{qg}^{(0)} + P_{gg}^{(0)}$$

at the LO approximation, i.e., after expanding Eq. (12) up to $\mathcal{O}(a_s)$. Only the off-diagonal entries change if dividing $F_L$ by $a_s C_{L,g}^{(1)}$; see Eqs. (41)-(45) in Ref. [22]. For NLO kernels we refer to [18].

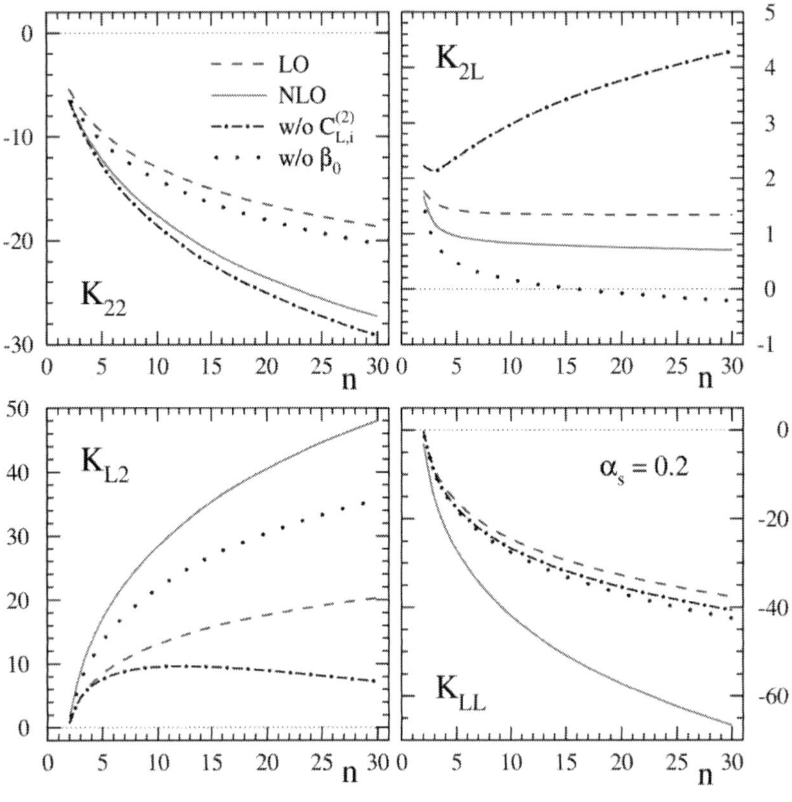


**Figure 4.** Physical anomalous dimensions $K_{AB}(n)$ in LO and NLO for $\{F_A, F_B\} = \{F_2, F_L/(a_s C_{L,q}^{(1)})\}$ assuming $\alpha_s = 0.2$ and $n_f = 3$. The dash-dotted and dotted lines show the NLO results where all contributions from $C_{L,i}^{(2)}$ and $\beta_0$, respectively, to $K_{AB}^{(1)}$ have been omitted. A global factor of $\alpha_s/4\pi$ has been ignored in the perturbative expansion, i.e., $K_{AB} = K_{AB}^{(0)} + (a_s/4\pi) K_{AB}^{(1)} + \mathcal{O}(a_s^2)$ is displayed.

In Fig. 4 we illustrate the $n$ dependence of the LO and NLO singlet kernels $K_{AB}$ for the evolution of $\{F_A, F_B\} = \{F_2, F_L/(a_s C_{L,q}^{(1)})\}$ assuming $\alpha_s = 0.2$ and $n_f = 3$. As can be seen, NLO

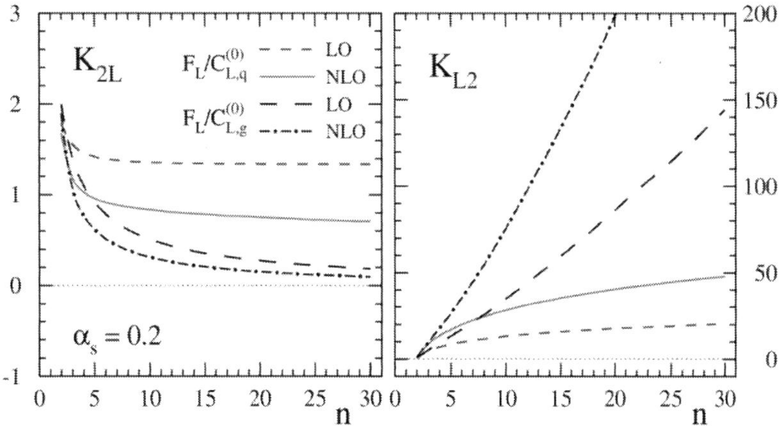


**Figure 5.** As in Fig. 4 but now comparing the LO and NLO off-diagonal kernels $K_{2L}$ and $K_{L2}$ for $\{F_A, F_B\} = \{F_2, F_L/(a_s C^{(1)}_{L,q})\}$ and $\{F_2, F_L/(a_s C^{(1)}_{L,g})\}$; see text.

corrections are sizable for all singlet kernels, in particular, when compared with the perturbative expansion of the singlet splitting functions $P_{kl}(n)$ in (6); see Figs. 1 and 2 in Ref. [11]. This is, however, not too surprising given that the known large higher order QCD corrections to the Wilson coefficients $C_{L,g}$ and $C_{L,q}$ [27] are absorbed into the physical anomalous dimensions $K_{AB}$ for the evolution of the DIS structure functions $F_2$ and $F_L$. The impact of contributions from the NLO coefficients $C^{(2)}_{L,g}$ and $C^{(2)}_{L,q}$ on the results obtained for $K_{AB}$ is illustrated by the dash-dotted lines in Fig. 4. Another source for large corrections are the terms proportional to $\beta_0$ in the NLO corrections as can be inferred from the dotted lines; note that $K^{(1)}_{L2}$ and $K^{(1)}_{LL}$ include terms proportional to $\beta_0 C^{(2)}_{L,g}$ and $\beta_0 C^{(2)}_{L,q}$. In Sec. 2.4 we will demonstrate how the differences between the LO and NLO kernels become apparent in the scale evolution of $F_{2,L}(x, Q^2)$.

Figure 5 compares the LO and NLO off-diagonal kernels $K_{2L}$ and $K_{L2}$ for $\{F_A, F_B\} = \{F_2, F_L/(a_s C^{(1)}_{L,q})\}$ and $\{F_2, F_L/(a_s C^{(1)}_{L,g})\}$. The most noticeable difference is the strong rise with $n$ for the kernel $K_{L2}$ governing the evolution of $\{F_2, F_L/(a_s C^{(1)}_{L,g})\}$. At LO accuracy, this is readily understood by inspecting the $n \to \infty$ limit which yields, see Eq. (41) in [22], $K^{(0)}_{L2} \sim n \ln n$, recalling that asymptotically $C^{(1)}_{L,q} \sim 1/n$, $C^{(1)}_{L,g} \sim 1/n^2$, $P^{(0)}_{qq} \sim \ln n$, $P^{(0)}_{qg} \sim 1/n$, $P^{(0)}_{gq} \sim 1/n$, and $P^{(0)}_{gg} \sim \ln n$. The NLO kernel $K_{L2}$ exhibits an even stronger rise with $n$. In the same way one obtains, for instance, that $K^{(0)}_{L2}$ governing the evolution of $\{F_2, F_L/(a_s C^{(1)}_{L,q})\}$ only grows like $\ln n$, see Eq. (14).

Despite this peculiar $n$ dependence and the differences between the singlet kernels shown in Fig. 5, both sets of observables, $\{F_2, F_L/(a_s C^{(1)}_{L,q})\}$ and $\{F_2, F_L/(a_s C^{(1)}_{L,g})\}$ can be used interchangeably in an analysis at LO and NLO accuracy. Results for the QCD scale evolution are identical, and one does not encounter any numerical instabilities related to the inverse Mellin transform, which we perform along a contour as described in Ref. [23]. In fact, it is easy to see that the eigenvalues

$$\lambda_\pm = \frac{1}{2}\left[K^{(0)}_{22} + K^{(0)}_{LL} \pm \sqrt{(K^{(0)}_{22} - K^{(0)}_{LL})^2 + 4K^{(0)}_{2L}K^{(0)}_{L2}}\,\right], \tag{15}$$

Journal of Physics: Conference Series **651** (2015) 012011 doi:10.1088/1742-6596/651/1/012011

which appear when solving the matrix valued evolution equation (9), are identical for both sets of kernels and also agree with the corresponding eigenvalues for the matrix of singlet anomalous dimensions $P_{kl}^{(0)}$; see also the discussions in Sec. 2.4

### *2.3. Example II: $F_2$ and $dF_2/dt$*

Of future phenomenological interest could be also the pair of observables $\{F_2, dF_2/dt\}$, in particular, in the absence of precise data for $F_L$. Determining experimentally the $t$ or $Q^2$ slope of $F_2$ is, of course, also challenging.

Defining $F_D \equiv dF_2/dt$, we obtain the following physical evolution kernels

$$\begin{aligned} K_{22}^{(0)} &= 0\,, \quad K_{2D}^{(0)} = 2\,, \\ K_{D2}^{(0)} &= \frac{1}{2}\left[P_{gq}^{(0)} P_{qg}^{(0)} - P_{gg}^{(0)} P_{qq}^{(0)}\right]\,, \\ K_{2D}^{(0)} &= P_{\rm gg}^{(0)} + P_{\rm qq}^{(0)} \end{aligned} \tag{16}$$

at LO to be used in Eq. (9); for NLO kernels see [18]. The kernels $K_{AB}$ in (16) exhibit more moderate higher order corrections, mainly through terms proportional to $\beta_{0,1}$, than those listed in Sec. 2.2. This shall become apparent in the next Section when we discuss results for the scale dependence of both $\{F_2, dF_2/dt\}$ and $\{F_2, F_L\}$.

### *2.4. Numerical Studies*

In this Section we apply the methodology based on physical anomalous dimensions as outlined above and compare with the results obtained in the conventional framework of scale-dependent quark and gluon densities and coefficient functions. Due to the lack of precise enough data for $F_L$ or $dF_2/dt$ we will adopt the following realistic "toy" initial conditions for the standard DGLAP evolution of PDFs at a scale $Q_0 = \sqrt{2}\,\text{GeV}$ [23]

$$\begin{aligned} xu_v(x, Q_0^2) &= 5.1072\, x^{0.8}\,(1-x)^3, \\ xd_v(x, Q_0^2) &= 3.06432\, x^{0.8}\,(1-x)^4, \\ x\bar{u}(x, Q_0^2) &= (1-x)\, x\, \bar{d}(x, Q_0^2), \\ x\bar{d}(x, Q_0^2) &= 0.1939875\, x^{-0.1}\,(1-x)^5, \\ x\bar{s}(x, Q_0^2) &= 0.2\, x\, [\bar{u}(x, Q_0^2) + \bar{d}(x, Q_0^2)], \\ xg(x, Q_0^2) &= 1.7\, x^{-0.1}\,(1-x)^5. \end{aligned} \tag{17}$$

for all our numerical studies. The value of the strong coupling $\alpha_s$ at $Q_0$ is taken to be 0.35. For our purposes we can ignore the complications due to heavy flavor thresholds and set $n_f = 3$ throughout. We use this set of PDFs to compute the flavor singlet parts of $F_2$, $F_L$, and $dF_2/dt$ at the input scale $Q_0$ using Eq. (11). For studies of DIS in the small $x$ region, say $x \lesssim 10^{-3}$, in which we are mainly interested in, the flavor singlet parts are expected to dominate over NS contributions and, hence, shall be a good proxy for the full DIS structure functions. Results at scales $Q > Q_0$ are obtained by either solving the RGEs for PDFs or by evolving the input structure functions directly adopting Eq. (9). For the solution in terms of PDFs we adopt from now on the standard choice $\mu = Q$.

For completeness and to facilitate the discussions below, let us quickly review the solution of the matrix-valued RGEs such as Eqs. (5) and (9). While one can truncate the QCD beta function and the anomalous dimensions consistently at any given order in $a_s$, there exists no unique solution to beyond the LO accuracy. The matrix-valued nature of (9) only allows for iterations around the LO solution, which at order $a_s^k$ can differ in various ways by formally higher-order terms of $\mathcal{O}(a_s^{l>k})$.

To this end, we employ the standard truncated solution in Mellin moment space, which can be found, for instance, in Ref. [24], see also [23], and reads

$$\Gamma_i(n,Q) = L_i(n,a_s,a_0)\Gamma_i(n,Q_0) \tag{18}$$

where the evolution operator up to NLO is defined as

$$L_i(n,a_s,a_0) = L_i^{(0)}(n,a_s,a_0) + L_i^{(1)}(n,a_s,a_0) \tag{19}$$

with

$$\begin{aligned} L_i^{(0)}(n,a_s,a_0) &= \left(\frac{a_s}{a_0}\right)^{-\frac{\lambda_-}{\beta_0}} e_- + (+) \leftrightarrow (-) \\ L_i^{(1)}(n,a_s,a_0) &= \left(\frac{a_s}{a_0}\right)^{-\frac{\lambda_-}{\beta_0}} \Bigg[(a_0 - a_s)e_- R_i^{(1)} e_- \\ &\quad - \left(a_0 - a_s\left(\frac{a_s}{a_0}\right)^{\frac{\lambda_- - \lambda_+}{\beta_0}}\right)\beta_0 \frac{e_- R_i^{(1)} e_+}{\lambda_+ - \lambda_- - \beta_0}\Bigg] + \quad (+) \leftrightarrow (-)\,. \end{aligned} \tag{20}$$

Here, $a_0 = a_s(Q_0)$, $\Gamma_P = \binom{\Sigma}{g}$, and $\Gamma_K = \binom{F_A}{F_B}$, i.e., the index $i = P$ refers to the coupled RGE for the quark singlet and gluon and $i = K$ to the RGE for the pair $\{F_A, F_B\}$ of DIS structure functions in (9). For $i = K$ one has

$$\left(R_K^{(0)}\right)_{AB} = \frac{1}{\beta_0} K_{AB}^{(0)}\,, \quad \left(R_K^{(1)}\right)_{AB} = \frac{1}{\beta_0} K_{AB}^{(1)} - \frac{\beta_1}{\beta_0^2} K_{AB}^{(0)}\,, \tag{21}$$

with a corresponding definition for $i = P$ in terms of the $2 \times 2$ matrices of singlet splitting functions $P^{(0)}$ and $P^{(1)}$. $\lambda_\pm$ denote the eigenvalues given in Eq. (15) and $e_\pm$ the projection operators onto the corresponding eigenspaces; see Refs. [23, 24].

As has been mentioned already at the end of Sec. 2.2, the eigenvalues $\lambda_\pm(n)$ are identical when computed for the kernels $K_{AB}$ and $P_{kl}$. This in turn implies that as long as, say, $F_2$ and $F_L$ are calculated at $\mu = Q_0$ with LO accuracy, their scale evolution based on physical anomalous dimensions reproduces exactly the conventional results obtained with the help of scale-dependent PDFs.

Figure 6 shows our results for the scale dependence of the DIS structure functions $F_2$ and $F_L$. The input functions at $Q_0 = \sqrt{2}\,\mathrm{GeV}$ are shown as dotted lines. While LO results are identical, starting from NLO accuracy the comparison between the two methods of scale evolution becomes more subtle, and results *seemingly* differ significantly as can be inferred from the middle panels of Fig. 6.

The origin of the differences between $F_{2,L}(x,Q^2)$ computed based on Wilson coefficients and scale-dependent PDFs and physical anomalous dimensions can be readily understood from terms which are formally beyond NLO accuracy. For instance, upon inserting the NLO Wilson coefficients (4) and the truncated NLO solution (18)-(20) into Eq. (11), $F_2$ at $\mathcal{O}(a_s)$ contains spurious terms of both $\mathcal{O}(a_0 a_s)$ and $\mathcal{O}(a_s^2)$. Since $F_L$ starts one order higher in $a_s$, similar terms are less important here. On the other hand, when we evolve $F_{2,L}$ with the help of physical anomalous dimensions we first compute, due to the lack of data, the input at $a_0$ based on Eq. (11), which then enters the RGE solution (18)-(20). Again, this leads to terms beyond

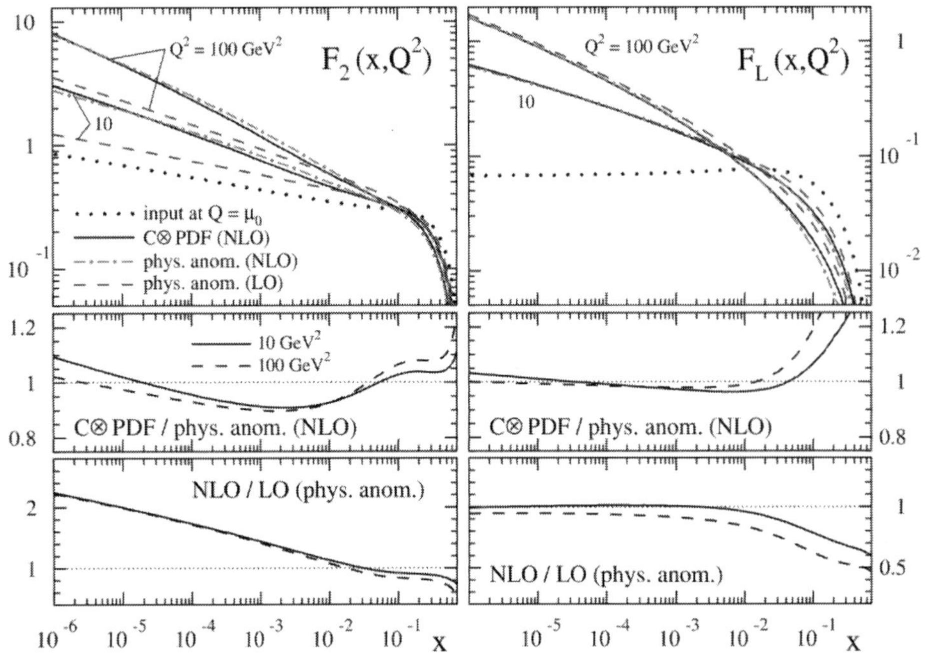


**Figure 6.** Scale dependence of the DIS structure functions $F_2$ (left) and $F_L$ (right) at NLO accuracy obtained with physical anomalous dimensions (dash-dotted lines) and in the standard way through a convolution of PDFs and Wilson coefficients (solid lines). The dashed and dotted lines show the results obtained at LO accuracy and the input at $Q = Q_0 = \sqrt{2}\,\mathrm{GeV}$, respectively. The middle panels give the ratios of the two different methods to evolve $F_{2,L}$ at NLO, and the lower panels illustrate the size of NLO corrections when physical anomalous dimensions are being used; see text.

NLO. In case of $F_2$ they are now of the order $\mathcal{O}(a_0 a_s)$ and $\mathcal{O}(a_0^2)$, i.e., even more relevant than in case of PDFs since $a_0 > a_s$.

To test if the entire difference between the two evolution methods shown in Fig. 6 is caused by these formally higher order contributions, one can easily remove all $\mathcal{O}(a_s^2)$, $\mathcal{O}(a_0 a_s)$, and $\mathcal{O}(a_0^2)$ contributions from our results. Indeed, the scale evolution based on physical anomalous dimensions and the calculation of $F_{2,L}$ from PDFs then yields exactly the same results also at NLO accuracy. We note that this way of computing properly truncated physical observables from scale-dependent PDFs beyond the LO accuracy has been put forward some time ago in Ref. [28, 29] but was not pursued any further in practical calculation.

Another interesting aspect to notice from Fig. 6 are the sizable NLO corrections illustrated in lower panels, in particular, for $F_2$ in the small $x$ region. For this comparison, LO results refer to the same input structure functions $F_{2,L}$ as used to obtain the NLO results but now evolved at LO accuracy, i.e., by truncating the evolution operator in Eqs. (18)-(20) at $L_K^{(0)}$. At first sight the large corrections appear to be surprising given that global PDF fits in general lead to acceptable fits of DIS data even at LO accuracy [13]. However, this is usually achieved by exploiting the freedom to have different sets of PDFs at LO and, say, NLO accuracy. The framework based on physical anomalous dimensions does not provide this option as the input for the scale evolution is, in principle, fully determined by experimental data, and only the value for the strong coupling can be adjusted at any given order. In this sense it provides a much more stringent test of the underlying framework and perhaps a better sensitivity to, for instance, the possible presence of non-linear effects to the scale evolution in the kinematic regime dominated by small $x$ gluons.

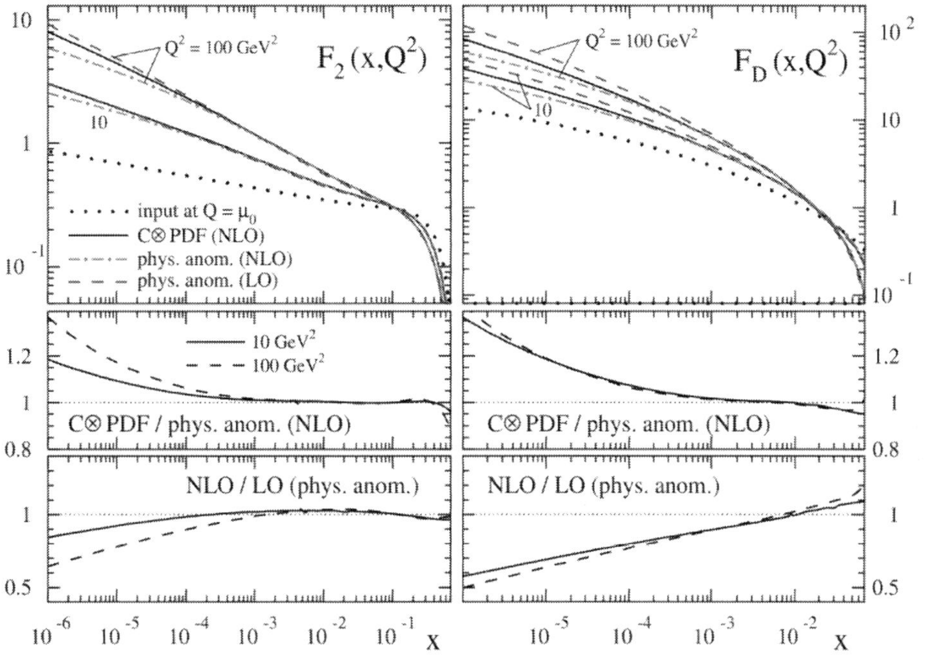


**Figure 7.** Same as in Fig. 7 but now for the pair of observables $F_2$ and $F_D \equiv dF_2/dt$.

In Fig. 7 we show the corresponding results for the scale dependence of the DIS structure function $F_2$ and its slope $F_D = dF_2/dt$. Again, any differences between the scale evolution performed with physical anomalous dimensions and based on PDFs are caused by formally higher order terms $\mathcal{O}(a_s^2)$, $\mathcal{O}(a_0 a_s)$, and $\mathcal{O}(a_0^2)$, which can be removed with the same recipe as above. As for $\{F_2, F_L\}$, NLO corrections are sizable in the small $x$ region due to numerically large contributions to $K_{D2}^{(1)}$ and $K_{DD}^{(1)}$ from the QCD beta function.

### *2.5. Summary*

We have presented a phenomenological study of the QCD scale evolution of deep-inelastic structure functions within the framework of physical anomalous dimensions. The method is free of ambiguities from choosing a specific factorization scheme and scale as it does not require the introduction of parton distribution functions. Explicit results for the physical evolution kernels needed to evolve the structure functions $F_2$, its $Q^2$ slope, and $F_L$ have been presented up to next-to-leading order accuracy.

It was shown that any differences with results obtained in the conventional framework of scale-dependent quark and gluon densities can be attributed to the truncation of the perturbative series at a given order in the strong coupling. At next-to-leading order accuracy the numerical impact of these formally higher order terms is far from being negligible but, if desired, such contributions can be systematically removed.

A particular strength of performing the QCD scale evolution based on physical anomalous dimensions rather than auxiliary quantities such as parton densities is that the required initial conditions are completely fixed by data and cannot be tuned freely in each order of perturbation theory. Apart from a possible adjustment of the strong coupling, this leads to easily testable predictions for the scale dependence of structure functions and also clearly exposes the relevance of higher order QCD corrections in describing deep-inelastic scattering data. Next-to-leading order corrections have been demonstrated to be numerically sizable, which is not too surprising given that the physical evolution kernels absorb all known large higher order QCD corrections to the hard scattering Wilson coefficients.

Once high precision deep-inelastic scattering data from future electron-ion colliders become available, an interesting application of our results will be to unambiguously quantify the size and relevance of non-linear saturation effects caused by an abundance of gluons with small momentum fractions. To this end, one needs to observe deviations from the scale evolution governed by the physical anomalous dimensions discussed in this work. The method of physical anomalous dimensions can be also used for a theoretically clean extraction of the strong coupling and is readily generalized to other processes such polarized deep-inelastic scattering or inclusive one-hadron production.

## 3. $F_2$ and $F_L$ at small $x$ using collinearly-improved BFKL resummation

### *3.1. Structure functions within the BFKL framework*

At small $x$ and center-of-mass energy $s = Q^2/x$, we can apply high energy factorization and write the structure functions $F_I$, $I = 2, L$ as

$$F_I(x, Q^2) \;=\; \int \frac{d^2\mathbf{q}_\perp}{\pi\, q^2} \int \frac{d^2\mathbf{p}_\perp}{\pi\, p^2} \Phi_I\left(q, Q^2\right) \Phi_P\left(p, Q_0^2\right) \mathcal{F}\left(x, q, p\right). \tag{22}$$

where $q \equiv \sqrt{\mathbf{q}_\perp^2)}$ . $\Phi_P$ is the non-perturbative proton impact factor which we model using

$$\Phi_P\left(p, Q_0^2\right) \;=\; \frac{\mathcal{C}}{\Gamma(\delta)} \left(\frac{p^2}{Q_0^2}\right)^{\delta} e^{-\frac{p^2}{Q_0^2}}, \tag{23}$$

where we have introduced two free parameters and a normalization. $\Phi_I$ is the impact factor associated to the photon which we treat at leading-order (LO), *i.e.*

$$\int \frac{d^2\mathbf{q}_\perp}{\pi q^2} \Phi_I\left(q, Q^2\right) \left(\frac{q^2}{Q^2}\right)^{\gamma-1} \;=\; \frac{\alpha_s(\mu^2)}{2\pi} \sum_{q=1}^{n_f} e_q^2\, c_I(\nu)\,, \tag{24}$$

where

$$c_I(\nu) \equiv \frac{\pi^2}{4} \frac{\Omega_I(\nu)}{(\nu + \nu^3)} \operatorname{sech}(\pi\nu) \tanh\left(\pi\nu\right) \tag{25}$$

$\nu = i(1/2 - \gamma)$, $\Omega_2 = (11 + 12\nu^2)/8$, $\Omega_L = \nu^2 + 1/4$, and the strong coupling $\alpha_s$ is fixed at the renormalization scale $\mu^2$. In the present work we will also use the kinematically improved impact factors proposed in [30, 31], which include part of the higher order corrections by considering exact gluon kinematics. Its implementation requires to replace the functions $c_I(\nu)$ by $\tilde{c}_I(\gamma, \omega)$ where

$$\tilde{c}_L(\gamma, \omega) = \frac{4\Gamma(\gamma{+}\xi{+}1)\Gamma(1{+}\gamma)\left[\left(\psi(\gamma{+}\xi) - \psi(\gamma)\right)\left(3\omega^2 - \xi^2 + 1\right) - 6\omega\xi\right]}{\xi\Gamma(1+\omega)(\xi^4 - 5\xi^2 + 4)} \tag{26}$$

and $\tilde{c}_2 = \tilde{c}_L + \tilde{c}_T$, with

$$\tilde{c}_T(\gamma,\omega) = \frac{\Gamma(\gamma+\xi)\Gamma(\gamma)}{\xi\Gamma(1+\omega)(\xi^4 - 5\xi^2 + 4)}\Big\{ - 2\xi\omega\left(\xi^2 + 3^2 + 6\omega + 11\right)$$
$$+ \left[\psi(\gamma+\xi) - \psi(\gamma)\right]\left[\xi^4 - 10\xi^2 + 3\omega^2\left(\omega^2 + 2\omega + 4\right) - 2\omega\left(\xi^2 - 1\right) + 9\right]\Big\}. \tag{27}$$

$\psi(\gamma)$ is the logarithmic derivative of the Euler Gamma function and $\xi = 1 - 2\gamma + \omega$, while $\omega$ is the Mellin variable conjugate to $x$ in the definition of the gluon Green function $\mathcal{F}$, see Eq. (28) below. The main difference between these impact factors is that the LO ones roughly double the value of their kinematically improved counterparts in the region with small $|\nu|$, while being very similar for $|\nu| \geq 1$.

The gluon Green function can be written in the form

$$\mathcal{F}\left(x,q,p\right) = \frac{1}{\pi}\int\frac{d\omega}{2\pi i}\int\frac{d\gamma}{2\pi i}\;\frac{1}{q^2}\left(\frac{q^2}{p^2}\right)^{\gamma} x^{-\omega}\frac{1}{\omega - \bar{\alpha}_s\hat{\mathcal{K}}\left(\gamma\right)}, \tag{28}$$

with $\bar{\alpha}_s = \alpha_s N_c/\pi$. The collinearly improved BFKL kernel as introduced in eq. (28) is an operator consisting of a diagonal (scale invariant) piece $\hat{\chi}(\gamma)$ with eigenvalue

$$\chi(\gamma) \;=\; \bar{\alpha}_s\chi_0(\gamma) + \bar{\alpha}_s^2\chi_1(\gamma) - \frac{1}{2}\bar{\alpha}_s^2\chi_0{}'(\gamma)\chi_0(\gamma) + \chi_{\rm RG}(\bar{\alpha}_s,\gamma,a,b)\,, \tag{29}$$

where $\chi_0(\gamma) = 2\psi(1) - \psi(\gamma) - \psi(1-\gamma)$, $a = \frac{5}{12}\frac{\beta_0}{N_c} - \frac{13}{36}\frac{n_f}{N_c^3} - \frac{55}{36}$ and $b = -\frac{1}{8}\frac{\beta_0}{N_c} - \frac{n_f}{6N_c^3} - \frac{11}{12}$, plus a term $\hat{\chi}_{\rm RC}(\gamma)$ proportional to $\beta_0$ which contains the running coupling corrections of the NLO kernel [32]:

$$\hat{\chi}_{\rm RC}(\gamma) \;=\; \bar{\alpha}_s^2\frac{\beta_0}{8N_c}\left(\chi_0(\gamma)\overrightarrow{\partial}_\gamma - \overleftarrow{\partial}_\gamma\chi_0(\gamma) + 2\log(\mu^2)\right). \tag{30}$$

The precise form of the NLO kernel $\chi_1$ can be found in [19, 33]. The resummation of collinear logarithms of order $\bar{\alpha}_s^3$ and beyond is realized by the term [19, 34, 35]

$$\begin{aligned}\chi_{\rm RG}(\bar{\alpha}_s,\gamma,a,b) \;&=\; \bar{\alpha}_s(1+a\bar{\alpha}_s)\left(\psi(\gamma) - \psi(\gamma - b\bar{\alpha}_s)\right)\\ &-\frac{\bar{\alpha}_s^2}{2}\psi''(1-\gamma) - b\bar{\alpha}_s^2\frac{\pi^2}{\sin^2(\pi\gamma)} + \frac{1}{2}\sum_{m=0}^{\infty}\Bigg(\gamma - 1 - m + b\bar{\alpha}_s\\ &-\frac{2\bar{\alpha}_s(1+a\bar{\alpha}_s)}{1-\gamma+m} + \sqrt{(\gamma-1-m+b\bar{\alpha}_s)^2 + 4\bar{\alpha}_s(1+a\bar{\alpha}_s)}\Bigg).\end{aligned} \tag{31}$$

Our final expression for the structure functions reads

$$F_I(x,Q^2) \propto \int d\nu\, x^{-\chi\left(\frac{1}{2}+i\nu\right)}\Gamma\left(\delta - \frac{1}{2} - i\nu\right)\left[1 + \frac{\bar{\alpha}_s^2\beta_0\chi_0\left(\frac{1}{2}+i\nu\right)}{8N_c}\log\left(\frac{1}{x}\right)\right.$$
$$\left.\times\left(i(\pi\coth(\pi\nu) - 2\pi\tanh(\pi\nu) - M_I(\nu)) - \psi\left(\delta - \frac{1}{2} - i\nu\right)\right)\right]\left(\frac{Q^2}{Q_0^2}\right)^{\frac{1}{2}+i\nu} c_I(\nu), \tag{32}$$

where $M_2$ and $M_L$ can be found in [19]. For the kinematical improved version of $F_I$ we replace $c_I(\nu)$ by $\tilde{c}_I(1/2 + i\nu, \chi(1/2 + i\nu))$.

In Eq. (33) the scale of the running of the coupling has been set to $\mu^2 = QQ_0$. Building on the work of [36] we found in [19] that in order to obtain a good description of the $Q^2$ dependence of the effective intercept of $F_2$, $\lambda$, for $x < 10^{-2}$, it is very useful to operate with non-Abelian physical renormalization schemes using the Brodsky-Lepage-Mackenzie (BLM) optimal scale setting [37] with the momentum space (MOM) physical renormalization scheme [38]. For technical details on our precise implementation we refer the reader to [19] (see also [39] for a review on the subject and [40] for a related work). More qualitatively, in these schemes the pieces of the NLO BFKL kernel proportional to $\beta_0$ are absorbed in a new definition of the running coupling in order to remove the infrared renormalon ambiguity. Once this is done, the residual scheme dependence in this framework is very small. We also found it convenient [19] to introduce, in order to describe the data with small $Q^2$, an analytic parametrization of the running coupling in the infrared proposed in [41].

### *3.2. Comparison to DIS experimental data*

In the following we compare our results with the experimental data for $F_2$ and $F_L$. Let us first compare the result obtained in [19] for the logarithmic derivative $d\log F_2/d\log(1/x)$ using Eq. (33) with a LO photon impact factor and our new calculation using the kinematically improved one. In Fig. 8 we present our results with the values of our best fits for both types of impact factors and compare them with the H1-ZEUS combined data [12] for $x < 10^{-2}$. The values of the parameters defining the proton impact factor in (23) and the position of the (regularized) Landau pole (we use $n_f = 4$) for the strong coupling are $\delta = 8.4$, $Q_0 = 0.28$ GeV, $\Lambda = 0.21$ GeV for the LO order case and $\delta = 6.5$, $Q_0 = 0.28$ GeV, $\Lambda = 0.21$ GeV for the kinematically improved (note that the normalization $\mathcal{C}$ does not contribute to this quantity).

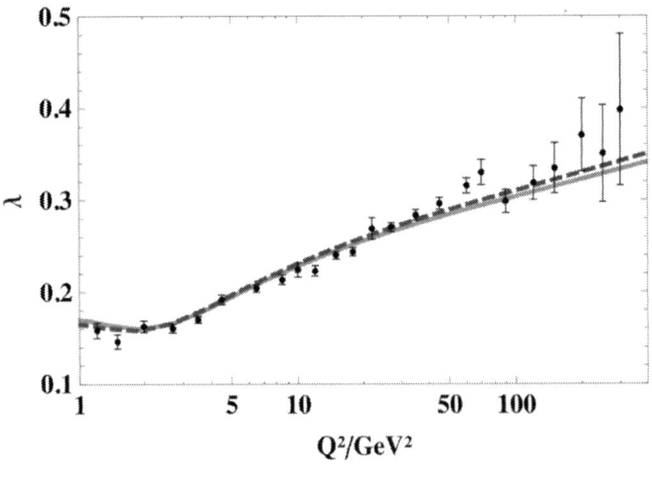


**Figure 8.** Fit to $\lambda$ for $F_2$ with the LO photon impact factor (solid line) and the kinematically improved one (dashed line). The data set has been extracted from [12].

The LO impact factor generates lower values than the kinematically improved one in the high $Q^2$ region and slightly higher ones when $Q^2 \lesssim 2\,\text{GeV}^2$. It is interesting to see how the approach presented here allows for a good description of the data in a very wide range of $Q^2$, not only for high values, where the experimental uncertainties are larger, but also in the non-perturbative regions due to our treatment of the running of the coupling.

Encouraged by these positive results we now turn to investigate more differential distributions. We select data with fixed values of $x$ and compare the $Q^2$ dependence of our theoretical predictions with them, now fixing the normalization for the LO impact factor to $\mathcal{C} = 1.50$ and 2.39 for the kinematically improved. Our results are presented in Fig. 9. The equivalent

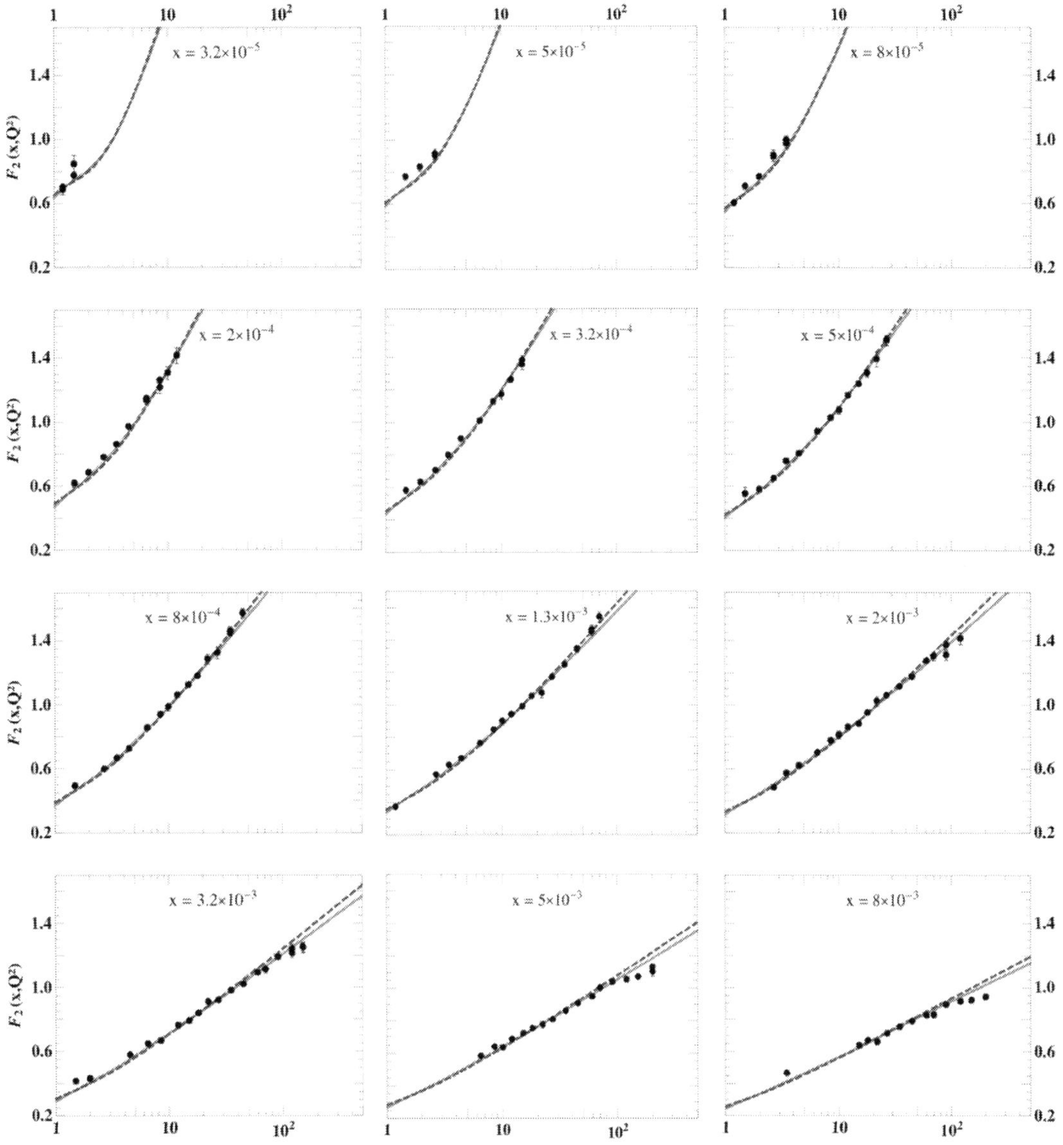


**Figure 9.** Study of the dependence of $F_2(x, Q^2)$ on $Q^2$ using the LO photon impact factor (solid lines) and the kinematically improved one (dashed lines). $Q^2$ runs from 1.2 to 200 GeV$^2$.

comparison to data, this time fixing $Q^2$ and looking into the evolution in the $x$ variable, is shown in Fig. 10. We observe that our predictions give a very accurate description of the data for both types of impact factors.

Let us remark that the values for the parameters in this fit are in syntony with the theoretical expectations for the proton impact factor since $Q_0$ is very similar to the confinement scale and

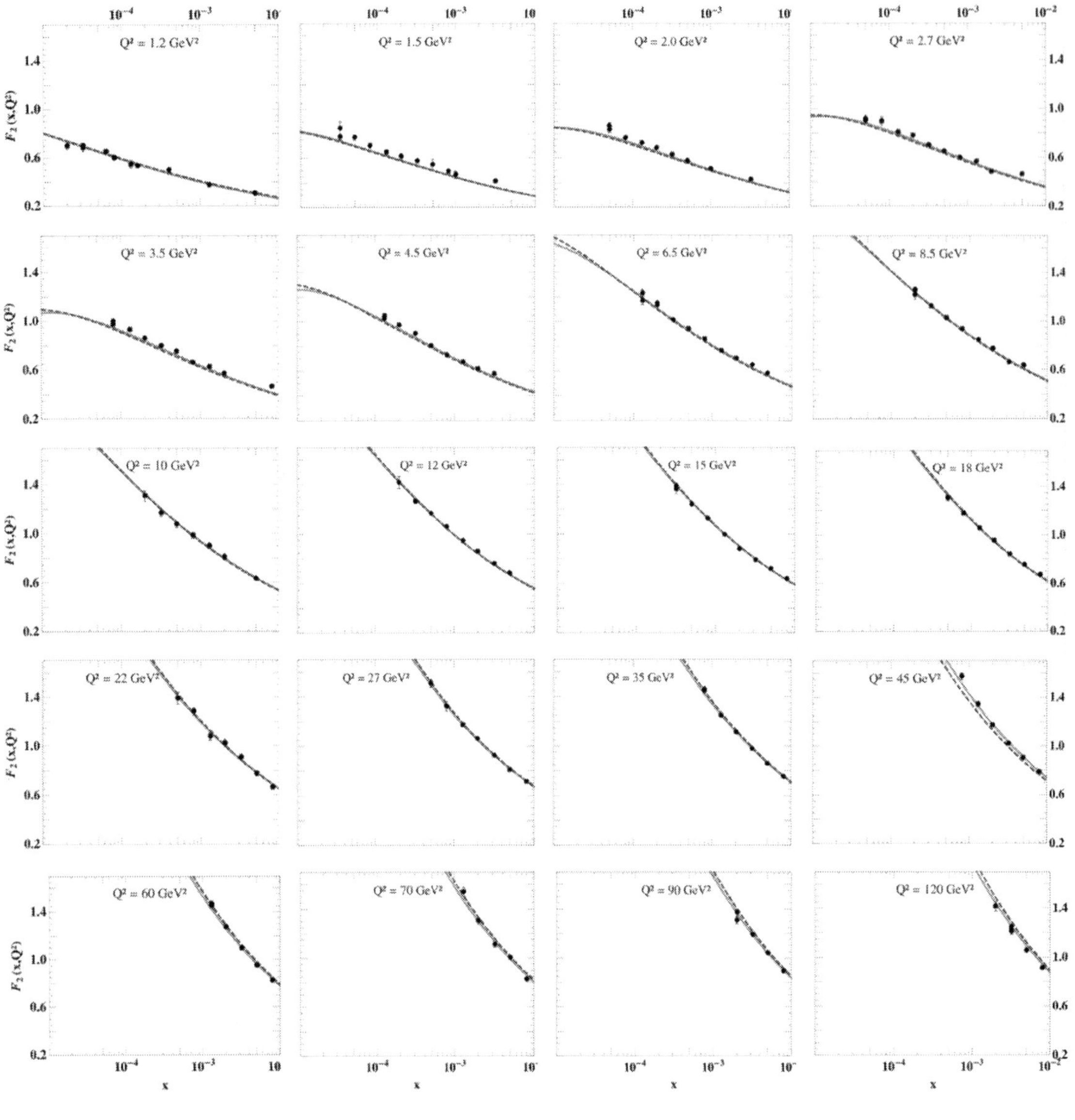


**Figure 10.** Study of the dependence of $F_2(x, Q^2)$ on $x$ using the LO photon impact factor (solid lines) and the kinematically improved one (dashed lines). $Q^2$ runs from 1.2 to 120 GeV$^2$.

the value of $\delta$ sets the maximal contribution from the impact factor also in that region. This is reasonable given that the proton has a large transverse size.

The longitudinal structure function is an interesting observable which is very sensitive to the gluon content of the proton. We will now present our predictions for $F_L$ using the best values for the parameters previously obtained in the fit of $F_2$. We will see that the agreement with the data is very good. First, $Q^2$ is fixed and the $x$ dependence is investigated in Fig. 11. The experimental data have been taken from [43]. To present the $Q^2$ dependence it is convenient to calculate, for each bin in $Q^2$, the average value of $x$, see Fig. 12. In some sense this is a similar plot to the one previously presented for $\lambda$ in the $F_2$ analysis and we can see that the effect of using different types of impact factors is to generate a global shift in the normalization. Again we note that we have an accurate description of the transition from high to low $Q^2$, which was one of the main targets of our work.

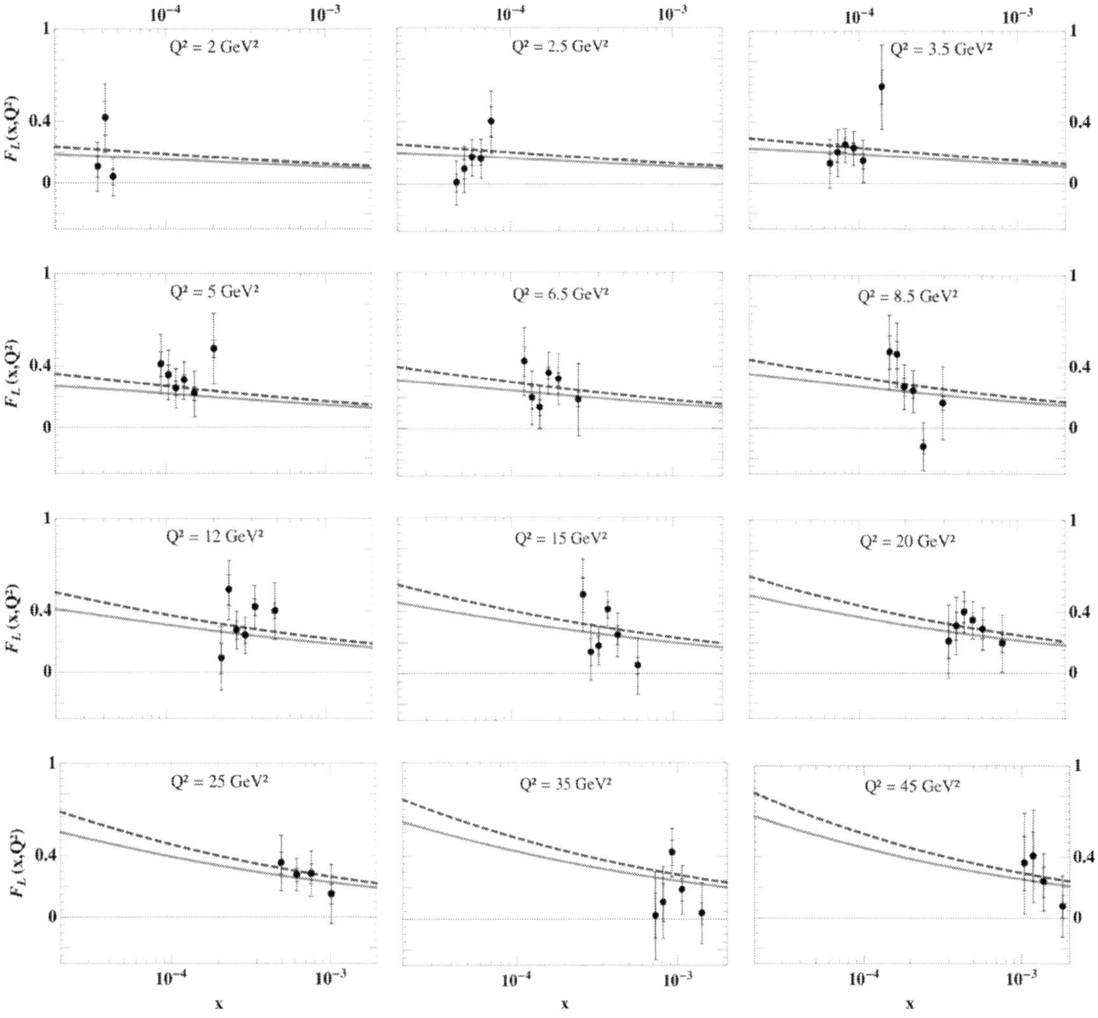


**Figure 11.** Fit to $F_L$ with the LO photon impact factor (solid lines) and the improved one (dashed lines). The experimental data are taken from [43].

### *3.3. Predictions for future colliders*

While our predictions for the structure functions are in agreement with the data from the HERA collider experiments H1 and ZEUS, these observables are too inclusive to provide unambiguous evidence for BFKL evolution (for other recent studies in this context see [42]). Comparable in quality fits can be obtained by both DGLAP evolution and saturation models, see *e.g.* [43, 44]. In order to distinguish among different parton evolution pictures new collider experiments are needed, such as the proposed Electron-Ion-Collider (EIC) at BNL/JLab (USA) [1] and the Large Hadron Electron Collider (LHeC) at CERN (Switzerland) [45], which will be able to measure both $F_2$ and $F_L$ at unprecedented small values of Bjorken $x$. In Fig. 13 we present two studies with our predictions for $F_2$ and $F_L$ down to values of $x = 10^{-6}$.

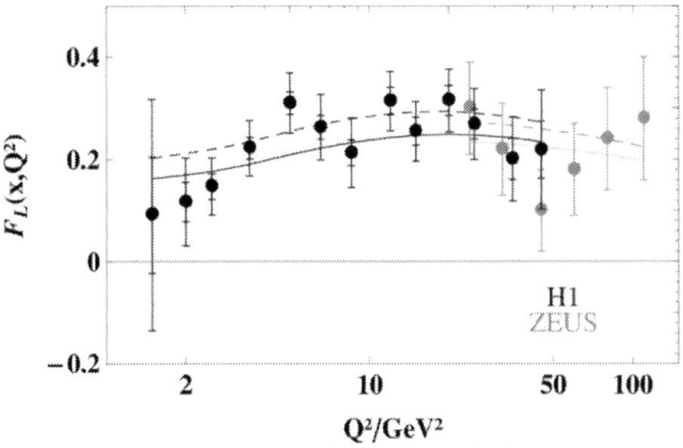


**Figure 12.** The proton structure function $F_L$ as a function of $Q^2$. The average $x$ values for each $Q^2$ of the H1 data (black) are given in Figure 13 of [43]. ZEUS data are taken from [44]. The solid line represents our calculation with the LO photon impact factor and the dashed line using the kinematically improved one.

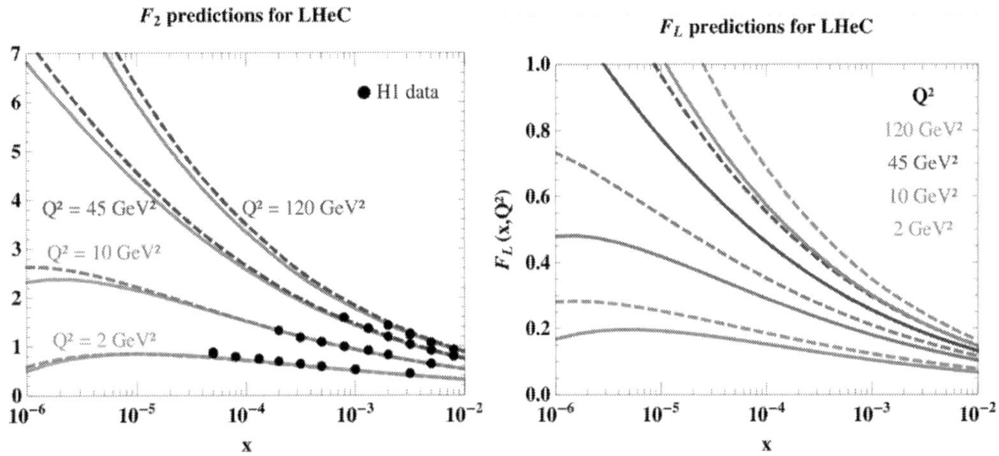


**Figure 13.** Predictions for $F_2$ (left) and $F_L$ (right) for LHeC. On the left plot, the curve with $Q^2 = 10\text{GeV}^2$ can be compared with Figure 4.13 of [45]. Simulated measurements for $F_L$ in the kinematic range plotted here (right) can be found in Figure 3.7 of the same reference.

### *3.4. Conclusions*

We have presented an application of the BFKL resummation program to the description of the $x$ and $Q^2$ dependence of structure functions as extracted from Deep Inelastic Scattering data at HERA. We have also provided some predictions for these observables at future colliders. In order to obtain the correct dependence on the virtuality of the photon at high values of the scattering energy, we have included in the BFKL kernel the main collinear contributions to all orders. We have also used optimal renormalization and an analytic running coupling in the infrared in order to accurately describe the regions of low $Q^2$.

## 4. Summary

DIS scattering experiments allow to explore the proton in terms of its QCD content, *i.e.*, quarks and gluons. At small values of $x$, the description in terms of linear DGLAP evolution and parton distribution functions is expected to break down and non-linear effects, associated with high gluon densities, are believed to set in. To pin down such effects in DIS, new collider experiments are needed, which allow to measure both structure functions $F_2$ and $F_L$ with high accuracy for both electron-proton and electron-nucleus scattering. DGLAP evolution formulated in terms of physical evolution kernels allows for a direct evolution of structure function doublets. Apart from removing scale- and scheme dependence in the description of structure functions, it further reduces the number of free parameters, used in the parametrization of non-perturbative initial conditions. BFKL evolution describes on the other hand directly evolution in $x$ and hence drives the system into the non-linear regime, promising higher sensitivity to non-linear effects. Being less explored than DGLAP evolution, we took a first step towards such applications by confronting BFKL evolution with the combined HERA data. In particular we demonstrated that NLO BFKL evolution is capable to describe the combined HERA data, if the NLO BFKL kernel is supplemented with collinear resummation and optimal scale setting for the QCD running coupling is used.

## References

[1] D. Boer *et al.*, arXiv:1108.1713 [nucl-th]; A. Accardi, J. L. Albacete, M. Anselmino, N. Armesto, E. C. Aschenauer, A. Bacchetta, D. Boer and W. Brooks *et al.*, "*Electron Ion Collider: The Next QCD Frontier - Understanding the glue that binds us all,*" arXiv:1212.1701 [nucl-ex];
[2] See, e.g., J. C. Collins, D. E. Soper, and G. F. Sterman, Adv. Ser. Direct. High Energy Phys. **5**, 1 (1988).
[3] R. P. Feynman, Phys. Rev. Lett. **23**, 1415 (1969).
[4] W. Furmanski and R. Petronzio, Z. Phys. C **11**, 293 (1982).
[5] W. L. van Neerven and E. B. Zijlstra, Phys. Lett. B **272**, 127 (1991); Phys. Lett. B **273**, 476 (1991); Nucl. Phys. B **383**, 525 (1992).
[6] S. Moch and J. A. M. Vermaseren, Nucl. Phys. B **573**, 853 (2000).
[7] D. I. Kazakov and A. V. Kotikov, Nucl. Phys. B **307**, 721 (1988) [Erratum-ibid. B **345**, 299 (1990)]; D. I. Kazakov, A. V. Kotikov, G. Parente, O. A. Sampayo, and J. Sanchez Guillen, Phys. Rev. Lett. **65**, 1535 (1990) [Erratum-ibid. **65**, 2921 (1990)]; J. Sanchez Guillen, J. Miramontes, M. Miramontes, G. Parente, and O. A. Sampayo, Nucl. Phys. B **353**, 337 (1991).
[8] J. Blumlein, Prog. Part. Nucl. Phys. **69**, 28 (2013).
[9] D. J. Gross and F. Wilczek, Phys. Rev. D **9**, 980 (1974); H. Georgi and H. D. Politzer, Phys. Rev. D **9**, 416 (1974); V. N. Gribov and L. N. Lipatov, Sov. J. Nucl. Phys. **15**, 438 (1972) [Yad. Fiz. **15**, 781 (1972)]; G. Altarelli and G. Parisi, Nucl. Phys. B **126**, 298 (1977); Y. L. Dokshitzer, Sov. Phys. JETP **46**, 641 (1977) [Zh. Eksp. Teor. Fiz. **73**, 1216 (1977)]; K. J. Kim and K. Schilcher, Phys. Rev. D **17**, 2800 (1978).
[10] E. G. Floratos, D. A. Ross, and C. T. Sachrajda, Nucl. Phys. B **129**, 66 (1977) [Erratum-ibid. B **139**, 545 (1978)]; G. Curci, W. Furmanski, and R. Petronzio, Nucl. Phys. B **175**, 27 (1980); W. Furmanski and R. Petronzio, Phys. Lett. B **97**, 437 (1980); A. Gonzalez-Arroyo, C. Lopez, and F. J. Yndurain, Nucl. Phys. B **153**, 161 (1979); A. Gonzalez-Arroyo and C. Lopez, Nucl. Phys. B **166**, 429 (1980); E. G. Floratos, R. Lacaze, and C. Kounnas, Phys. Lett. B **98**, 89 (1981); Phys. Lett. B **98**, 285 (1981); Nucl. Phys. B **192**, 417 (1981).
[11] S. Moch, J. A. M. Vermaseren, and A. Vogt, Nucl. Phys. B **688**, 101 (2004); Nucl. Phys. B **691**, 129 (2004).
[12] F. D. Aaron *et al.* [H1 and ZEUS Collaboration], JHEP **1001**, 109 (2010).
[13] For recent PDF analyses, see: H. -L. Lai *et al.*, Phys. Rev. D **82**, 074024 (2010); J. Gao *et al.*, arXiv:1302.6246 [hep-ph]; A. D. Martin, W. J. Stirling, R. S. Thorne, and G. Watt, Eur. Phys. J. C **63**, 189 (2009); R. D. Ball *et al.* [The NNPDF Collaboration], Nucl. Phys. B **874**, 36 (2013); S. Alekhin, J. Blumlein, S. Klein, and S. Moch, Phys. Rev. D **81**, 014032 (2010).
[14] L. N. Lipatov, Sov. J. Nucl. Phys. **23** (1976) 338, E. A. Kuraev, L. N. Lipatov, V. S. Fadin, Phys. Lett. B **60** (1975) 50, Sov. Phys. JETP **44** (1976) 443, Sov. Phys. JETP **45** (1977) 199. I. I. Balitsky, L. N. Lipatov, Sov. J. Nucl. Phys. **28** (1978) 822.
[15] I. Balitsky, Nucl. Phys. B **463** (1996) 99 [hep-ph/9509348]; Y. V. Kovchegov, Phys. Rev. D **55** (1997) 5445 [hep-ph/9701229].
[16] J. Jalilian-Marian, A. Kovner, A. Leonidov and H. Weigert, Phys. Rev. D **59**, 014014 (1998) [hep-ph/9706377]; Phys. Rev. D **59**, 034007 (1999) [Erratum-ibid. D **59**, 099903 (1999)] [hep-ph/9807462]; J. Jalilian-Marian, A. Kovner and H. Weigert, Phys. Rev. D **59**, 014015 (1998) [hep-ph/9709432]; H. Weigert, Nucl. Phys. A **703**, 823 (2002) [hep-ph/0004044]; E. Iancu, A. Leonidov and L. D. McLerran, Nucl. Phys. A **692**,

583 (2001) [hep-ph/0011241]; E. Iancu, K. Itakura and L. McLerran, Nucl. Phys. A **708**, 327 (2002) [hep-ph/0203137].
[17] J. L. Abelleira Fernandez *et al.*, arXiv:1211.4831 [hep-ex].
[18] M. Hentschinski and M. Stratmann, arXiv:1311.2825 [hep-ph].
[19] M. Hentschinski, A. Sabio Vera and C. Salas, Phys. Rev. Lett. **110** (2013) 041601 [arXiv:1209.1353 [hep-ph]].
[20] M. Hentschinski, A. Sabio Vera and C. Salas, Phys. Rev. D **87** (2013) 076005 [arXiv:1301.5283 [hep-ph]].
[21] S. Catani, Z. Phys. C **75**, 665 (1997).
[22] J. Blumlein, V. Ravindran, and W. L. van Neerven, Nucl. Phys. B **586**, 349 (2000).
[23] A. Vogt, Comput. Phys. Commun. **170**, 65 (2005).
[24] M. Gluck, E. Reya, and A. Vogt, Z. Phys. C **48**, 471 (1990).
[25] J. Blumlein, Comput. Phys. Commun. **133**, 76 (2000).
[26] J. Blumlein and S. Kurth, Phys. Rev. D **60**, 014018 (1999).
[27] S. Moch, J. A. M. Vermaseren and A. Vogt, Phys. Lett. B **606**, 123 (2005) [hep-ph/0411112].
[28] M. Gluck, K. Grassie, and E. Reya, Phys. Rev. D **30**, 1447 (1984).
[29] M. Gluck, E. Reya, and A. Vogt, Phys. Rev. D **46**, 1973 (1992).
[30] J. Kwiecinski, A. D. Martin and A. M. Stasto, Phys. Rev. D **56**, 3991 (1997) [hep-ph/9703445].
[31] A. Bialas, H. Navelet and R. B. Peschanski, Nucl. Phys. B **603**, 218 (2001) [hep-ph/0101179].
[32] A. Sabio Vera, F. Schwennsen, Phys. Rev. D **77** (2008) 014001 [arXiv:0708.0549 [hep-ph]], Nucl. Phys. B **776** (2007) 170 [hep-ph/0702158 [HEP-PH]], A. Sabio Vera, Nucl. Phys. B **746** (2006) 1 [hep-ph/0602250], F. Caporale, A. Papa, A. Sabio Vera, Eur. Phys. J. C **53** (2008) 525 [arXiv:0707.4100 [hep-ph]].
[33] V. S. Fadin, L. N. Lipatov, Phys. Lett. B **429** (1998) 127, M. Ciafaloni, G. Camici, Phys. Lett. B **430** (1998) 349.
[34] G. P. Salam, JHEP **9807** (1998) 019 [hep-ph/9806482].
[35] A. Sabio Vera, Nucl. Phys. B **722** (2005) 65 [hep-ph/0505128].
[36] S. J. Brodsky, V. S. Fadin, V. T. Kim, L. N. Lipatov and G. B. Pivovarov, JETP Lett. **70** (1999) 155 [hep-ph/9901229], JETP Lett. **76** (2002) 249 [Pisma Zh. Eksp. Teor. Fiz. **76** (2002) 306] [hep-ph/0207297].
[37] S. J. Brodsky, G. P. Lepage and P. B. Mackenzie, Phys. Rev. D **28** (1983) 228.
[38] W. Celmaster and R. J. Gonsalves, Phys. Rev. D **20** (1979) 1420, Phys. Rev. Lett. **42** (1979) 1435, P. Pascual and R. Tarrach, Nucl. Phys. B **174** (1980) 123 [Erratum-ibid. B **181** (1981) 546].
[39] S. J. Brodsky and L. Di Giustino, Phys. Rev. D **86** (2012) 085026 [arXiv:1107.0338 [hep-ph]].
[40] M. Angioni, G. Chachamis, J. D. Madrigal and A. Sabio Vera, Phys. Rev. Lett. **107** (2011) 191601 [arXiv:1106.6172 [hep-th]].
[41] B. R. Webber, JHEP **9810** (1998) 012. [hep-ph/9805484].
[42] H. Kowalski, L. N. Lipatov, D. A. Ross and G. Watt, Eur. Phys. J. C **70** (2010) 983 [arXiv:1005.0355 [hep-ph]], H. Kowalski, L. N. Lipatov and D. A. Ross, arXiv:1109.0432 [hep-ph], arXiv:1205.6713 [hep-ph].
[43] F. D. Aaron *et al.* [ H1 Collaboration], Eur. Phys. J. C **71**, 1579 (2011) [arXiv:1012.4355 [hep-ex]].
[44] S. Chekanov *et al.* [ZEUS Collaboration], Phys. Lett. B **682**, 8 (2009) [arXiv:0904.1092 [hep-ex]].
[45] J. L. Abelleira Fernandez *et al.* [LHeC Study Group Collaboration], J. Phys. G **39**, 075001 (2012) [arXiv:1206.2913 [physics.acc-ph]].

# Differences and analogies between quantum chromodynamics and ferromagnets

**C P Hofmann**[1]

[1] Facultad de Ciencias, Universidad de Colima, Bernal Díaz del Castillo 340, Colima C.P. 28045, Mexico

E-mail: christoph.peter.hofmann@gmail.com

**Abstract.** The low-energy physics of quantum chromodynamics (QCD) and ferromagnets is dominated by Goldstone bosons. While the effective theory of QCD - chiral perturbation theory - is well established in the particle physics community, the systematic studies of ferromagnetic systems within the effective Lagrangian framework are not well-known. We analyze the low-temperature properties of ferromagnets in one, two and three space dimensions up to three-loop order in the effective expansion, i.e., beyond the accuracy of any previous results obtained with conventional condensed matter methods. In particular, in the nonrelativistic domain, the effective method perfectly works in one space dimension.

## 1. Motivation

The effective field theory of quantum chromodynamics (QCD), chiral perturbation theory, is very-well established in particle physics. The method was originally devised for zero temperature [1, 2, 3], but has been extended to finite temperature soon after [4, 5]. There are many excellent outlines of the method available – the interested reader may want to consult Refs. [6, 7, 8, 9, 10]. In the present overview, we are interested in the description of relativistic and nonrelativistic systems at nonzero temperature. In fact, the low-temperature properties of QCD have been systematically analyzed in a series of papers a long time ago [4, 5, 11]. One important result of these studies is the three-loop formula for the order parameter, i.e., the quark condensate, which takes the following form [11],

$$\frac{\langle\bar{q}q\rangle(T,m_q)}{\langle 0|\bar{q}q|0\rangle} = 1 - \beta_0 T^2 - \beta_1 \mathbf{T^4} - \beta_2 \mathbf{T^6 \ln T} + \mathcal{O}(\mathbf{T^8}), \tag{1}$$

where $\langle 0|\bar{q}q|0\rangle$ is the quark condensate at zero temperature. Note that the boldfaced terms are those contributions that result from the interaction among the Goldstone bosons, while the leading temperature-dependent term corresponds to free Goldstone bosons.

The effective field theory method, however, is not restricted to the relativistic domain. Indeed, the effective Lagrangian technique has been transferred to nonrelativistic systems in Ref. [12]. In particular, the effective Lagrangian for the ferromagnet was established in that reference. In analogy to the low-temperature expansion of the quark condensate and other thermodynamic quantities, in this talk, we will be interested in the behavior of the (spontaneous) magnetization and other observables of ferromagnets in one, two and three space dimensions – detailed information can be found in the original papers [13, 14, 15, 16, 17, 18]. We will present a three-loop analysis of the partition function in each case and point our both differences and analogies between QCD and ferromagnets. One of our main themes is to determine at what order in the low-temperature expansion, the interaction among the Goldstone bosons (the spin waves or magnons in the case of the ferromagnet, the pseudoscalar mesons in the case of QCD) does manifest itself.

## 2. From the underlying theory to the effective field theory

### *2.1. Symmetries, Goldstone theorem and order parameters*

In the chiral limit, i.e. when the light quark masses $m_u, m_d, m_s$ are set to zero, the QCD Lagrangian,

$$\mathcal{L}_{\rm QCD} = -\frac{1}{2g^2}{\rm tr}_c G_{\mu\nu}G^{\mu\nu} + \bar{q}i\gamma^\mu D_\mu q - \bar{q}mq, \tag{2}$$

is invariant under the chiral transformation $SU(3)_R\times SU(3)_L$. The QCD vacuum, on the other hand, is only invariant under $SU(3)_V$: the chiral symmetry is spontaneously broken, and the quark condensate serves as an order parameter. Goldstone's theorem then predicts eight pseudoscalar mesons in the low-energy spectrum of QCD: three pions, four kaons and the $\eta$-particle. In the real world we rather have $m_u, m_d, m_s \neq 0$, i.e., the chiral symmetry is only approximate, such that there emerge pseudo-Goldstone bosons: the pseudoscalar mesons are not massless, but their masses are still small compared to the mass of, e.g., the proton.

An analogous situation arises in the context of ferro- and antiferromagnets that are described by the Heisenberg model on the underlying or microscopic level,

$$\mathcal{H}_0 = -J\sum_{n.n.}\vec{S}_m\cdot\vec{S}_n, \qquad J = const. \tag{3}$$

Depending on the sign of the exchange integral $J$, a ferromagnetic ($J > 0$) or antiferromagnetic ($J < 0$) ground state is preferred. But again, whereas the Hamiltonian $\mathcal{H}_0$ is invariant under the group G = O(3), describing rotations in the space of the spin variables, the ground state is only invariant under the subgroup H = O(2), describing rotations around the third axis. In the case of the three-dimensional ferromagnet, the spontaneous magnetization $\Sigma$ serves as an order parameter. Below the Curie temperature, we are in the broken phase ($\Sigma \neq 0$), above the Curie temperature, the system is in the symmetric phase ($\Sigma = 0$). The Curie temperature is thus the analog of the chiral phase transition temperature in QCD.

Goldstone's theorem for nonrelativistic systems [19, 20, 21] also implies that there are low-energy excitations in the spectrum whose energy tends to zero for large wavelengths: $\omega \to 0$ for $|\vec{k}| \to 0$. However, in the case of the ferromagnet, although we have two magnon fields, we only have one magnon particle that follows a quadratic dispersion relation [12, 22]. These magnons or spin waves can be viewed as collective fluctuations of the spins.

The Heisenberg model can be extended by incorporating an external magnetic field $\vec{H}$ that couples to the total spin vector,

$$\mathcal{H} = -J\sum_{n.n.}\vec{S}_m\cdot\vec{S}_n - \mu\sum_n\vec{S}_n\cdot\vec{H} = \mathcal{H}_0 + \mathcal{H}_{sb}. \tag{4}$$

Much like the quark masses in the QCD Lagrangian explicitly break chiral symmetry, the magnetic field is a symmetry breaking parameter that explicitly breaks the spin rotation symmetry O(3) of the Heisenberg Hamiltonian. For small quark masses and for weak magnetic fields, these quantities represent perturbations of the spontaneously broken symmetry.

### *2.2. Relativistic and nonrelativistic effective Lagrangians*

The construction of effective Lagrangians is based on an analysis of the symmetries in the underlying theory (QCD Lagrangian, Heisenberg Hamiltonian), in particular of the spontaneously broken symmetry [1, 23, 24, 25]. The degrees of freedom in the effective Lagrangian are the Goldstone bosons that dominate the low-energy properties of the system. One then proceeds to write down the most general possible Lagrangian that includes all terms consistent with the symmetries identified in the underlying theory, and then calculates observables with this Lagrangian to the desired order of perturbation theory. This corresponds to an expansion of observables in powers of momenta (or temperature), i.e., to a derivative expansion of the effective Lagrangian. In chiral perturbation theory, Lorentz-invariance implies that the derivative expansion consists of a series

$$\mathcal{L}_{eff} = \mathcal{L}_{eff}(\pi, \partial\pi, \partial^2\pi, \ldots, m) = \mathcal{L}^2_{eff} + \mathcal{L}^4_{eff} + \mathcal{L}^6_{eff} + \ldots, \tag{5}$$

where the leading term (order $p^2$) is given by

$$\mathcal{L}^2_{eff} = \tfrac{1}{4}F_\pi^2 \,\mathrm{tr}\,\{\partial_\mu U \partial^\mu U^\dagger + \chi(U + U^\dagger)\}, \qquad \chi = 2Bm. \tag{6}$$

Note that the Goldstone boson fields, i.e. the pseudoscalar mesons, are collected in the matrix $\pi(x)$ that is related to $U(x)$ by

$$U = \exp[2i\pi(x)/F_\pi]. \tag{7}$$

At leading order $\mathcal{O}(p^2)$, we have two effective constants: $F_\pi$ (pion decay constant) and $B$.

The next-to-leading order ($\mathcal{O}(p^4)$) chiral effective Lagrangian,

$$\mathcal{L}^4_{eff} = L_1 \,\mathrm{tr}\,\{D_\mu U^\dagger D^\mu U\}^2 + \ldots + L_5 \,\mathrm{tr}\,\{D_\mu U^\dagger D^\mu U(\chi^\dagger U + U^\dagger \chi)\} + \ldots, \tag{8}$$

already involves ten effective constants ($L_1, \ldots$) that parametrize the physics of the underlying theory (QCD). Note that these effective constants have to be determined by experiment (e.g. pion-pion scattering) or in numerical simulations – symmetry does not fix their actual values. It is important to point out, however, that symmetry does unambiguously determine the derivative structure of the various terms in the effective Lagrangian.

In the nonrelativistic domain, charge densities are allowed to pick up nonzero expectation values, since Lorentz-invariance no longer represents a symmetry of the system. This is precisely what happens in the case of the ferromagnet, where the spontaneous magnetization $\Sigma$ shows up as a leading-order effective constant of a term that involves one time derivative only and thus dominates the low-energy dynamics [12],

$$\mathcal{L}^F_{eff} = \Sigma \frac{\partial_0 U^1 U^2 - \partial_0 U^2 U^1}{1 + U^3} + \Sigma H^i U^i - \tfrac{1}{2}F^2 D_r U^i D_r U^i. \tag{9}$$

Note that the spontaneous magnetization also occurs in the term that involves the magnetic field $\vec{H} = (0, 0, H)$. The spin waves correspond to the fluctuations of the magnetization vector $\vec{U}$, defined by

$$\vec{U} = (U^1, U^2, U^3) = (\pi^1, \pi^2, \sqrt{1 - \pi^a \pi^a}), \quad a = 1, 2, \qquad |\vec{U}| = 1. \tag{10}$$

Note the difference with respect to chiral perturbation theory, where the leading-order effective Lagrangian reads

$$\mathcal{L}^{QCD}_{eff} = \tfrac{1}{4}F_\pi^2 \,\mathrm{tr}\,\{\partial_\mu U \partial^\mu U^\dagger + 2Bm(U + U^\dagger)\}, \quad U = \exp[2i\pi(x)/F_\pi]. \tag{11}$$

Accordingly, the dispersion relation of pseudoscalar mesons takes a linear (relativistic) form

$$\omega = \sqrt{c^2 \vec{k}^2 + c^4 M^2}, \qquad M^2 = (m_u + m_d)B, \tag{12}$$

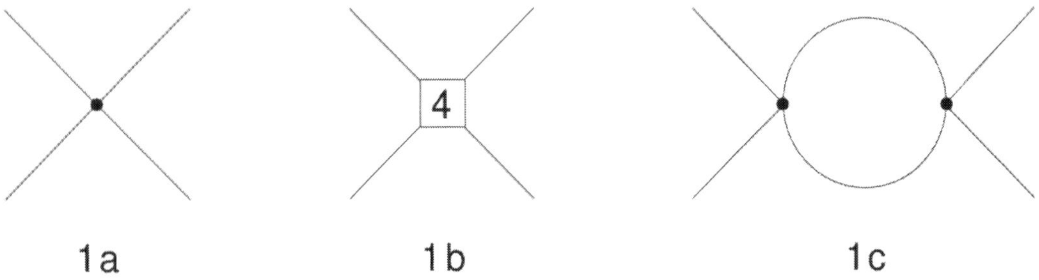


**Figure 1.** *Loops in chiral perturbation theory are suppressed by two powers of momentum. The ultraviolet divergences of the one-loop graph 1c can be absorbed into next-to-leading effective constants contained in the tree graph 1b.*

while for ferromagnetic magnons it is quadratic,

$$\omega = \gamma \vec{k}^2 + \mu H, \qquad \gamma = F^2/\Sigma. \tag{13}$$

The constant $F$ in the leading-order effective Lagrangian of the ferromagnet is thus related to the helicity modulus $\gamma = F^2/\Sigma$.

An essential point for the effective framework to be consistent is that the ultraviolet divergences that occur in loop graphs can systematically be absorbed by effective constants, order by order in the low-energy expansion. In chiral perturbation theory, the pion-pion scattering graphs up to next-to-leading order are shown in Fig. 1. While the tree graph 1a of order $p^2$ is finite, the one-loop graph 1c of order $p^4$ is singular in the ultraviolet. However, these divergences can be absorbed into coupling constants of order $p^4$ showing up in graph 1b. The crucial point is that these two graphs do "communicate", because Goldstone-boson loops in 3+1 dimensions are suppressed by two powers of momentum: graphs 1b and 1c are of the same momentum order $p^4$.

It is important to point out that the suppression of loops depends both on the dispersion relation of the system and on the space dimension $d_s$. Consider a diagram with a single Goldstone boson loop. The corresponding integrals are

$$\int \frac{\mathrm{d}E \, \mathrm{d}^{d_s} p}{E^2 - \vec{p}^2} \propto \mathcal{P}^{d_s - 1}, \qquad \int \frac{\mathrm{d}E \, \mathrm{d}^{d_s} p}{E - \vec{p}^2} \propto \mathcal{P}^{d_s}. \tag{14}$$

In a Lorentz-invariant framework we thus have

- $d$=3+1: Loops are suppressed by two powers of momentum
- $d$=2+1: Loops are suppressed by one power of momentum

On the other hand, in a nonrelativistic setting, like for the ferromagnet that exhibits a quadratic dispersion relation, we conclude

- $d$=3+1: Loops are suppressed by three powers of momentum
- $d$=2+1: Loops are suppressed by two powers of momentum
- $d$=1+1: Loops are suppressed by one power of momentum

Hence there is an analogy between ferromagnets in two space dimensions and QCD: in the effective field theory expansion, loops are suppressed by two powers of momentum in either case. Accordingly, the structure of the low-temperature series, as we discuss in the next section,

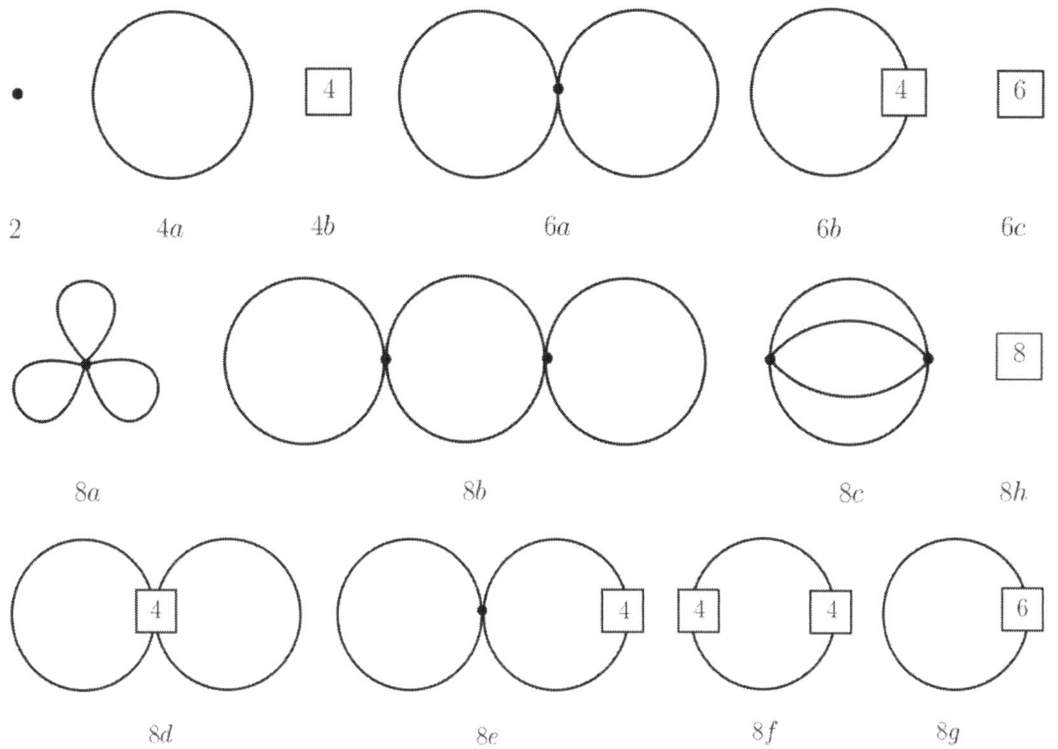


**Figure 2.** *Feynman graphs: QCD partition function and partition function of the $d_s$=2 ferromagnet up to three-loop order $p^8$ in the perturbative expansion. Note that loop graphs are suppressed by two powers of momentum. In the case of the ferromagnet, the T-independent graphs 4b, 6c, 8h are absent [13].*

exhibit common features. But there is also an important difference: whereas the effective loop expansion fails in 1+1 space dimensions in the case of Lorentz-invariant theories, it is perfectly consistent in the nonrelativistic domain: in particular, ferromagnetic spin-chains are accessible by effective field theory.

## 3. Low-temperature expansion of the partition function

### *3.1. Analogies between QCD and ferromagnets in two space dimensions*

The relevant Feynman graphs corresponding to the low-temperature expansion of the partition function, both for QCD or ferromagnets in two space dimensions, are depicted in Fig. 2. While the next-to-leading order chiral effective Lagrangian $\mathcal{L}^4_{eff}$ was given in Eq. (8), the subleading pieces in the effective Lagrangian for the ferromagnet are

$$\begin{aligned}
\mathcal{L}^2_{eff} &= \Sigma \frac{\varepsilon_{ab} \partial_0 U^a U^b}{1+U^3} + \Sigma \mu H U^3 - \tfrac{1}{2} F^2 \partial_r U^i \partial_r U^i , \\
\mathcal{L}^4_{eff} &= l_1 (\partial_r U^i \partial_r U^i)^2 + l_2 (\partial_r U^i \partial_s U^i)^2 + l_3 \Delta U^i \Delta U^i , \\
\mathcal{L}^6_{eff} &= c_1 U^i \Delta^3 U^i , \\
\mathcal{L}^8_{eff} &= d_1 U^i \Delta^4 U^i .
\end{aligned} \tag{15}$$

Note that the higher-order pieces $\mathcal{L}^6_{eff}$ and $\mathcal{L}^8_{eff}$ are irrelevant for the Goldstone boson interaction according to Fig. 2: they only occur in one-loop or tree graphs. Note that the relation between momentum and temperature depends on the dispersion relation: in chiral perturbation theory we have $p \propto E \propto T$, whereas for the ferromagnet it reads $p^2 \propto E \propto T$.

The evaluation of the partition function for ferromagnets in two space dimensions and QCD is discussed in detail in Refs. [11, 15, 16]. Here we quote the results for the pressure,

$$P_{\rm QCD} = b_0 T^4 + \mathbf{b_1 T^6} + \mathbf{b_2 T^8 \ln T} + \mathcal{O}(\mathbf{T^{10}}), \tag{16}$$

$$P_{Ferro2D} = \hat{a}_0 T^2 + \hat{a}_1 T^3 + \mathbf{\hat{a}_2^A T^4} + \mathbf{\hat{a}_2^B T^4 \ln T} + \mathcal{O}(\mathbf{T^5}), \tag{17}$$

and the order parameter: while the low-temperature expansion of the quark condensate is

$$\frac{\langle \bar{q}q \rangle (T, m_q)}{\langle 0 | \bar{q}q | 0 \rangle} = 1 - \beta_0 T^2 - \mathbf{\beta_1 T^4} - \mathbf{\beta_2 T^6 \ln T} + \mathcal{O}(\mathbf{T^8}), \tag{18}$$

for the magnetization we obtain

$$\frac{\Sigma(T,H)}{\Sigma} = 1 - \hat{\alpha}_0 T - \hat{\alpha}_1 T^2 - \mathbf{\hat{\alpha}_2^A T^3} - \mathbf{\hat{\alpha}_2^B T^3 \ln T} + \mathcal{O}(\mathbf{T^4}). \tag{19}$$

Note that all interaction contributions in the above expressions are boldfaced. Again there is this analogy between QCD and ferromagnets in two space dimensions: due to the identical loop suppression, the structure of temperature powers is analogous. Also, so-called chiral logarithms arise in the renormalization of loop graphs, and logarithms in the low-temperature series emerge at three-loop order. More precisely, as in chiral perturbation theory, the cateye-graph for ferromagnets in two space dimensions is logarithmically divergent,

$$T^{d_s+2} (\mu H)^{\frac{d_s-2}{2}} \Bigg\{ \sum_{n=1}^{\infty} \frac{e^{-\mu H n/T}}{n^{\frac{d_s+2}{2}}} \Bigg\}^2 \Gamma(1 - \frac{d_s}{2}). \tag{20}$$

This expression is dimensionally regularized in the space dimension $d_s$. In two space dimensions, the cateye-graph 8c exhibits an ultraviolet divergence. This singularity can be absorbed into the next-to-leading order low-energy constants $l_1$ and $l_2$ from $\mathcal{L}^4_{eff}$ contained in the two-loop graphs 8d and 8e. This logarithmic renormalization of the effective constants leads to chiral logarithms – $\log(H/\mu)$ in the case of the ferromagnet, and $\log(M/\mu)$ in the case of QCD – where $\mu$ is the renormalization scale. Moreover, in the limit $\vec{H} \to 0$ (much like in the chiral limit $m_q \to 0$), the cateye graph gives rise to logarithmic terms $\propto T^n \, lnT$ in the low-temperature series for the pressure and the order parameter at the three-loop level, both for QCD ($n$=8,6) and ferromagnetic films ($n$=4,3).

There is, however, an important difference at the two-loop level as far as the Goldstone boson interaction is concerned. While the two-loop graph 6a (see Fig. 2) does contribute in the case of QCD, the same graph turns out to be zero for the ferromagnet (in any space dimension) because of space rotation invariance or parity. As a consequence, the terms proportional to $T^3$ ($T^2$) in the low-temperature series for the pressure (magnetization) for the ferromagnet in two space dimensions are exclusively determined by free magnons. On the other hand, the analogous terms in the low-temperature expansion of the QCD pressure and the quark condensate ($T^6$ and $T^4$, respectively) are affected by the interaction.

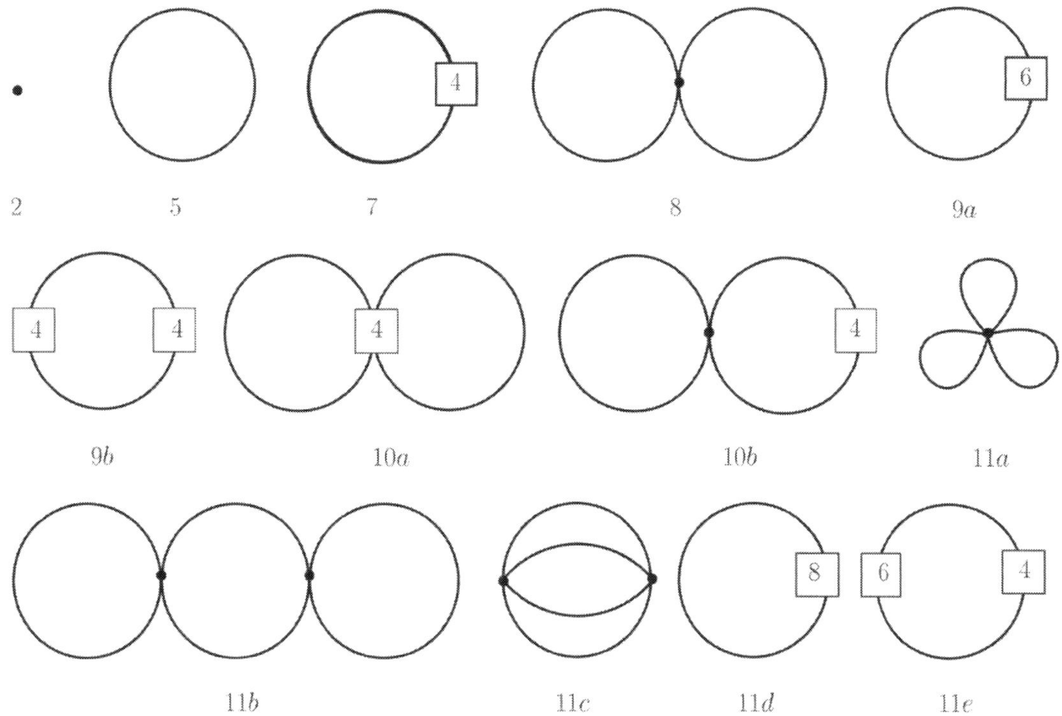


**Figure 3.** *Feynman graphs: Partition function of the $d_s$=3 ferromagnet up to three-loop order $p^{11}$ in the perturbative expansion. Note that loop graphs are suppressed by three powers of momentum.*

*3.2. Ferromagnets in odd space dimensions are different*

Let us proceed with ferromagnets in three space dimensions [13, 14]. The relevant Feynman graphs for the partition function are shown in Fig. 3. Loops are now suppressed by three powers of momentum, such that the structure of the temperature powers and the "communication" among the graphs is rather different. In particular, the three-loop graphs (11a-c) are no longer of the same order as the two-loop graphs with insertions from $\mathcal{L}^4_{eff}$ (10a-b). This seems to indicate a serious problem, as one might think that potential divergences in the three-loop graphs can no longer be absorbed into next-to-leading order effective constants of the two-loop graphs. However, as we have pointed out, the cateye graph is not singular in three space dimensions. Also, there are no logarithmic terms in the temperature in these series at three-loop order.

The leading contribution in the low-temperature expansion of the partition function is given by the one-loop graph 5 of order $p^5$, giving rise to a $T^{5/2}$-term in the pressure. In contrast to QCD or ferromagnets in two space dimensions, we thus also have half-integer powers of the temperature. The full three-loop results for the pressure and the spontaneous magnetization turn out to be [14]

$$P = a_0 T^{\frac{5}{2}} + a_1 T^{\frac{7}{2}} + a_2 T^{\frac{9}{2}} + \mathbf{a_3 T^5} + \mathbf{a_4 T^{\frac{11}{2}}} + \mathcal{O}(\mathbf{T^6}), \tag{21}$$

and

$$\frac{\Sigma(T)}{\Sigma} = 1 - \alpha_0 T^{\frac{3}{2}} - \alpha_1 T^{\frac{5}{2}} - \alpha_2 T^{\frac{7}{2}} - \boldsymbol{\alpha_3} \mathbf{T^4} - \boldsymbol{\alpha_4} \mathbf{T^{\frac{9}{2}}} + \mathcal{O}(\mathbf{T^5}), \tag{22}$$

respectively. Again we have boldfaced all terms that are related to the Goldstone boson interaction. The coefficients in these series depend on the low-energy effective constants. In the case of the spontaneous magnetization they are explicitly given by

$$\begin{aligned}
\alpha_0 &= \frac{1}{8\pi^{\frac{3}{2}}\Sigma\gamma^{\frac{3}{2}}}\,\zeta(\tfrac{3}{2}),\\
\alpha_1 &= \frac{15\,l_3}{16\pi^{\frac{3}{2}}\Sigma^2\gamma^{\frac{7}{2}}}\,\zeta(\tfrac{5}{2}),\\
\alpha_2 &= \frac{105}{32\pi^{\frac{3}{2}}\Sigma^2\gamma^{\frac{9}{2}}}\left(\frac{9l_3^2}{2\Sigma\gamma}-c_1\right)\zeta(\tfrac{7}{2}),\\
\alpha_3 &= \frac{3(8l_1+6l_2+5l_3)}{64\pi^3\Sigma^3\gamma^5}\,\zeta(\tfrac{5}{2})\zeta(\tfrac{3}{2}),\\
\alpha_4 &= \frac{945}{64\pi^{\frac{3}{2}}\Sigma^2\gamma^{\frac{11}{2}}}\left(d_1-\frac{11l_3c_1}{\Sigma\gamma}\right)\zeta(\tfrac{9}{2})-\frac{1}{2\Sigma^3\gamma^{\frac{9}{2}}}\,j_2,
\end{aligned} \tag{23}$$

where the constant $j_2$ stems from the cateye 11c graph and takes the value

$$j_2 = -0.00008\,. \tag{24}$$

Before the pioneering work by Dyson [26] and Zittartz [27], it was quite unclear at which order in the low-$T$ series the spin-wave interaction shows up in the spontaneous magnetization or in the other thermodynamic quantities. Here not only have we rederived this result in a more elegant manner, but also have we extended the calculation up to three-loop order. Interestingly, the two-loop graphs 10a-b with insertions from the next-to-leading order Lagrangian $\mathcal{L}^4_{eff}$ represent the leading contribution from the spin-wave interaction, while the three-loop graphs 11a-c only contribute at next-to-leading order. Again, this is a conseqeuence of the peculiar momentum counting in the nonrelativistic domain: in three space dimensions, loop are suppressed by three powers of momentum.

Let us have a closer look at the scales involved in QCD and the present case of the ferromagnet in three space dimensions. The series for the respective order parameters, the quark condensate and the magnetization, are

$$\frac{\langle\bar{q}q\rangle(T,m_q)}{\langle 0|\bar{q}q|0\rangle} = 1-\beta_0T^2-\beta_1\mathbf{T^4}-\beta_2\mathbf{T^6\,lnT}+\mathcal{O}(\mathbf{T^8}), \tag{25}$$

and

$$\frac{\Sigma(T,H)}{\Sigma} = 1-\alpha_0T^{\frac{3}{2}}-\alpha_1T^{\frac{5}{2}}-\alpha_2T^{\frac{7}{2}}-\alpha_3\mathbf{T^4}-\alpha_4\mathbf{T^{\frac{9}{2}}}+\mathcal{O}(\mathbf{T^5}). \tag{26}$$

In the chiral limit ($m_q \to 0$) and in zero magnetic field, the leading coefficients read

$$\begin{aligned}
\beta_0 &= \frac{1}{8F_\pi^2},\\
\alpha_0 &= \frac{\zeta(\frac{3}{2})}{8\pi^{\frac{3}{2}}\Sigma\gamma^{\frac{3}{2}}},
\end{aligned} \tag{27}$$

such that the corresponding temperature scales differ in more than ten orders of magnitude,

$$\Lambda^T_{\mathrm{QCD}} = \sqrt{8}F_\pi \approx 250\mathrm{MeV}, \qquad \Lambda^T_{\mathrm{F}} = \alpha_0^{-2/3} \approx 10\mathrm{meV}. \tag{28}$$

For temperatures small compared to $\Lambda^T_{\mathrm{QCD}}$ and $\Lambda^T_{\mathrm{F}}$, the low-temperature series are perfectly valid. As we discuss below, the situation in one and two space dimensions is more subtle, because of the Mermin-Wagner theorem [28].

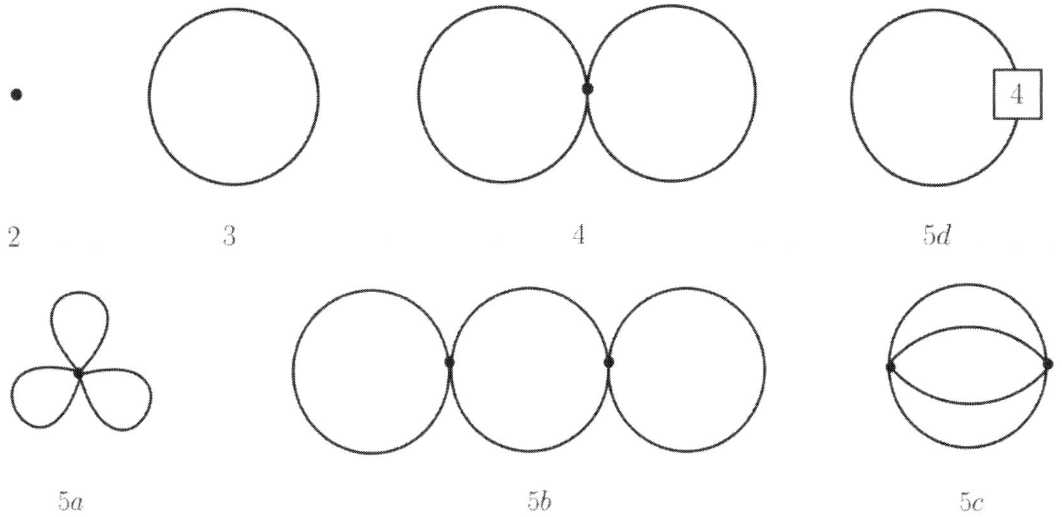


**Figure 4.** *Feynman graphs: Partition function of the $d_s = 1$ ferromagnet up to three-loop order $p^5$ in the perturbative expansion. Note that loop graphs are suppressed by one power of momentum.*

Finally, we turn to ferromagnetic spin chains [17, 18]. Remember that in a Lorentz-invariant setting, the effective Lagrangian method does not consistently work in one space dimension. However, ferromagnetic loops in one space dimension are still suppressed by one power of momentum. The relevant Feynman graphs, again up to three-loop order in the perturbative expansion of the partition function, are depicted in Fig. 4. Remarkably, in one space dimension only very few effective constants are required. In particular, while $\mathcal{L}^6_{eff}$ and even $\mathcal{L}^8_{eff}$ contribute up to three-loop order in three and two space dimensions, in the case of ferromagnetic spin chains only $\mathcal{L}^4_{eff}$ is relevant – as a matter of fact, it only occurs in the one-loop graph 5d, such that the spin-wave interaction is not affected at all by subleading pieces of the effective Lagrangian.

For the pressure we obtain,

$$P = \tilde{a}_0 T^{\frac{3}{2}} + \tilde{\mathbf{a}}_1 \mathbf{T}^{\frac{5}{2}} + \mathcal{O}(\mathbf{T}^3), \tag{29}$$

while the magnetization amounts to

$$\frac{\Sigma(T,H)}{\Sigma} = 1 - \tilde{\alpha}_0 T^{\frac{1}{2}} - \tilde{\alpha}_1 \mathbf{T}^{\frac{3}{2}} + \mathcal{O}(\mathbf{T}^2). \tag{30}$$

Much like in the case of the ferromagnet in three space dimensions, there are no logarithmic terms in the temperature in these series, and we also have half-integer powers of $T$. Note that the spin-wave interaction already sets in at next-to-leading order. The corresponding coefficient $\tilde{a}_1$ turns out to be positive, such that the spin-wave interaction in the pressure is repulsive. Remarkably, this is a rigorous statement following from the leading effective Lagrangian $\mathcal{L}^2_{eff}$ alone – it is not affected by any higher-order effective constants up to the three-loop level we have considered in the present analysis.

At the end of this overview, we like to point out some subtleties that arise at finite temperature in space dimensions lower or equal than two. Although the spin symmetry O(3) of the underlying Heisenberg model is spontaneously broken at zero temperature, both for ferromagnetic films and for ferromagnetic spin chains, matters are quite different at nonzero temperature. Following the Mermin-Wagner theorem [28], there is no spontaneous symmetry breaking in space dimensions

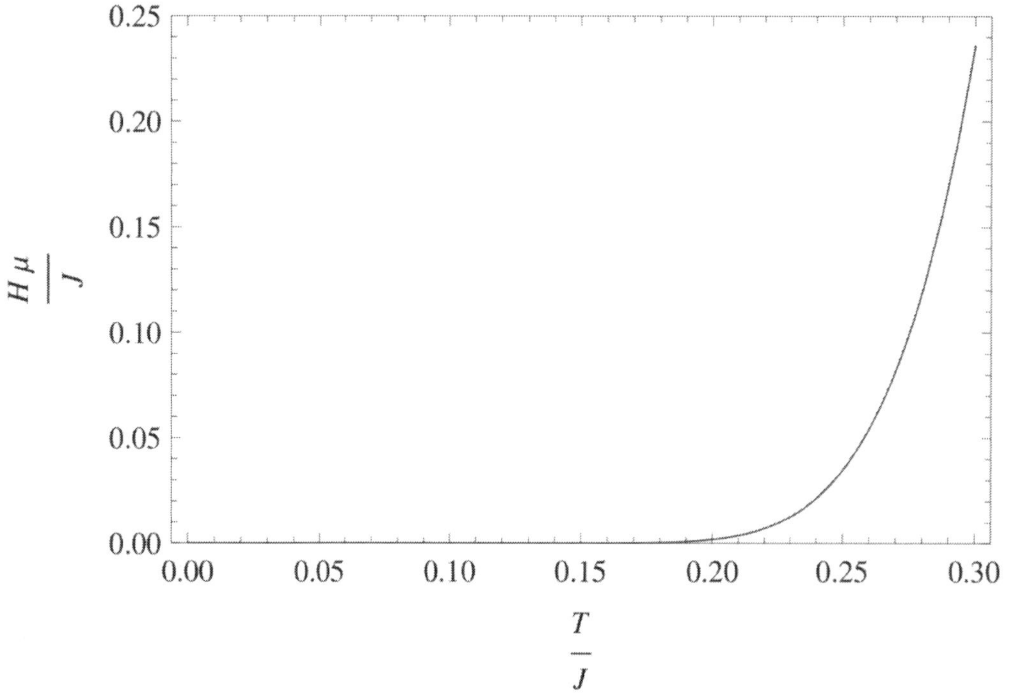


**Figure 5.** *Domain of validity for the ferromagnet in two space dimensions: in the parameter regime above the curve the effective series are safe.*

two and one in the isotropic Heisenberg model. Indeed, the low-energy spectrum of both systems is characterized by a nonperturbatively generated energy gap. Accordingly, the correlation length of magnons is no longer infinite (as for a true Goldstone bosons), but finite. In the case of ferromagnets in two dimensions, the correlation length is still exponentially large [29],

$$\xi_{np} = C_\xi a S^{-\frac{1}{2}} \sqrt{\frac{T}{JS^2}} \exp\Big[\frac{2\pi J S^2}{T}\Big] \qquad (d_s = 2), \tag{31}$$

where $a$ is the distance between two neighboring sites on the square lattice, and $C_\xi \approx 0.05$ is a dimensionless constant. For ferromagnetic spin chains, on the other hand, the correlation length is only proportional to the inverse temperature [30],

$$\xi_{np} = a\, C^{(0)}_\xi \frac{JS^2}{T} \Big[1 + C^{(1)}_\xi \frac{1}{\pi} \sqrt{\frac{T}{JS^3}} + \mathcal{O}(T)\Big] \qquad (d_s = 1). \tag{32}$$

Here $a$ is the distance between two neighboring sites of the spin chain and $C^{(0)}_\xi$ and $C^{(1)}_\xi$ are dimensionless constants.

As we have discussed in much detail in the original publications [15, 16, 17, 18], this then implies that the external magnetic field cannot be switched off completely, because we would then leave the validity domain of the effective low-temperature expansion presented here. Still, the excluded regions are rather tiny, in particular in the case of ferromagnets in two space dimensions. We have illustrated this in two plots: Fig. 5 refers to ferromagnetic films, while Fig. 6 corresponds to ferromagnetic spin chains. In either case, in the parameter regime above

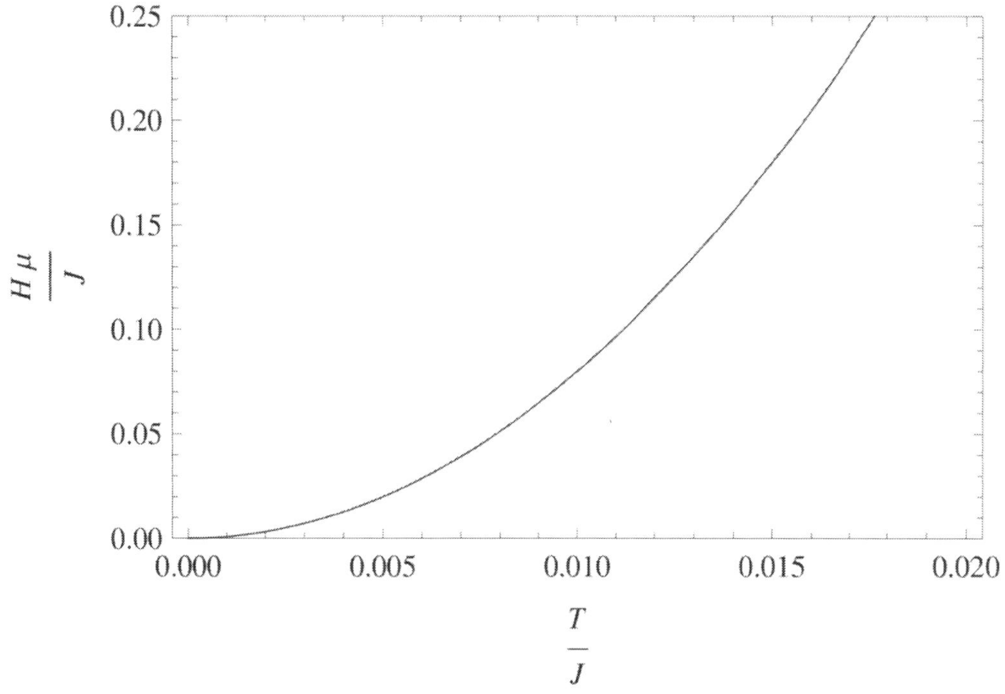


**Figure 6.** *Domain of validity for the ferromagnetic spin chain: in the parameter regime above the curve the effective series are safe.*

the plotted curve, the effective expansion works perfectly. Note however, that the horizontal axis lies outside the allowed domain: the magnetic field cannot be switched off completely. This is different from ferromagnets in three space dimensions or QCD: there, the chiral limit or the limit of zero external magnetic field does not pose any problems: the effective series are valid as they stand.

## 4. Conclusions

The low-energy dynamics of systems with a spontaneously broken symmetry is dominated by the corresponding Goldstone bosons and a systematic effective field theory can be constructed. While the effective field theory for QCD (chiral perturbation theory) has been established a long time ago, the technique has been transferred and applied to nonrelativistic systems only recently. In the present overview we have focused on the properties of QCD and ferromagnets in one, two and three space dimensions at finite temperature. In particular, we have discussed the partition function and various observables derived from there, including order parameters, up to the three-loop level.

The structure of these low-temperature series is an immediate consequence of the symmetries of the underlying theory, i.e., the QCD Lagrangian and the Heisenberg Hamiltonian, respectively. Although ferromagnetic magnons obey a quadratic dispersion relation, there is an interesting analogy between ferromagnets in two space dimensions and QCD: in either case, Goldstone boson loops are suppressed by two powers of momentum in the effective loop expansion. Accordingly, the next-to-leading order effective constants undergo logarithmic renormalization and the ultraviolet divergences in three-loop graphs are absorbed into effective constants of

$\mathcal{L}^4_{eff}$ arising in two-loop graphs. Moreover, in the low-temperature expansion of the various thermodynamic quantities, terms logarithmic in the temperature emerge at three-loop order, related to the Goldstone boson interaction.

On the other hand, ferromagnets in one or three space dimensions are rather different. In the latter case, loops are suppressed by three powers of momentum, in the former case, loops are still suppressed by one momentum power. It is quite remarkable that the systematic effective field theory method works in one space dimension in the nonrelativistic domain – in a Lorentz-invariant setting, the method does not work consistently in one space dimension, since loops are no longer suppressed. In the low-temperature series of the various thermodynamic quantities, no logarithmic terms in the temperature show up in the case of ferromagnets in one and three space dimension. Instead, we also have half-integer powers of the temperature. Furthermore, the lower the space dimension, the less higher-order pieces of the effective Lagrangian and the less low-energy effective constants are needed up to the three-loop level.

The scales involved in QCD and in ferromagnets differ in about ten orders of magnitude. Still, the effective field theory captures the low-temperature properties of both systems in a systematic way. What is important is not the absolute scale, but the question whether or not the temperature is low compared to the specific scale involved.

It is important to point out that many papers have been devoted in the past to the problem of how the spin-wave interaction in ferromagnets manifests itself in the partition function. In particular, in three space dimensions there are more than a hundred papers available, that are based on different methods such as spin-wave theory, Schwinger boson mean field theory, Green's functions, etc. We like to stress that many of these results are not only in contradiction with each other, but also in contradiction with the effective field theory results presented here. The systematic effective field theory approach hence proves to be superior to conventional condensed matter techniques, as it addresses the problem from a model-independent point of view.

## Acknowledgments
The author would like to thank U.-J. Wiese and H. Leutwyler for stimulating discussions.

## References
[1] Weinberg S 1979 *Physica* A **96** 327
[2] Gasser J and Leutwyler H 1984 *Ann. Phys. (NY)* **158** 142
[3] Gasser J and Leutwyler H 1985 *Nucl. Phys.* B **250** 465
[4] Gasser J and Leutwyler H 1987 *Phys. Lett.* B **184** 83
[5] Gasser J and Leutwyler H 1987 *Phys. Lett.* B **188** 477
[6] Leutwyler H 1995 *Hadron Physics 94 – Topics on the Structure and Interaction of Hadronic Systems*, ed. V E Herscovitz, C A Z Vasconcellos and E Ferreira (Singapore: World Scientific) p 1
[7] Scherer S 2003 *Adv. Nucl. Phys.* **27** 277
[8] Goity J L 2001*Czech. J. Phys.* **51** B35
[9] Burgess C P 2000 *Phys. Rep.* **330** 193
[10] Brauner T 2010 *Symmetry* **2** 609
[11] Gerber P and Leutwyler H 1989 *Nucl. Phys.* B **321** 387
[12] Leutwyler H 1994 *Phys. Rev.* D **49** 3033
[13] Hofmann C P 2002 *Phys. Rev.* B **65** 094430
[14] Hofmann C P 2011 *Phys. Rev.* B **84** 064414
[15] Hofmann C P 2012 *Phys. Rev.* B **86** 054409
[16] Hofmann C P 2012 *Phys. Rev.* B **86** 184409
[17] Hofmann C P 2013 *Phys. Rev.* B **87** 184420
[18] Hofmann C P 2013 Thermodynamics of ferromagnetic spin chains in a magnetic field: Impact of the spin-wave interaction *preprint* cond-mat/1306.0600
[19] Lange R V 1965 *Phys. Rev. Lett.* **14** 3; 1966 *Phys. Rev.* **146** 301
[20] Guralnik G S, Hagen C R and Kibble T W B 1968 *Advances in Particle Physics* vol 2, ed. R L Cool and R E Marshak (New York: Wiley) p 567

Journal of Physics: Conference Series **651** (2015) 012012 doi:10.1088/1742-6596/651/1/012012

[21] Nielsen H B and Chadha S 1976 *Nucl. Phys.* B **105** 445
[22] Hofmann C P 1999 *Phys. Rev.* B **60** 388
[23] Coleman S, Wess J and Zumino B 1969 *Phys. Rev.* **177** 2239
[24] Callan C G, Coleman S, Wess J and Zumino B 1969 *Phys. Rev.* **177** 2247
[25] Leutwyler H 1994 *Ann. Phys. (NY)* **235** 165
[26] Dyson F J 1956 *Phys. Rev.* **102** 1217; 1956 **102** 1230
[27] Zittartz J 1965 *Z. Phys.* **184** 506
[28] Mermin N D and Wagner H 1966 *Phys. Rev. Lett.* **17** 1133
[29] Kopietz P and Chakravarty S 1989 *Phys. Rev.* B **40** 4858
[30] Kopietz P 1989 *Phys. Rev.* B **40** 5194

Journal of Physics: Conference Series **651** (2015) 012013 doi:10.1088/1742-6596/651/1/012013

# Hadronic decay asymmetries of the Top quark.

**C. G. Honorato**[1], **F. Larios**[2] **and M.A. Pérez**[1]

[1]Departamento de Física, CINVESTAV, A.P. 14- 40, 0 000, México D.F., México

[2] Departamento de Física Aplicada, CINVESTAV-Mérida, A.P. 3, 310 Mérida, ucatán, México

E-mail: `larios@mda.cinvestav.mx`

**Abstract.** We study some New Physics effects on the hadronic decays of the Top quark, like $t \to b\bar{b}c$, and on a forward-backward-like CP even asymmetry $A_t$ constructed in such a way that it is zero in the SM. We  nd that an anomalous right-handed contribution of the effective $tbW$ vertex may induce an asymmetry $A_t$ of the order of 20%.

## 1. Introduction

The $A_{FB}$ asymmetry measured at the Tevatron [1] gives us a strong hint that more new physics (  P) effects involving the Top quark will be observed at the L  C[2].  odels that attempt to explain $A_{FB}$ usually invoke  avor changing effects. The effects of  avor changing neutral currents (FC  C) in processes involving the top quark have been studied even before the actual discovery of this quark [3]. While these processes are highly suppressed in the Standard  odel (S  ), several FC  C top-quark decays may be enhanced by several orders of magnitude in scenarios beyond the S   and some of them falling within the reach of the L  C [4]. FC  C top-quark processes may thus serve as a window to test new physics effects. Recently, the FC  C contributions to the decay $t \to bbc$ have been used to identify deviations from the S   predictions[5]. In particular, it was pointed that four observables, two CP-even and two CP-odd, that are formulated in such a way that they are zero in the S  , may produce measurable values at the L  C and thus may be used to identify new physics effects. In this report, we want to focus on a forward-backward-like CP even asymmetry proposed in Ref. [5] and compute the predictions from different  ew Physics scenarios. In particular, we are interested in 1) the contributions from an effective $tbW$ vertex, and in 2) the contributions induced by the new heavy gauge bosons introduced to explain the $A_{FB}$ asymmetry measured at the Tevatron[6]. We find that a right-handed ($f_2^R$) $tbW$ coupling may induce a CP-even asymmetry $A_t$ of the order of 20  . On the other hand, a new light $W'$ boson, with mass of order 180-300 GeV, and with pure right-handed $tdW'$ couplings may also contribute to $A_t$ with the same order of magnitude[7].

In any three body decay ($M \to m_1, m_2, m_3$) one can define three invariant masses $m_{ij}^2 = (p_i + p_j)^2$, that satisfy the constraint $m_{12}^2 + m_{23}^2 + m_{13}^2 = M^2 + m_1^2 + m_2^2 + m_3^2$. Only two are independent variables and we can define two asymmetries depending which $i, j$ pairs we choose. The asymmetry proposed in Ref. [5] for the decay $t \to bbc$ depends on $\rho^2 \equiv (p_b + p_{\bar{b}})^2$.  owever, the S   amplitude squared in the limit $m_b = 0$ depends on a function of $(p_b + p_c)^2$ divided by the $W$-propagator which depends on $(p_{\bar{b}} + p_c)^2$. Therefore, we have chosen $(p_b + p_c)^2$ instead of $\rho^2$ to parametrize the asymmetries of interest.

Journal of Physics: Conference Series **651** (2015) 012013 doi:10.1088/1742-6596/651/1/012013

## 2. Asymmetry in the hadronic decay of top

Let us start by writing the tree level amplitude for the hadronic decay $t \to bW^+ \to bbc$ of the top quark. For the sake of generality, we will use the notation $t \to bW^+ \to bdu$ for the initial formula. Even though our main interest is in $t \to bW^+ \to bbc$, other decay modes like $t \to dW^+ \to dbc$, or $t \to sW^+ \to ssc$ could be addressed in a similar manner. Besides $u$, $d$ and $s$, we consider $c$ to be massless. We allow the $b$ quark mass to be nonzero. As one would expect, effects from non-zero $m_b$ are very small and we usually set $m_b = 0$ in our calculations, but we have kept $m_b$ terms in the initial equations as they can be useful for future studies. For the propagator of the intermediate $W$ boson we have first taken the general expression given in Ref. [8] and then we have simplified it by taking $m_c = 0$. The S amplitude is:

$$\begin{aligned} \mathcal{M}(t \to bud) &= u_b \frac{ig}{\sqrt{2}} \gamma_\mu P_L u_t u_u \frac{ig}{\sqrt{2}} V_{ud} \gamma_\nu P_L v_d \left( -g^{\mu\nu} + (1 + i r_W) \frac{q^\mu q^\nu}{m_W^2} \right) G_T^{-1}, \qquad (1) \\ G_T &= q^2 - m_W^2 + i q^2 r_W , \end{aligned}$$

with $r_W \equiv \Gamma_W / m_W$ the width-to-mass ratio of the $W$ boson. By momentum conservation $p_t = p_b + p_u + p_d$, and we define $q \equiv p_u + p_d = p_t - p_b$, $t \equiv p_b + p_u$ and $\rho = p_b + p_d$.

Let us now consider the particular case $t \to bbc$ with $m_b$ non-zero. After summing (averaging) over final (initial) spins, and neglecting terms of order higher than $m_b^2$ the amplitude squared is given by

$$\begin{aligned} \overline{|\mathcal{M}_{\mathrm{SM}}|^2}(t \to bbc) &= \frac{3}{2} g^4 |V_{cb}|^2 |V_{tb}|^2 M_2(q^2, t^2) |G_T|^{-2}, \qquad (2) \\ M_2(q^2, t^2) &= t^2 (m_t^2 - t^2) + m_b^2 \left( t^2 (2 - \frac{m_t^2}{m_W^2}) + m_t^2 q^2 (m_t^2 - q^2) \frac{1 + r_W^2}{4 m_W^4} - m_t^2 \right). \end{aligned}$$

otice that $M_2^{SM}$ depends on $t^2$ in the numerator and on $q^2$ in the denominator. One can choose to apply the constraint $q^2 + t^2 + \rho^2 = m_t^2$ and change the $t^2$ dependence for a $\rho^2$ dependence as is done in Ref. [5]. As expected, our asymmetry based on $t^2$ is equivalent to the asymmetry based on $\rho^2$. As we will see, the definition of the asymmetry involves finding a particular integration limit that satisfies a cubic equation. We will find that it is easier to solve the equation based on $t^2$.

The total width for a three body decay can be written as[9]:

$$\Gamma = \frac{1}{(2\pi)^3 32 m_t^3} \int dm_{12}^2 \int dm_{23}^2 \; \overline{|\mathcal{M}|^2}. \qquad (3)$$

Let us now assign $p_1 = p_{\bar{b}}$, $p_2 = p_c$ and $p_3 = p_b$, then $m_{12}^2 = q^2$ and $m_{23}^2 = t^2$. Furthermore, let us rewrite the integral in terms of dimensionless variables $x \equiv q^2/m_t^2$ and $y \equiv t^2/m_t^2$ ($m_b \to 0$, $\hat{m}_W = m_W/m_t$):

$$\Gamma^{SM}(t \to bbc) = \frac{m_t}{2^8 \pi^3} \int_0^1 dx \int_{y_{min}}^{y_{max}} dy \; \frac{3}{2} g^4 |V_{cb}|^2 |V_{tb}|^2 \; f_2^{SM}(x, y). \qquad (4)$$

Where $f_2^{SM}$ is defined as $f_2^{SM}(x,y) = (y - y^2)/[(x - \hat{m}_W^2)^2 + x^2 r_W^2]$ and the integration limits for $m_b \to 0$ are simply $y_{min} = 0$ and $y_{max} = 1 - x$. For $m_b$ nonzero we refer to[7].

Our goal is to define an asymmetry that is zero in the S but not necessarily so for other models. We will split the integral in $y$ into two equal parts by finding $y_{min} < t_x < y_{max}$ which must satisfy the cubic equation $4t_x^3 - 6t_x^2 + b_x = 0$ [7], where $b_x$ is defined by $b_x = (1-x)^2(1+2x)$. We obtain a rather simple analytical expression for $t_x$. As mentioned before, choosing the variable $\rho^2$ instead of $t^2$ leads to a more complicated cubic equation. otice that the authors in Ref. [5] decided to use a numerical method to obtain the solution in their study. There are three solutions to this equation. The following one satisfies that $0 \le t_x \le 1$ for $0 \le x \le 1$:

$$t_x = \frac{1}{2}\mathcal{R}e\left[1-(1+i\sqrt{3})z_x^{1/3}\right] , \tag{5}$$

where $z_x = 1 - b_x + i\sqrt{b_x(2-b_x)}$. The general solution to $t_x$ (for $m_b$ nonzero) can be found in Ref. [7].

The forward-backward-like asymmetry is defined as

$$A_t = \frac{\int_0^1 dx \int_{t_x}^{y_{max}} dy\overline{|\mathcal{M}|^2} - \int_0^1 dx \int_{y_{min}}^{t_x} dy\overline{|\mathcal{M}|^2}}{\int_0^1 dx \int_{t_x}^{y_{max}} dy\overline{|\mathcal{M}|^2} + \int_0^1 dx \int_{y_{min}}^{t_x} dy\overline{|\mathcal{M}|^2}} . \tag{6}$$

otice that if P effects do not change significantly the value for $BR(t \to bbc)$ in the S , we can indeed make an approximation and use the S value for the denominator in $A_t$.

The asymmetry defined in [5] that is based on $\rho^2$ yields similar results. For instance, the coupling constant $X_{LR}^V$ (expected to be of order 1) associated to a four fermion vector operator yields $A_\rho = 0.0393|X_{LR}^V|^2$ (see Eq. (24) in [5]). If we consider this coupling and calculate its contribution to the $A_t$ asymmetry we obtain a value of $0.041|X_{LR}^V|^2$.

## 3. Effective $tbW$ effects on $A_t$

The first example of P effects that we want to consider is based on the effective $tbW$ vertex[10]:

$$\begin{aligned}\mathcal{L}_{tbW} &= \frac{g}{\sqrt{2}} W_\mu^- \, b\gamma^\mu \left(f_1^L P_L + f_1^R P_R\right) t \\ &- \frac{g}{\sqrt{2}M_W} \partial_\nu W_\mu^- \, b\,\sigma^{\mu\nu} \left(f_2^L P_L + f_2^R P_R\right) t \quad + \ h.c. ,\end{aligned} \tag{7}$$

In the S (tree level) the coupling constants are $f_1^L = V_{tb} \simeq 1$ and $f_1^R = f_2^R = f_2^L = 0$. The amplitude squared given by this vertex can be found in [7].

The asymmetry in terms of the effective couplings is[7]:

$$\begin{aligned}A_t &= 0.01 f_1^L f_1^R + 0.04 f_1^L f_2^L + 0.60 f_1^L f_2^R \\ &- 0.44(f_1^R)^2 - 1.23(f_2^L)^2 + 0.78(f_2^R)^2 + 0.01 f_2^L f_2^R .\end{aligned} \tag{8}$$

In principle, we expect the coe cients to assume values somewhat (maybe much) less than one. To estimate how large can $A_t$ become due to effects from the general $tbW$ vertex we will consider the bounds presented in one recent study based on $b \to s\gamma$ measurements[11]: $|f_1^R| \le 2.5 \times 10^{-3}$, $|f_2^L| \le 1.3 \times 10^{-3}$ and $|f_2^R| \le 0.57$. The potential contributions to the asymmetry are $A_t \le 3 \times 10^{-5}$, $A_t \le 5 \times 10^{-5}$ and $A_t \le 0.34$ respectively.

We can define an asymmetry in terms of the $x = q^2/m_t^2$ variable. owever, the sensitivity of $A_q$ to P effects is much lower than the sensitivity of $A_t$[7].

## 4. Asymmetry from models associated with $A_{FB}$

There are a good number of models proposed in the literature that attempt to explain the FB asymmetry of $tt$ production at the Tevatron[12]. In many cases the production process is modified by P effects that are based on FC C couplings[13].

Let us first refer to three examples presented in Ref. [6]. In particular, we want to consider the cases where there is a heavy boson involved with a mass of 2 TeV that can contribute to the $qq \to tt$ process via a t-channel diagram. One is a neutral vector $Z'$, another is a charged vector $W'$, and the third one is a neutral $SU(2)$ scalar $S$. It is possible that these new bosons contribute to the $t \to bbc$ decay mode that we are interested. For instance, the top quark could make a transition $t \to bW'^+ \to bbc$ (i.e. ust as the S decay but with $W$ being replaced by $W'$). nlike the S $W$ boson, the $W'$ can couple to both left and right chiralities and the amplitude squared will have a different form than Eq. (2). Another case is a heavy neutral scalar $H'$ with strong avor violating couplings that contributes via the process $t \to cH' \to cbb$ (or maybe $t \to cH' \to cbs$). Similarly the process induced by the scalar could also stem from a heavy vector boson $Z'$. For more details we refer to Ref. [6]. Table 1 shows the contributions to $A_t$ from each case.

| model | couplings | $A_t$ |
|---|---|---|
| $e\, q_d\gamma^\mu(f_LP_L + f_RP_R)tW'^-_\mu$ | $f_L = 2,\ f_R = 20$ | $5\times 10^{-3}$ |
| $y_{ij}\, q_iq_jH'$ | $y_{tc} = 15,\ y_{bb} = 1$ | 0.05 |
| $ef_R\, c_L\gamma^\mu t_R Z'_\mu$ | $f_R = 15$ | $-0.09$ |
| $\sqrt{2}g_s\frac{\kappa_{tcg}}{\Lambda}\, b_L\sigma^{\mu\nu}T^at_RG_{\mu\nu}$ | $\frac{\kappa_{tcg}}{\Lambda} \leq \frac{0.06}{\text{TeV}}$ | $-0.07$ |

**Table 1.** Asymmetries predicted by FC interactions. The first three involve new heavy bosons with mass $m_V = 2$ TeV that can give rise to a $tt$ FB asymmetry $A_{FB} = 0.2$[6]. The last row is for an effective dimension 5 $tcg$ operator[14] (limits from Tevatron measurements[15]).

Another possible FC C (this one is not used to explain $A_{FB}$) that we would like to consider involves the gluon field via a dimension 5 $tcg$ tensor coupling[14]. This coupling could be generated at loop level and could induce a new single top production process $gc \to t \to bW^+$ where the top quark is produced without any additional particles. Table 1 shows the contribution to $A_t$ from this coupling.

For the results shown so far it has become apparent that there are several P scenarios, that are of interest for the Top quark research program and that could also yield a non-zero $A_t$. From the results in Table 1 the heavy $W'$ has no significant contribution to $A_t$. owever, a light $W'$ with pure right handed FV couplings $tdW'$ [16] could indeed give significant contributions[7]. This mode has a negligible width in the S ($\sim 10^{-8}$ GeV), so the mere observance of this decay would signal P effects. owever, the possibility of such a light $W'$ boson is severely constrained by recent L C $ttj$ data [17].

**Acknowledgments** We thank RedFAE, Conacyt and S I for support.

[1] . Peters [ATLAS and CDF and CMS and D0 Collaborations], ar iv:1211.6028 [hep-ex].
[2] E. L. Berger, Q. -H. Cao, C. -R. Chen and H. hang, ar iv:120 .48 [hep-ph]
[3] . L. Diaz-Cruz, R. Martinez, M. A. Perez and A. Rosado, Phys.Rev.D**41**, 8 1 (1 0) G. Eilam, . L. Hewett and A. Soni, Phys. Rev. D**44**, 14 3 (1 1) [Erratum-ibid. D**59**, 03 01 (1 )].
[4] F. Larios, R. Martinez and M. A. Perez, Int. . Mod. Phys. A**21**, 34 3 (2006) and refrences therein.
[5] . iers, T. nighton, D. London, M. Russell, A. Szynkman, . Webster, Phys. Rev. D**84**, 0 4018 (2011).
[6] Q. H. Cao, D. Mc een, . L. Rosner, G. Shaughnessy and C. E. M. Wagner, Phys. Rev. D**81**, 114004 (2010).
[ ] F. Larios and M. A. Perez, Phys. Rev. D **85**, 01 503 (2012).
[8] D. Atwood *et al*, Phys. Rev. D**49**, 28 (1 4).
[ ] . Nakamura et al. (Particle Data Group), . Phys. G 3 , 0 5021 (2010).
[10] C. R. Chen, F. Larios and C. P. uan, Phys.Lett. B**631**, 126 (2005).
[11] . Drobnak, S. Fa fer and . F. amenik, Nucl. Phys. B**855**, 82 (2012).
[12] T. Aaltonen *et al.* [CDF Collaboration], Phys. Rev. D **83**, 112003 (2011) V. M. Abazov *et al.* [D0 Collaboration], ar iv:110 .4 5 [hep-ex].
[13] . Shu, T. M. P. Tait and . Wang, Phys. Rev. D**81**, 034012 (2010) M. Cvetic, . Halverson and P. Langacker, HEP **1211**, 064 (2012).
[14] E. Malkawi and T. M. P. Tait, Phys. Rev. D **54**, 5 58 (1 6).
[15] V. M. Abazov *et al.* [D0 Collaboration], Phys. Lett. B 6 3, 81 (2010).
[16] E. L. Berger, Q. -H. Cao, C. -R. Chen, . -H. u and H. hang, ar iv:1111.3641 [hep-ph].
[1 ] D. Duffty, . Sullivan and H. hang, Phys. Rev. D **85**, 0 402 (2012).

Journal of Physics: Conference Series **651** (2015) 012014 doi:10.1088/1742-6596/651/1/012014

# Top quark electromagnetic dipole moments

**Antonio O. Bouzas and F. Larios**

Departamento de Física Aplicada, CINVESTAV-Mérida, A.P. 73, 97310 Mérida, Yucatán, México

E-mail: `larios@mda.cinvestav.mx`

**Abstract.** The magnetic and electric dipole moments of the top quark are constrained indirectly by the $\mathrm{Br}(B \to X_s\gamma)$ and the $A_{CP}(B \to X_s\gamma)$ measurements. They can also be tested by top quark production and decay processes. The recent measurement of $t\bar{t}\gamma$ production by CDF are used to set direct constraints. The $B \to X_s\gamma$ measurements by themselves define an allowed parameter region that sets up stringent constraints on both dipole moments. The measurement by CDF has a $\sim 37\%$ error that is too large to set any competitive bounds, for which a much lower 5% error would be required. For the LHC it is found that with its higher energy the same measurement could indeed further constrain the allowed parameter region given by the $B \to X_s\gamma$ measurement [1]. In addition, the proposed LHeC experiment (electron-proton) could provide even more stringent constraints than the LHC via the $t\bar{t}$ photoproduction channel [2].

## 1. Introduction

Currently, the anomalous magnetic and electric dipole moments (MDM and EDM) of the top quark get the most stringent constraints from the $\mathrm{Br}(B \to X_s\gamma)$. We make an evaluation of these constraints, where in addition to the $\mathrm{Br}(B \to X_s\gamma)$ we consider a CP asymmetry for this process that indeed sets the strongest bounds on the EDM of the top quark. Our bounds are more stringent than reported previously [3]. The CDF collaboration has reported a measurement of $t\bar{t}\gamma$ production with $6fb^{-1}$ of data [4]. (Some preliminary study has also been done for the LHC [5].) This process has been considered as a probe of the dipole moments of the top quark by Baur et. al [6]. Their overall conclusion was that at the energy of the Tevatron it would not be possible to set bounds as stringent as those from the $B \to X_s\gamma$ measurements. However, at the energy of the LHC the bounds would become as stringent. The reason for this is that since the dipole coupling is proportional to the momentum of the photon there is more relative contribution (compared to the QED coupling) as the energy of the collider increases. In this work, we take the experimental result by [4] and make an estimate of the bounds on the MDM and EDM, where indeed we corroborate that $t\bar{t}\gamma$ at the Tevatron is far from competing with $B \to X_s\gamma$. But on the other hand, we also find that the LHC could in principle set significant *direct* bounds that would further improve what we already have from the *indirect* bounds from $B \to X_s\gamma$.

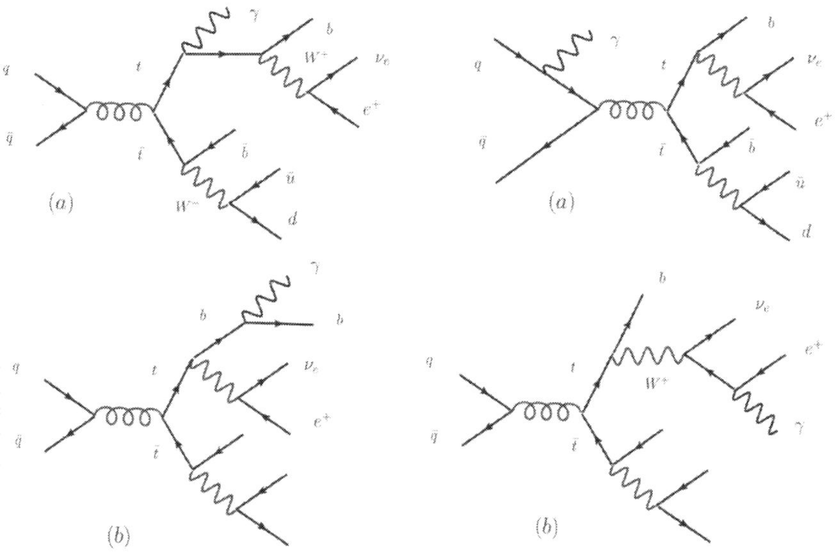

**Figure 1.** Feynman diagrams for $tt\gamma$ production at the Tevatron. In diagrams (a) the photon is radiated along with the on-shell top quarks, in diagrams (b) the photon is radiated by the decay products of one of the top quarks.

## 2. The MDM and EDM of the Top quark.

Following [7], we define the effective $t\bar{t}\gamma$ Lagrangian

$$\mathcal{L}_{t\bar{t}\gamma} = e\bar{t}\left(Q_t\gamma_\mu A^\mu + \frac{1}{4m_t}\sigma_{\mu\nu}F^{\mu\nu}(\kappa + i\tilde{\kappa}\gamma_5)\right)t, \tag{1}$$

where the CP even $\kappa$ and CP odd $\tilde{\kappa}$ terms are related to the anomalous MDM and EDM of the top quark, respectively. This Lagrangian is also defined in Refs. [6] and [3]; comparing with their effective Lagrangian (notice a relative minus sign in the charge term) we obtain the following relations:

$$\begin{aligned} \kappa &= -F_{2V}^{\gamma} = \frac{2m_t}{e}\mu_t = Q_t a_t\,, \\ \tilde{\kappa} &= F_{2A}^{\gamma} = \frac{2m_t}{e}d_t\,. \end{aligned} \tag{2}$$

Where $a_t = (g_t - 2)/2$ is the anomalous MDM in terms of the gyromagnetic factor $g_t$. The factors $F_{2V}^{\gamma}$ and $F_{2A}^{\gamma}$ are used in [6]. The factors $\mu_t$ and $d_t$ are used in [3]. The SM prediction for $a_t$ is $a_t^{\rm SM} = 0.02$ [8], which translates to $\kappa^{\rm SM} = 0.013$. The bounds for $\kappa$ that we will obtain will be about two orders of magnitude greater, therefore the SM prediction will not be considered in our calculations. On the other hand, the CP violating EDM factor $d_t$ is strongly suppresed in the SM: $d_t^{\rm SM} < 10^{-30} e\,cm$ ($\tilde{\kappa} < 1.75 \times 10^{-14}$)[9]. The EDM is thus a very good probe of new physics. There are models with vector like multiplets that predict values as high as $10^{-19} e\,cm$ ($\tilde{\kappa} < 1.75 \times 10^{-3}$) [10].

## 3. Limits from $t\bar{t}\gamma$ at the Tevatron

The CDF collaboration has reported a cross–section measurement of top–quark pair production with an additional photon that carries at least 10 GeV of transverse energy, $\sigma_{t\bar{t}\gamma} = 0.18 \pm 0.08$

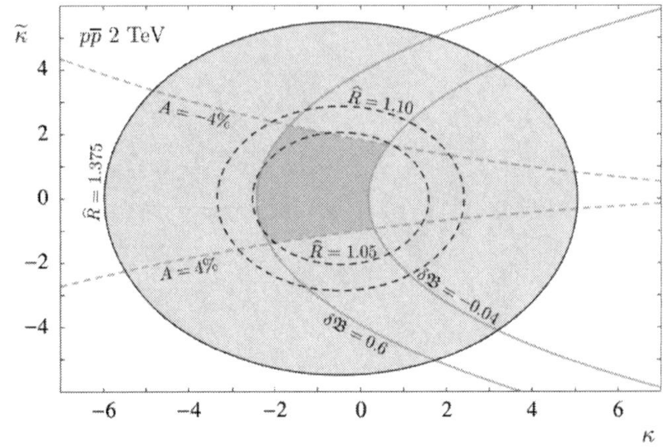


**Figure 2.** The allowed parameter space for the anomalous magnetic and electric dipole moments of the top quark. Gray solid lines: region allowed by the experimental results for the branching ratio for $B \to X_s\gamma$. Gray dashed lines: region allowed by the experimental results for the $CP$ asymmetry for $B \to X_s\gamma$. Black solid line: region allowed by the CDF measurement of $\hat{R}$ for $t\bar{t}\gamma$ production at $\sqrt{s} = 2$ TeV. Black dashed lines: regions allowed by the values of $\hat{R}$ indicated in the figure.

pb [4]. In addition, using events with the same selection criteria as for the $t\bar{t}\gamma$ candidates, but without the photon, they also perform a measurement of the $t\bar{t}$ production cross section. In this way they determine the ratio

$$R^{\text{exp}} \equiv \frac{\sigma_{t\bar{t}\gamma}}{\sigma_{t\bar{t}}} = (0.024 \pm 0.009)\text{pb} , \tag{3}$$

in which systematic uncertainties are eliminated. The experimental result (3) is in excellent agreement with the SM prediction $R^{\text{SM}} = 0.024 \pm 0.005$ [4].

Since the main result by CDF is given in terms of $\sigma_{t\bar{t}\gamma}/\sigma_{t\bar{t}}$, we shall consider that ratio as a function of the MDM $\kappa$ and the EDM $\tilde{\kappa}$ and then use the CDF measurement to set bounds on those parameters. In the radiative production process two modes are predominant: (1) $t\bar{t}$ produced along with the radiated photon followed by the decay of the top pair, which is indeed $t\bar{t}\gamma$ production proper (see Figure 1 (a)), and (2) $t\bar{t}$ produced on-shell with one of them decaying radiatively (see Figure 1 (b)). The first mode may involve initial–state radiation if the initial partons are charged. The second mode may involve final–state radiation from the $b$ jets, the intermediate $W$ boson or the $W$ decay products.

In order to quantify the impact of the top–quark MDM and EDM on the cross section, we focus our attention on a normalized ratio $\hat{R} \equiv R/R^{\text{SM}} = \sigma_{t\bar{t}\gamma}/\sigma^{\text{SM}}_{t\bar{t}\gamma}$. In this way, the CDF result (3) can be translated to $\hat{R}^{\text{exp}} = 1 \pm 0.375$. We compute the cross sections for $p\bar{p} \to t\bar{t} \to bW^+\bar{b}W^-\gamma \to$ FS at leading order at the Tevatron energy, and the same processes with $pp$ initial state at LHC energies. We choose semileptonic final states, as done in the CDF measurement, but consider also a simplified process with final state $b\bar{b}WW$ as a cross check of our results.

The expression for $\hat{R}$ must have in general the quadratic form $\hat{R} = 1 + a_1\kappa + a_2\kappa^2 + b_2\tilde{\kappa}^2$. By computing $\hat{R}$ for several values of $\kappa$ and $\tilde{\kappa}$ we can obtain the coefficients $a_i$ in $\hat{R}$ at the desired energy. Then we use a relation of the form

$$\hat{R}_1 \leq \hat{R} = 1 + a_1\kappa + a_2\kappa^2 + a_3\tilde{\kappa}^2 \leq \hat{R}_2 \tag{4}$$

to find the allowed parameter region for $(\kappa,\tilde{\kappa})$ at that energy. In the case of the CDF measurement (3), we set $\widehat{R}_{1,2} = 1 \pm 0.375$ to define the allowed region at the $1\sigma$ level. As for the numerical coefficients we obtain $a_1 = -0.002, -0.008, -0.009$ for the Tevatron, LHC (7 TeV) and LHC (14 TeV). Similarly, $a_2 = 0.011, 0.055, 0.088$ and $a_3 = 0.011, 0.055, 0.089$. The measurement of $\widehat{R}$ at the Tevatron by the CDF collaboration sets limits on $(\kappa,\tilde{\kappa})$ through (4).

At the Tevatron energy $\sqrt{s} = 2$ TeV, the production of $t\bar{t}$ and $t\bar{t}\gamma$ receives its dominant contribution from $u\bar{u}$ initial states, but we took into account also the smaller contributions from initial $d\bar{d}$ and $gg$. For the numerical computation we considered the semileptonic process $p\bar{p} \to t\bar{t} \to b\bar{b}qq'\ell\nu_\ell\gamma$ with three lepton flavors, where the final photon can originate from any initial, intermediate or final charged particle. In analogy with the measurement reported by CDF [4], we applied cuts in the transverse energy of the photon, missing transverse energy and pseudorapidity of the final particles given by

$$E_T^\gamma > 10\text{GeV},\ \not{E}_T > 20\text{GeV}, |\eta_q| < 3.6, |\eta_b| < 2, |\eta_\gamma| < 1, |\eta_\ell| < 1. \tag{5}$$

With those cuts we obtain a SM cross section $\sigma^{\text{SM}}_{t\bar{t}\gamma} = 0.07261$ pb at $\sqrt{s} = 2$ TeV, in agreement with the leading-order result reported in [4]. In order to increase the sensitivity of the process to the dipole moments of the intermediate top quark it is necessary to reduce the background from photons originating in final-state charged particles. For that purpose we impose a lower bound on the distance from the photon to the charged particles in the $\eta$–$\phi$ plane, $\Delta = \sqrt{(\Delta\eta)^2 + (\Delta\phi)^2}$, $\Delta_{\gamma,\text{ch}} > 0.4$, which plays the same role as the analogous cuts introduced in the actual measurement [4]. With the cut on $\Delta$ the SM cross section at 2 TeV is $\sigma^{\text{SM}}_{t\bar{t}\gamma} = 0.0193$ pb.

## 4. Limits from $B \to X_s\gamma$

We separate the SM value $C_7^{\text{SM}}(m_b) = -0.31$ from the NP contributions:

$$C_7(m_b) = -0.31 + 0.67\,\delta C_7(\mu_h) + 0.09\,\delta C_8(\mu_h) + \cdots, \tag{6}$$

We set $\delta C_8(\mu_h) = 0$ and keep $\delta C_7(\mu_h)$ which is given in terms of $\kappa$ and $\tilde{\kappa}$ in Ref. [7]. A numerical expression for the branching ratio $\mathcal{B}(B \to X_s\gamma)$ in terms of the coefficients $C_7(\mu_h)$:

$$\begin{aligned}\delta\mathcal{B}(B \to X_s\gamma) &\equiv \mathcal{B}(B \to X_s\gamma) - \mathcal{B}^{\text{SM}}(B \to X_s\gamma) = 10^{-4}\times \\ &\times \left(\text{Re}(-7.184\,\delta C_7 - 2.225\,\delta C_8 + 2.454\,\delta C_7\,\delta C_8^*) + 4.743\,|\delta C_7|^2 + 0.789\,|\delta C_8|^2\right),\end{aligned}$$

Using the predicted value for $\mathcal{B}^{\text{SM}}(B \to X_s\gamma)$ and a measured value for $\mathcal{B}(B \to X_s\gamma)$ we get the relation:

$$\delta\mathcal{B}(B \to X_s\gamma) = (3.43 \pm 0.22) - (3.15 \pm 0.23) = -0.234\kappa + 0.005\kappa^2 + 0.039\tilde{\kappa}^2,$$

The CP asymmetry

$$A_{\text{CP}}(B \to X_s\gamma) = \frac{\Gamma(\bar{B} \to X_s\gamma) - \Gamma(B \to X_{\bar{s}}\gamma)}{\Gamma(\bar{B} \to X_s\gamma) + \Gamma(B \to X_{\bar{s}}\gamma)} \tag{7}$$

was first proposed in [15]. Its latest experimental value is quoted in [13] as $A_{\text{CP}}^{\text{Exp.}}(B \to X_s\gamma) = (-0.8 \pm 2.9)\%$. Concerning the SM prediction, the most recent study is given in Ref. [16]. With the definition $C_7(m_b)/C_7^{\text{SM}}(m_b) = r_7 e^{i\theta_7}$ we get

$$A_{\text{CP}}[\%] = a_7\frac{\sin(\theta_7)}{r_7} + 1.07\frac{\cos(\theta_7)}{r_7} + \frac{0.03}{r_7^2}\,.$$

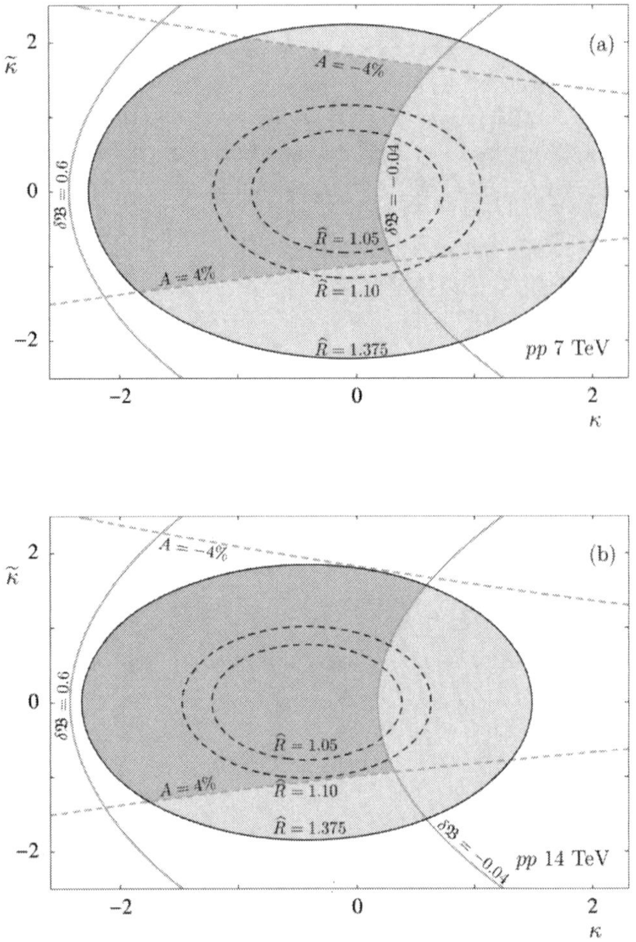


**Figure 3.** Gray lines as in previous figure. Black solid and dashed lines delimit the regions allowed by hypothetical measurements of $\widehat{R}$ for $t\bar{t}\gamma$ production at the LHC. In (a) $\sqrt{s} = 7$ TeV. In (b) $\sqrt{s} = 14$ TeV.

We can write $r_7 e^{i\theta_7} = C_7(m_b)/C_7^{\rm SM}(m_b) = 1 - 0.0705\kappa - i0.1962\widetilde{\kappa}$. Then,

$$A_{\rm CP}[\%] = \frac{1.1 - 0.075\kappa - 3.08\widetilde{\kappa}}{(1-0.07\kappa)^2 + 0.0385\widetilde{\kappa}^2}.$$

## 5. Allowed parameter space for $\kappa$ and $\widetilde{\kappa}$

We can now show the allowed region in the $\kappa$ vs. $\widetilde{\kappa}$ plane. For the asymmetry we require $|A_{\rm CP}(X_s\gamma)| < 4\%$, to be consistent with the uncertainty of the SM prediction as well as the experimental error. The region of allowed values for $(\kappa, \widetilde{\kappa})$ is shown in figures 2 and 3 by the gray dashed lines. For the branching ratio, the region allowed at the $1\sigma$ level is $-0.038 < \delta\mathcal{B}(B \to X_s\gamma) < 0.598$. That region is delimited in figures 2 and 3 by gray solid lines. Roughly speaking the MDM is bounded to be $-2 < \kappa < 1$ which translated to the $m_t\mu_t = \kappa e/2 = 0.15\kappa$ term used in [3] means that $-0.3 < m_t\mu_t < 0.15$. Our limits are significantly more stringent than reported by [3].

At the $1\sigma$ level the allowed region for $(\kappa, \tilde{\kappa})$ is bounded by the inequalities $0.625 < \widehat{R} < 1.375$. The lower value turns out to be unattainable, so it does not set any bound. The region delimited by $\widehat{R} = 1.375$ is shown in figure 2 by the black solid line. The black dashed lines in that figure show the regions that would be delimited by hypothetical measurements $\widehat{R} = 1 \pm 0.1$ and $1 \pm 0.05$. We see from the figure that, as expected from the analysis in [6], the bounds set by the Tevatron measurement of $\widehat{R}$ are much less constraining than those arising from the asymmetry and branching ratio for $B \to X_s\gamma$. This is so even in the hypothetical case of an experimental result $\widehat{R} = 1 \pm 0.1$ with a 10% measurement error. Only a 5% measurement uncertainty could yield bounds of the same order of magnitude at most.

We have also performed the same computation for hypothetical measurements of $\widehat{R}$ for $t\bar{t}\gamma$ production in $pp$ collisions at the LHC, both at $\sqrt{s} = 7$ TeV and $\sqrt{s} = 14$ TeV. In this case the dominant contribution to the production process comes from $gg$ initial states, but we also took into account the smaller contributions from the initial states $u\bar{u}$ and $d\bar{d}$. The results are shown in figure 3 (a) for the lower LHC energy and in figure 3 (b) for the higher one. As seen in the figure, the hypothetical experimental results at the LHC would remove significant portions of the region of the $(\kappa, \tilde{\kappa})$ plane allowed by the measurements of the branching ratio and $CP$ asymmetry of $B \to X_s\gamma$. Whereas this is true already at $\sqrt{s} = 7$ TeV, the constraints set by a measurement of $\widehat{R}$ at $\sqrt{s} = 14$ TeV with an experimental uncertainty smaller than, say, 30% would lead to strikingly tighter bounds on $(\kappa, \tilde{\kappa})$ than those currently available.

Our overall conclusions agree with those given by Baur et. al [6] even though their analysis was not based on the ratio $R$ reported by CDF but on the change in the $p_T$ distribution of the photon.

## Acknowledgments
We thank RedFAE, Conacyt and SNI for support.

## References
[1] A. O. Bouzas and F. Larios, *Phys. Rev.* D**87**, 074015 (2013).
[2] A. O. Bouzas and F. Larios, *Phys. Rev.* D**88**, 094007 (2013).
[3] J. F. Kamenik, M. Papucci and A. Weiler, *Phys. Rev.* D**85**, 071501 (2012).
[4] T. Aaltonen *et al.* [CDF Collaboration], *Phys. Rev.* D**84**, 031104 (2011).
[5] J. Erdmann, arXiv:1206.5696 [hep-ex].
[6] U. Baur, A. Juste, L. H. Orr and D. Rainwater, *Phys. Rev.* D**71**, 054013 (2005).
[7] J. L. Hewett and T. G. Rizzo, *Phys. Rev.* D**49**, 319 (1994).
[8] W. Bernreuther, et al, *Phys. Rev. Lett.* **95**, 261802 (2005).
[9] A. Soni and R. M. Xu, *Phys. Rev. Lett.* 69, 33 (1992); F. Hoogeveen, *Nucl. Phys.* B341, 322 (1990); M. E. Pospelov and I. B. Khriplovich, Yad. Fiz. 53, 1030 (1991) [*Sov. J. Nucl. Phys.* 53, 638 (1991)].
[10] T. Ibrahim and P. Nath, *Phys. Rev.* D**82**, 055001 (2010).
[11] E. Lunghi and J. Matias, *JHEP* **0704**, 058 (2007).
[12] M. Misiak, et al, *Phys. Rev. Lett.* **98**, 022002 (2007).
[13] The Heavy Flavor Averaging Group, http://www.slac.stanford.edu/xorg/hfag.
[14] T. Hurth and F. Mahmoudi, arXiv:1211.6453 [hep-ph].
[15] A. L. Kagan and M. Neubert, *Phys. Rev.* D**58**, 094012 (1998).
[16] M. Benzke, S. J. Lee, M. Neubert and G. Paz, *Phys. Rev. Lett.* **106**, 141801 (2011).

# Pseudoscalar Fields in Torsionful Geometries of the Early Universe, the Baryon Asymmetry and Majorana Neutrino Mass Generation

**Nick E. Mavromatos**

Theoretical Particle Physics and Cosmology Group, Department of Physics, King's College London, Strand, London WC2R 2LS, UK and
Theory Division, Physics Department, CERN CH-1211 Geneva 23, Switzerland

E-mail: `Nikolaos.Mavromatos@cern.ch`

**Abstract.** We discuss here a specific field-theory model, inspired from string theory, in which the generation of a matter-antimatter asymmetry in the Cosmos is due to the propagation of fermions in a non-trivial, spherically asymmetric (and hence Lorentz violating) gravitational background that may characterise the epochs of the early universe. The background induces different dispersion relations, hence populations, between fermions and antifermions, and thus CPT Violation (CPTV) already in thermal equilibrium. Species populations may freeze out leading to leptogenesis and baryogenesis. More specifically, after reviewing some generic models of background-induced CPTV in early epochs of the Universe, we consider a string-inspired scenario, in which the CPTV is associated with a cosmological background with torsion provided by the Kalb-Ramond (KR) antisymemtric tensor field of the string gravitational multiplet. In a four-dimensional space time this field is dual to a pseudoscalar "axion-like" field. The thermalising processes in this model are (right-handed) Majorana neutrino-antineutrino oscillations, which are induced in the presence of the KR axion background. These processes freeze out at a (high) temperature $T_c \gg m$, where $m$ is the Majorana neutrino mass, at which the KR background goes to zero or is diminished significantly, through appropriate phase transitions of the (string) universe. An additional, but equally important, rôle, of the KR field is that its quantum fluctuations and mixing with an ordinary axion, which couples to the Majorana neutrinos via appropriate Yukawa couplings, can also lead to the generation of a Majorana neutrino mass through quantum anomalies. This provides a novel way for generating neutrino masses, independent of the traditional seesaw mechanism.

## 1. Introduction

One of the most important issues of fundamental physics, relates to an understanding of the magnitude of the observed baryon asymmetry $n_B - n_{\overline{B}}$ (where $B$ denotes baryon, $\overline{B}$ denotes antibaryon, $n_B$ is the number density of baryons and $n_{\overline{B}}$ the number density of antibaryons in the universe). The universe is overwhelmingly made up of matter rather than anti-matter. According to the standard Big Bang theory, matter and antimatter have been created in equal amounts in the early universe. However, the observed charge-parity (CP) violation in particle physics [1], prompted A. Sakharov [2] to conjecture that for baryon asymmetry in the universe (BAU) we need: (i) Baryon number violation to allow for states with $\Delta B \neq 0$ starting from states with $\Delta B = 0$ where $\Delta B$ is the change in baryon number. (ii) If C or CP conjugate processes to a scattering process were allowed with the same amplitude then baryon asymmetry would disappear. Hence C and CP need to be broken. (iii) Chemical equilibrium does not permit asymmetries. Consequently Sakharov required that chemical equilibrium does not hold during an epoch in the early universe. Hence non-equilibrium physics in the early universe together with baryon number (B), charge (C) and charge-parity (CP) violating interactions/decays of anti-particles, may result in the observed BAU. In fact there are two types of non-equilibrium processes in the early universe that can produce this asymmetry: the first type concerns processes generating asymmetries between leptons and antileptons (*leptogenesis*), while the second produces asymmetries between baryons and antibaryons (*baryogenesis*). The near complete observed asymmetry today, is estimated in the Big-Bang theory [3] to imply:

$$\Delta n(T \sim 1\ \mathrm{GeV}) = \frac{n_B - n_{\overline{B}}}{n_B + n_{\overline{B}}} \sim \frac{n_B - n_{\overline{B}}}{s} = (8.4 - 8.9) \times 10^{-11} \tag{1}$$

at the early stages of the expansion, e.g. for times $t < 10^{-6}$ s and temperatures $T > 1$ GeV. In the above formula $s$ denotes the entropy density. Unfortunately, the observed CP violation within the Standard Model (SM) of particle physics (found to be $O(10^{-3})$ for the standard parameter $\epsilon$ in the neutral Kaon experiments [1]) induces an asymmetry much less than that in (1) [4]. There are several ideas that go beyond the SM (*e.g.* grand unified theories, supersymmetry, extra dimensional models *etc.*) which involve the decays of right-handed sterile neutrinos. For relevant important works on this see [5, 6, 7, 8, 9, 10, 11]. These ideas lead to extra sources for CP violation that could generate the observed BAU. Some degree of fine tuning and somewhat *ad hoc* assumptions are involved in such scenarios; so the quest for an understanding of the observed BAU still needs further investigation. An example of fine tuning is provided by the choice of the hierarchy of the right-handed Majorana neutrino masses. For instance, enhanced CP violation, necessary for BAU, can be achieved in models with three Majorana neutrinos, by assuming two of these neutrinos are nearly degenerate in mass.

The requirement of non-equilibrium is on less firm ground [12] than the other two requirements of Sakharov, *e.g.* if the non-equilibrium epoch occurred prior to inflation then its effects would be hugely diluted by inflation. A basic assumption in the scenario of Sakharov is that *CPT symmetry* [13] (where $T$ denotes time reversal operation) holds in the very early universe. *CPT symmetry* leads to the production of matter and antimatter in equal amounts. Such *CPT invariance* is a cornerstone of all known *local* effective *relativistic* field theories without gravity, and consequently of current particle-physics phenomenology. It should be noted that the necessity of non-equilibrium processes in CPT invariant theories can be dropped if the requirement of CPT invariance is relaxed [14]. This violation of CPT (denoted by CPTV) is the result of a breakdown of Lorentz symmetry (which might happen at ultrahigh energies [15]). For many models with CPTV, in the time-line of the expanding universe, CPTV generates first lepton asymmetries (*leptogenesis*); subsequently, through sphaleron processes [16] or baryon-lepton (B-L) number conserving processes in Grand Unified Theories (GUT), the lepton asymmetry can be communicated to the baryon sector to produce the observed BAU.

In order to obtain the observed BAU CPTV in the early universe may obviate the need for fine tuning the decay widths of extra sources of CP violation, such as sterile neutrinos and/or supersymmetry partners. Instead, one has to "tune" the background space-time, assuming a phase transition at an appropriate (high) temperature, after which the geometry of the universe assumes its canonical Robertson-Walker form. In this note we shall consider a simplified scenario [17]: the observed matter-antimatter asymmetry in the universe today is due to the coupling of right-handed Majorana neutrinos to a pseudoscalar background field that originates from the Kalb-Ramond (KR) antisymmetric field of an ancestor string theory. The low energy limit of this ancestor string theory describes the observable universe. The oscillations of Majorana neutrinos between themselves and their antiparticles offer a microscopic realisation

Journal of Physics: Conference Series **651** (2015) 012015 doi:10.1088/1742-6596/651/1/012015

of chemical equilibrium processes which freeze out at a particular (high) temperature $T_D$ -the universe is assumed to undergo a phase transition such that the background KR field goes either to zero or to a very small value, compatible with the absence today of any observed CPTV effect. Such right-handed neutrinos characterise simple of the extensions of the Standard Model, termed neutrino-minimal-Standard-Model ($\nu$MSM) [5], in the absence of supersymmetry or extra dimensions. $\nu$MSM can provide candidates for dark matter. However, there are delicate issues associated with the realisation of the baryogenesis scenarios in this model, give that for the range of masses of the right-handed neutrinos employed in the model (two degenerate ones, with mass of order GeV, and a light one (dark matter), with mass of order O(10) keV); the baryogenesis is supposed to take place via coherent oscillations between the degenerate right-handed neutrinos. Such coherent oscillations, though, may be destroyed in the high-temperature plasma of particles that characterises the early universe.

Our work provides a simple geometric scenario to avoid such dilemmas. We consider a model such as the $\nu$MSM, in a KR background which breaks Lorentz symmetry. The background couples to the right-handed neutrinos; a lepton asymmetry is induced by tuning the background. The crucial rôle of right-handed neutrinos for the realisation of our scenario [17], as sketched above, is compatible with the important rôle of the lightest of them as dark matter, envisaged in [5, 18]. Moreover in an era characterised by the apparent absence of supersymmetry signals in the large hadron collider (LHC) [19], the issue of the identification of the nature of the dark matter becomes even more pressing.

There is an additional significant rôle, for the KR axion field. Even if the background value of the field is zero in the present era, its quantum fluctuations, which survive today, may be responsible for giving the right-handed Majorana neutrinos their mass. This may happen through anomalous couplings of the KR field with the gravitational background and its mixing with an ordinary axion field, which couples via appropriate Yukawa couplings to the right-handed neutrinos [20]. In this way, by an appropriate choice of the axion-neutrino Yukawa couplings, one may generate masses for the three right-handed neutrinos. Such masses lie in the range envisaged in $\nu$MSM [5], so that the lightest of them (keV mass range) can play the rôle of a dark matter candidate. The ordinary axions in this model may provide additional dark matter candidates.

The structure of the talk is the following: in the next section 2 we shall review some models where background geometries do not respect rotational symmetry, and so violate Lorentz symmetry (LIV). The background can induce CPTV matter-antimatter asymmetries in thermal equilibrium in the early universe. In section 3 we shall discuss our specific string-inspired model where the KR axion field plays the rôle of torsion. Torsion provides a LIV geometry and matter-antimatter asymmetry is generated. We discuss right-handed neutrino-antineutrino oscillations of Pontercorvo type[21, 22]; the oscillations violate *both* CP and CPT. We also estimate the freeze-out temperature, which is the temperature at which the KR field switches off (or diminishes significantly) due to a phase transition of the string universe [17]. In section 4 we discuss the rôle of the quantum fluctuations of the KR field in providing Majorana masses for the right-handed neutrinos. Conclusions and an outlook appear in section 5.

## 2. Lorentz-Violating Geometries and Matter-Antimatter Asymmetry in the Universe

We shall briefly review some existing models of CPTV induced asymmetry between matter and antimatter in the early universe. These existing models can be contrasted with our approach in this article.

### 2.1. CPTV Models with Particle-Antiparticle Mass Difference

The simplest possibility [23] for inducing CPTV in the early universe is through particle-antiparticle mass differences $m \neq \overline{m}$. These would affect the particle phase-space distribution function $f(E, \mu)$,

$$f(E, \mu) = [\exp(E - \mu)/T) \pm 1]^{-1} \ , \quad E^2 = \vec{p}^2 + m^2 \ , \tag{2}$$

and antiparticle phase-space distribution function

$$f(\overline{E}, \bar{\mu}) = [\exp(\bar{E} - \bar{\mu})/T) \pm 1]^{-1} \ , \quad \bar{E}^2 = \vec{p}^2 + \bar{m}^2 \ , \tag{3}$$

with $\vec{p}$ being the 3−momentum. (Our convention will be that an overline over a quantity will refer to an antiparticle, + will correspond to Fermi statistics (fermions), whereas − will correspond to Bose statistics (bosons)).

Mass differences between particles and antiparticles, $\overline{m} - m \neq 0$, generate a matter-antimatter asymmetry in the relevant number densities $n$ and $\overline{n}$,

$$n - \overline{n} = g_{d.o.f.} \int \frac{d^3p}{(2\pi)^3} [f(E, \mu) - f(\overline{E}, \bar{\mu})], \tag{4}$$

where $g_{d.o.f.}$ denotes the number of degrees of freedom of the particle species under study.

In the case of spontaneous Lorentz violation [24] there is a vector field $A_\mu$ with a non-zero time-like expectation value which couples to a global current $J^\mu$ such as baryon number through an interaction lagrangian density

$$\mathcal{L} = \lambda A_\mu J^\mu. \tag{5}$$

This leads to $m \neq \bar{m}$ and $\mu \neq \bar{\mu}$. Alternatively, following [23] we can make the assumption that the dominant contributions to baryon asymmetry come from quark-antiquark mass differences, and that their masses "run" with the temperature i.e. $m \sim gT$ (with $g$ the QCD coupling constant). One can provide estimates for the induced baryon asymmetry on noting that the maximum quark-antiquark mass difference is bounded by the current experimental bound on the proton-antiproton mass difference, $\delta m_p (= |m_p - \overline{m}_p|)$, known to be less than $2 \cdot 10^{-9}$ GeV. Taking $n_\gamma \sim 0.24 T^3$ (the photon equilibrium density at temperature $T$) we have [23]:

$$\beta_T = \frac{n_B}{n_\gamma} = 8.4 \times 10^{-3} \frac{m_u \delta m_u + 15 m_d \delta m_d}{T^2} \ , \quad \delta m_q = |m_q - \overline{m}_q|. \tag{6}$$

Thus, $\beta_T$ is too small compared to the observed one. To reproduce the observed $\beta_{T=0} \sim 6 \cdot 10^{-10}$ one would need $\delta m_q (T = 100 \text{ GeV}) \sim 10^{-5} - 10^{-6}$ GeV $\gg \delta m_p$, which is somewhat unnatural.

However, active (*light*) neutrino-antineutrino mass differences alone may reproduce BAU; some phenomenological models in this direction have been discussed in [25], considering, for instance, particle-antiparticle mass differences for active neutrinos compatible with current oscillation data. This leads to the result

$$n_B = n_\nu - n_{\overline{\nu}} \simeq \frac{\mu_\nu T^2}{6}, \tag{7}$$

yielding $n_B/s \sim \frac{\mu_\nu}{T} \sim 10^{-11}$ at $T \sim 100$ GeV, in agreement with the observed BAU. (Here $s$, $n_\nu$, and $\mu_\nu$ are the entropy density, neutrino density and chemical potential respectively.)

### 2.2. CPTV-induced by Curvature effects in Background Geometry

In the literature the rôle of gravity has been explicitly considered within a local effective action framework which is essentially that of (5). A coupling to scalar curvature $R$ [26, 27, 28, 29] through a CP violating interaction Lagrangian $\mathcal{L}$:

$$\mathcal{L} = \frac{1}{M_*^2} \int d^4x \sqrt{-g}\, (\partial_\mu R)\, J^\mu \ , \tag{8}$$

where $M_*$ is a cut-off in the effective field theory and $J^\mu$ could be the current associated with baryon (B) number. There is an implicit choice of sign in front of this interaction, which has been fixed so as to ensure matter dominance. It has been shown that [26]

$$\frac{n_{B-L}}{s} = \frac{\dot{R}}{M_*^2 T_d}, \tag{9}$$

with $T_d$ the freeze-out temperature for $B-L$ interactions. The idea then is that this asymmetry can be converted to baryon number asymmetry provided the $B+L$ violating (but B-L conserving) electroweak sphaleron interaction has not frozen out. To leading order in $M_*^{-2}$ we have $R = 8\pi G(1-3w)\rho$ where $\rho$ is the energy density of matter and the equation of state is $p = w\rho$ where $p$ is pressure. For radiation $w = 1/3$ and so in the radiation dominated era of the Friedmann-Robertson-Walker cosmology $R = 0$. However $w$ is precisely 1/3 when $T^\mu_\mu = 0$. In general $T^\mu_\mu \propto \beta(g) F^{\mu\nu} F_{\mu\nu}$ where $\beta(g)$ is the beta function of the running gauge coupling $g$ in a $SU(N_c$ gauge theory with $N_c$ colours. This allows $w \neq 1/3$. Further issues in this approach can be found in [26, 27, 28, 29].

Another approach involves an axial vector current [30, 31, 32, 33] instead of $J_\mu$. The scenario is based on the well known fact that fermions in curved space-times exhibit a coupling of their spin to the curvature of the background space-time.The Dirac Lagrangian density of a fermion can be re-written as:

$$\mathcal{L} = \sqrt{-g}\,\overline{\psi}\left(i\gamma^a \partial_a - m + \gamma^a \gamma^5 B_a\right)\psi \ , \quad B^d = \epsilon^{abcd} e_{b\lambda}\left(\partial_a e^\lambda_{\ c} + \Gamma^\lambda_{\nu\mu} e^\nu_{\ c} e^\mu_{\ a}\right) \ , \tag{10}$$

in a standard notation, where $e^\mu_{\ a}$ are the vielbeins, $\Gamma^\mu_{\ \alpha\beta}$ is the Christoffel connection and Latin (Greek) letters denote tangent space (curved space-time) indices. The space-time curvature background has, therefore, the effect of inducing an "axial" background field $B_a$ which can be non-trivial in certain anisotropic space-time geometries, such as Bianchi-type cosmologies or axisymmetric Kerr black holes [30, 31, 32, 33]. For an application to particle-antiparticle asymmetry it is necessary for this axial field $B_a$ to be a constant in some local frame. The existence of such a frame has not been demonstrated. As before if it can be arranged that $B_a \neq 0$ for $a = 0$ then for constant $B_0$ CPT is broken: the dispersion relation of neutrinos in such backgrounds differs from that of antineutrinos. Explicitly, for the case of light-like $B_0 = |\vec{B}|$-background one has [34]:

$$(E \pm |\vec{B}|)^2 = (\vec{p} \pm \vec{B})^2 + m^2 \ , \tag{11}$$

and for pure time-like B-backgrounds, of interest to us in the next section 3 [34],

$$E^2 = m^2 + (B_0 \pm |\vec{p}|)^2, \tag{12}$$

where $m$ is the fermion mass and the $+$ $(-)$ signs refer to particles (antiparticles) (in the case of Majorana neutrinos these are helicity states). For small $m, B_0 << |\vec{p}|$ one may then obtain the (approximate) dispersion relations given in [31, 30]

$$E \simeq |\vec{p}| + \frac{m_{\rm eff}^2}{2|\vec{p}|} + B_0 \ , \quad \overline{E} \simeq |\vec{p}| + \frac{m_{\rm eff}^2}{2|\vec{p}|} - B_0 \ , \quad m_{\rm eff}^2 = m^2 + B_0^2 \ll |\vec{p}| \ , \tag{13}$$

which we shall make use of in what follows.

The relevant neutrino asymmetry emerges on following the same steps used when there was an explicit particle-antiparticle mass difference. As a consequence, for the pure-time like case considered above, and assuming a constant $B_0$, which will be of interest to us here, the following neutrino-antineutrino density difference is found from (13):

$$\Delta n_\nu \equiv n_\nu - n_{\overline{\nu}} \sim g^\star T^3 \left(\frac{B_0}{T}\right), \tag{14}$$

with $g^\star$ the number of degrees of freedom for the (relativistic) neutrino. An excess of particles over antiparticles is predicted only when $B_0 > 0$, which had to be assumed in the analysis of [30, 31, 32, 33]; we should note, however, that the sign of $B_0$ and its constancy have not been justified in this phenomenological approach (The above considerations concern the dispersion relations for any fermion, not only neutrinos. However, when one considers matter excitations from the vacuum, as relevant for leptogenesis, we need chiral fermions to get non trivial CPTV asymmetries in *populations* of particle and antiparticles, because $< \psi^\dagger \gamma^5 \psi > = - < \psi_L^\dagger \gamma^5 \psi_L > + < \psi_R^\dagger \gamma^5 \psi_R >$.). At temperatures $T < T_d$, with $T_d$ the decoupling temperature of the lepton-number violating processes, the ratio of the net Lepton number $\Delta L$ (neutrino asymmetry) to entropy density (which scales as $T^3$) remains constant,

$$\Delta L(T < T_d) = \frac{\Delta n_\nu}{s} \sim \frac{B_0}{T_d} . \tag{15}$$

This implies a lepton asymmetry (leptogenesis) which, by tuning $B_0$ (for a given decoupling temperature $T_d$, that depends on the details of the underlying Lepton-number violating processes) can lead to a $\Delta L$ of the phenomenologically right order $\Delta L \sim 10^{-10}$. The latter can then be communicated to the baryon sector to produce the observed BAU (baryogenesis) by a B-L conserving symmetry in the context of either Grand Unified Theories (GUT) [30], or sphaleron processes in the standard model.

In the following section we shall discuss a case of a background where the constancy of $B_0$ in the Robertson-Walker cosmological frame is guaranteed by construction. This case is inspired by string theory.

## 3. Kalb-Ramond (KR) Torsion Background, Majorana Neutrinos and Baryogenesis

In this section we will discuss the case of a constant $B^0$ "axial" field that appears due to the interaction of the fermion spin with a string-theory background geometry with *torsion*. This is a novel observation, which (as far as we are aware) was discussed for first time in [35]. In the presence of torsion the Christoffel symbol contains a part that is antisymmetric in its lower indices: ${\Gamma^\lambda}_{\mu\nu} \neq {\Gamma^\lambda}_{\nu\mu}$. Hence the last term of the right-hand side of the Eqn.(10) is *not* zero. Since the torsion term is of gravitational origin it couples universally to all fermion species. The effect of the coupling to neutrinos will be clarified below.

The massless gravitational multiplet in string theory contains the dilaton (spin 0, scalar), $\Phi$, the graviton (spin 2, symmetric tensor), $g_{\mu\nu}$, and the spin 1 antisymmetric tensor $B_{\mu\nu}$. The (Kalb-Ramond) field $B$ appears in the string effective action only through its totally antisymmetric field strength, $H_{\mu\nu\rho} = \partial_{[\mu} B_{\nu\rho]}$, where $[\ldots]$ denotes antisymmetrization of the indices within the brackets. The calculation of string amplitudes [36] shows that $H_{\mu\nu\rho}$ plays the role of *torsion* in a generalised connection $\overline{\Gamma}$:

$${\overline{\Gamma}^\lambda}_{\mu\nu} = {\Gamma^\lambda}_{\mu\nu} + e^{-2\Phi} H^\lambda_{\mu\nu} \equiv {\Gamma^\lambda}_{\mu\nu} + {T^\lambda}_{\mu\nu} . \tag{16}$$

${\Gamma^\lambda}_{\mu\nu} = {\Gamma^\lambda}_{\nu\mu}$ is the torsion-free Einstein-metric connection, and ${T^\lambda}_{\mu\nu} = -{T^\lambda}_{\nu\mu}$ is the *torsion.*

In ref. [37] exact solutions to the world-sheet conformal invariance conditions (to all orders in $\alpha'$) of the low energy effective action of strings have been presented. In four "large" (uncompactified) dimensions of the string, the antisymmetric tensor field strength can be written uniquely as

$$H_{\mu\nu\rho} = e^{2\Phi}\epsilon_{\mu\nu\rho\sigma}\partial^\sigma b(x) \tag{17}$$

with $\epsilon_{0123} = \sqrt{g}$ and $\epsilon^{\mu\nu\rho\sigma} = |g|^{-1}\epsilon_{\mu\nu\rho\sigma}$, with $g$ the metric determinant. The field $b(x)$ is a "pseudoscalar " *axion*-like field. The dilaton $\Phi$ and axion $b$ fields are fields that appear as Goldstone bosons of spontaneously broken scale symmetries of the string vacua, and so are exactly massless classically. In the effective string action such fields appear only through their derivatives.The exact solution of [37] in the *string frame* requires that both dilaton and axion fields are linear in the target time $X^0$, $\Phi(X^0) \sim X^0$, $b(X^0) \sim X^0$. This solution will shift the minima of all fields in the effective action which couple to the dilaton and axion by a space-time independent amount.

In the "physical" *Einstein frame* , relevant for cosmological observations, the temporal component of the metric is normalised to $g_{00} = +1$ by an appropriate change of the time coordinate. In this setting, the solution of [37] leads to a Friedmann-Robertson-Walker (FRW) metric, with scale factor $a(t) \sim t$, where $t$ is the FRW cosmic time. Moreover, the dilaton field $\Phi$ behaves as

$$\Phi(t) = -\ln t + \phi_0, \tag{18}$$

with $\phi_0$ a constant, and the axion field $b(x)$ is linear in $t$. There is an underlying world-sheet conformal field theory with central charge $c = 4 - 12Q^2 - \frac{6}{n+2} + c_I$ where $Q^2 (> 0)$ is the central-charge deficit and $c_I$ is the central charge associated with the world-sheet conformal field theory of the compact "internal" dimensions of the string model [37]. The condition of cancellation of the world-sheet ghosts that appear because of the fixing of reparametrisation invariance of world-sheet co-ordinates requires that $c = 26$. The solution for the axion field is

$$b(x) = \sqrt{2}e^{-\phi_0}\sqrt{Q^2}\frac{M_s}{\sqrt{n}}t\ , \tag{19}$$

where $M_s$ is the string mass scale and $n$ is a positive integer, associated with the level of the Kac-Moody algebra of the underlying world-sheet conformal field theory. For non-zero $Q^2$ there is an additional dark energy term in the effective target-space time action of the string [37] of the form $\int d^4x\sqrt{-g}e^{2\Phi}(-4Q^2)/\alpha'$. The linear axion field (19) *remains* a non-trivial solution *even* in the *static* space-time limit with a constant dilaton field [37]. In such a case space time is an Einstein universe with positive cosmological constant and constant positive curvature proportional to $6/(n+2)$.

For the solutions of [37], the covariant torsion tensor $e^{-2\Phi}H_{\mu\nu\rho}$ is *constant.* (This follows from (16) and (17) since the exponential dilaton factors cancel out in the relevant expressions.) Only the spatial components of the torsion are nonzero in this case,

$$T_{ijk} \sim \epsilon_{ijk}\dot{b} = \epsilon_{ijk}\sqrt{2Q^2}e^{-\phi_0}\frac{M_s}{\sqrt{n}}\ , \tag{20}$$

where the overdot denotes derivative with respect to $t$. As discussed in [35], in the framework of the target-space effective theory, the relevant Lagrangian terms for fermions (to lowest order in $\alpha'$) will be of the form (10), with the vector $B^0$ being associated with the spatial components of the constant torsion part

$$B^0 \sim \epsilon^{ijk}T_{ijk}\ , \tag{21}$$

where From (16), (17) and (10), we also observe that only the temporal component $B^0$ of the $B^d$ vector is nonzero. Note that the torsion-free gravitational part of the connection (for the

FRW or flat case) yields a vanishing contribution to $B^0$. From (10) and (20) then we obtain a constant $B^0$ of order

$$B^0 \sim \sqrt{2Q^2} e^{-\phi_0} \frac{M_s}{\sqrt{n}} \text{ GeV} > 0. \tag{22}$$

We follow the conventions of string theory for the sign of $B^0$ . From phenomenological considerations $M_s$ and $g_s^2/4\pi$ are taken to be larger than O($10^4$) GeV and about 1/20 respectively.

The particle-antiparticle asymmetry occurs already in thermal equilibrium, due to the background-induced difference in the dispersion relations between particles and antiparticles. Since the coupling of fermions to torsion is *universal*, the axion background would also couple to quarks and charged leptons. However, it is the right-handed neutrinos that play a crucial rôle and induce a phenomenologically viable leptogenesis, and then baryogenesis through sphaleron processes in the standard model or other B-L conserving processes. This is due to the fact that, as argued in [17], the right-handed Majorana neutrinos can oscillate between themselves and their antiparticles, unlike the charged fermions of the standard model. Such $B_0$-background-induced neutrino-antineutrino oscillations, which have been envisaged initially by Pontercorvo [21, 22], are induced by the mixing of neutrino and antineutrino states to produce mass eigenstates due to the constant 'environmental' field $B^0$ [32, 33]. To see this, we consider the Lagrangian for Majorana neutrinos in the presence of $B_a$, written in terms of two-component (Weyl) spinor fields $\psi, \psi^c$ (since a generic four-component right-handed Majorana spinor $\Psi$ may be written in our notation as $\Psi = \begin{pmatrix} \psi_R^c \\ \psi_R \end{pmatrix}$, where from now on we omit the left-handed suffix $R$):

$$\mathcal{L}_\nu = \sqrt{-g}\Big[\big(\psi^{c\dagger} \quad \psi^\dagger\big)\frac{i}{2}\gamma^0\gamma^\mu D_\mu \begin{pmatrix} \psi^c \\ \psi \end{pmatrix} - \big(\psi^{c\dagger} \quad \psi^\dagger\big)\begin{pmatrix} -B_0 & -m \\ -m & B_0 \end{pmatrix}\begin{pmatrix} \psi^c \\ \psi \end{pmatrix}, \tag{23}$$

where $D_\mu$ is the gravitational covariant derivative with respect to the torsion-free spin connection, and we assume for brevity that the neutrino has only lepton-number-violating Majorana-type masses. We shall discuss later on (*cf.* section 4) a way of generating dynamically such masses using quantum fluctuations of the KR torsion. For the moment we are interested in this "free" lagrangian, ignoring coupling it to the Lepton sector of the Standard Model. This is to be understood in the complete theory, e.g. in the $\nu$MSM model of [5].

At this stage we note that the energy eigenstates are appropriate linear combinations of the states $|\psi\rangle$ and $|\psi^c\rangle$. We observe from (23) that, in the presence of torsion, there are non-trivial and unequal diagonal lepton-number-conserving entries in the "mass" matrix $\mathcal{M}$ for $\psi$ and $\psi^c$:

$$\mathcal{M} = \begin{pmatrix} -B_0 & -m \\ -m & B_0 \end{pmatrix}.$$

This matrix is hermitean, so can be diagonalised by a unitary matrix, leading to two-component mass eigenstates $|\chi_{i,j}\rangle$ that are mixtures of the states $|\psi\rangle$ and $|\psi^c\rangle$ (and hence of the energy eigenstates):

$$\begin{aligned} |\chi_1\rangle &= \mathcal{N}^{-1}\{\Big(B_0 + \sqrt{B_0^2 + m^2}\Big)|\psi^c\rangle + m\,|\psi\rangle\}, \\ |\chi_2\rangle &= \mathcal{N}^{-1}\{-m\,|\psi^c\rangle + \Big(B_0 + \sqrt{B_0^2 + m^2}\Big)|\psi\rangle\}, \end{aligned} \tag{24}$$

where $\mathcal{N} \equiv \Big[2\Big(B_0^2 + m^2 + B_0\sqrt{B_0^2 + m^2}\Big)\Big]^{1/2}$, with eigenvalues $m_{1,2} = \mp\sqrt{B_0^2 + m^2}$. The reader should have already noticed that using this Weyl notation we treat the "antiparticle" $|\psi^c\rangle$ and "particle" $|\psi\rangle$ states as different "species". In this way, the energy differences between them induced by the KR background $B^0$ (13) for fixed momenta, may lead to oscillations among such states.

The above mixing can be expressed by writing the four-component mass-eigenstate neutrino spinor in terms of $\psi$ and $\psi^c$ using an angle $\theta$ [33]:

$$\nu \equiv \begin{pmatrix} \chi_1 \\ \chi_2 \end{pmatrix} = \begin{pmatrix} \cos\theta & \sin\theta \\ -\sin\theta & \cos\theta \end{pmatrix} \begin{pmatrix} \psi^c \\ \psi \end{pmatrix} \quad : \qquad \tan\theta \equiv \frac{m}{B_0 + \sqrt{B_0^2 + m^2}} \,. \tag{25}$$

It is readily seen that the four-component spinor $\nu$ is also Majorana, as it satisfies the Majorana condition $\nu^c = \nu$. We note that in the absence of torsion, $B_0 \to 0$, the mixing angle between the two-component spinors $\psi$ and $\psi^c$ is maximal: $\theta = \pi/4$, thereby reproducing the usual situation for Majorana spinors (which are mass eigenstates and thus of mixed chirality [22]), whereas it is non-maximal when $B_0 \neq 0$.

The mixing (24) enables us to understand the difference between the densities of fermions and antifermions mentioned earlier (15). The expectation values of the number operators of $\chi_i, i = 1, 2$ in the basis $|\psi\rangle$ and $|\psi^c\rangle$ are given by:

$$\begin{aligned} N_{\chi_1} &= <: \chi_1^\dagger \chi_1 :> = \cos^2\theta \; <: \psi^{c\dagger} \psi^c :> + \sin^2\theta \; <: \psi^\dagger \psi :>, \\ N_{\chi_2} &= <: \chi_2^\dagger \chi_2 :> = \sin^2\theta \; <: \psi^{c\dagger} \psi^c :> + \cos^2\theta \; <: \psi^\dagger \psi :>, \end{aligned} \tag{26}$$

where cross-terms do not contribute. We observe that, for general $\theta \neq \pi/4$, *i.e.*, $B_0 \neq 0$, as seen in (25), there is a difference between the populations of $\chi_1$ and $\chi_2$:

$$N_{\chi_1} - N_{\chi_2} = \cos 2\theta \Big( < n_{\psi^c} > - < n_\psi > \Big) \,, \tag{27}$$

where $< n_\psi > = <: \psi^\dagger \psi :> \neq < n_{\psi^c} > = <: \psi^{c\dagger} \psi^c :>$ are the corresponding number operators for the states $|\psi\rangle$ and $|\psi^c\rangle$.

This difference in the neutrino and antineutrino populations (15) is made possible by the presence of fermion-number-violating fermion-antifermion oscillations, whose probability was calculated in [33]:

$$\mathcal{P}(t) = |\langle \nu_1(t) | \nu_2(0) \rangle|^2 = \sin^2\theta \sin^2\Big( \frac{E_\nu - E_{\nu^c}}{2} \, t \Big) = \frac{m^2}{B_0^2 + m^2} \sin^2(B_0 \, t) \,, \tag{28}$$

where we used (13) with $\vec{B} = 0$, as in our specific background (22) and the definition of the mixing angle (25).

The time evolution of the system is calculated by expressing $|\psi\rangle$ and $|\psi^c\rangle$ as appropriate linear combinations of the eigenstates of the Hamiltonian. Using appropriate creation and annihilation operators acting on the vacuum state $|0\rangle$, we can then restrict first our attention only to positive-frequency (energy) states created from the vacuum. This determines the argument of the sinusoidal oscillation term $\sin^2\Big(\frac{E_\nu - E_{\nu^c}}{2} \, t\Big)$. In the case of relativistic neutrinos moving close to the speed of light, the oscillation length obtained from (28) is

$$L = \frac{\pi \, \hbar \, c}{|B_0|} = \frac{6.3 \times 10^{-14} \, \text{GeV}}{B_0} \, \text{cm}. \tag{29}$$

where we have reinstated $\hbar$ and $c$, and $B_0$ is measured in GeV.

Such oscillations among "particle/antiparticle" quantum states with positive energy thermalise the corresponding populations (2), (3). Similar asymmetries are created in the negative-frequency states, which contribute, together with the positive-frequency ones in the

Journal of Physics: Conference Series **651** (2015) 012015 doi:10.1088/1742-6596/651/1/012015

physical Majorana neutrino states. In this way thermalization of the physical states with different populations between particles and antiparticles, due to the background $B^0$, is obtained.

For oscillations to be effective at any given epoch in the early Universe, this length has to be less than the size of the Hubble horizon. We assume that a cosmological solution of the form discussed in [37], with a scale factor increasing linearly with time, is applicable some time after any earlier inflationary epoch. For a temperature

$$T_d \sim 10^9 \text{ GeV}, \tag{30}$$

the relevant Hubble horizon size $\sim 10^{-12}$ cm. On the other hand, we see from (15) that the correct order of magnitude for the lepton asymmetry $\sim 10^{-10}$ is obtained if

$$B_0 \sim 10^{-1} \text{ GeV}. \tag{31}$$

For this value of $B^0$, the oscillation length (29) $10^{-13}$ cm, which is within the Hubble horizon size $10^{-12}$ cm. This implies that Majorana neutrino/antineutrino oscillations occur sufficiently rapidly to establish chemical equilibrium and hence a lepton asymmetry. On the other hand, as already mentioned, charged leptons and quarks, although coupled to the KR torsion $H$, nevertheless do not exhibit such oscillations due to charge conservation.

At the temperature $T_d \simeq 10^9$ GeV the universe is assumed to undergo a *phase transition* [17] towards either a vanishing $B_0$ or at least a very small $B_0$ compatible with the current limits, $B_0 < 10^{-2}$ eV , of the relevant parameter of the standard model extension [38, 39, 40, 41]. In this scenario for leptogenesis no fine tuning for the width of the pertinent CP violating processes in the lepton sector is required, in contrast to the case of conventional leptogenesis [42, 43, 44, 5, 18]. However, the presence of right-handed neutrinos was essential, and this is consistent with the need for explaining the smallness of the active neutrino masses through see-saw mechanisms, or the rôle of sterile neutrinos as dark matter [5, 18].) The reader should note that the range of neutrino masses (Gev and keV) invoked in the latter works is consistent with the approximations leading to (29).

## 4. Kalb-Ramond Torsion Fluctuations, Anomalies and Majorana Mass Generation

Before concluding we would like to discuss another interesting aspect of the KR torsion: the generation of the masses of the right-handed Majorana neutrinos used above, e.g. in the range of GeV and keV as required in the $\nu MSM$ model [5]. So far we have discussed the rôle of background KR torsion. However, as we discussed above, at the temperature $T_d$ the universe may undergo a phase transition to a vanishing $B_0$. The quantum fluctuations of the torsion, however, survive. In this section we would like to make a suggestion [20] that links these fluctuations to a mechanism for dynamical generation of (chirality changing) Majorana mass terms for neutrinos.

To discuss quantum aspects of torsion we first notice that the KR H-torsion is a totally antisymmetric type of torsion coupled to fermions as (using for brevity differential form language):

$$S_\psi \ni -\frac{3}{4}\int S \wedge^\star J^5 \ ,$$

where $J^5_\mu = \overline{\psi}\gamma_\mu\gamma^5\psi$ is the axial fermion current. Here the fermions $\psi$ are generic and represent all sermonic excitations of the Standard Model plus right handed Majorana neutrinos. The totally antisymmetric part of the torsion $S =^\star T$, that is $S_d = \frac{1}{3!}\epsilon^{abc}_{\ \ \ d}T_{abc}$, where $T_{abc}$ is the contorsion which is proportional to $H_{abc} = \epsilon_{abcd}\partial^d b(x)$ in our case, with $b$ the KR axion field. Classically one has the Bianchi identity $d^\star S = 0$.

To discuss *quantum corrections* [20] we impose the constrain that quantum corrections should not affect this Bianchi identity, which allows for a definition of a *conserved torsion charge*

$Q = \int^\star S$. Implementing this constraint via a delta function in the relevant path integral $\delta(d^\star S)$ leads to the introduction of a Lagrange multiplier field $b$

$$\mathcal{Z} \ni \int D\mathbf{S}\, Db \exp\left[i \int \frac{3}{4\kappa^2}\mathbf{S}\wedge{}^\star\mathbf{S} - \frac{3}{4}\mathbf{S}\wedge{}^\star\mathbf{J}^5 + \left(\frac{3}{2\kappa^2}\right)^{1/2} b\, d^\star\mathbf{S}\right]$$
$$= \int Db \exp\left[-i\int \frac{1}{2}\mathbf{d}b\wedge{}^\star\mathbf{d}b + \frac{1}{f_b}\mathbf{d}b\wedge{}^\star\mathbf{J}^5 + \frac{1}{2f_b^2}\mathbf{J}^5\wedge\mathbf{J}^5\right], \tag{32}$$

where $f_b = (3\kappa^2/8)^{-1/2} = \frac{M_P}{\sqrt{3\pi}}$ and the non-propagating $\mathbf{S}$ field has been integrated out. Here we have used the same notation $b$ for the Lagrange multiplier field as the background KR axion field. This is for reasons of economy. The field $b$ in (32) denotes quantum fluctuations of the KR axion, and we assume a vanishing background for this field today. If one considers the quantum fluctuations about the background then the background terms are understood (but not explicitly given) in (32). The reader should notice that, as a result of this integration, the corresponding *effective* field theory contains a *non-renormalizable* repulsive four-fermion axial-current-current interaction. By partially integrating the term $db\wedge^\star J^5$ and using the (one-loop exact) *chiral anomaly equation*

$$\nabla_\mu J^{5\mu} = \frac{e^2}{8\pi^2}F^{\mu\nu}\widetilde{F}_{\mu\nu} - \frac{1}{192\pi^2}R^{\mu\nu\rho\sigma}\widetilde{R}_{\mu\nu\rho\sigma}\,, \tag{33}$$

where $\mathbf{F}$ denotes field strength of gauge fields, and $R$ is the four-dimensional space time gravitational curvature, we obtain an effective "axion-like" coupling for the KR axion with the gauge sector

$$S_{\rm eff} \ni -\frac{e^2}{8\pi^2 f_b}\int b(x)F^{\mu\nu}\widetilde{F}_{\mu\nu} + \frac{1}{192\pi^2 f_b}\int b(x)R^{\mu\nu\rho\sigma}\widetilde{R}_{\mu\nu\rho\sigma}, \tag{34}$$

where the $(\widetilde{A}_{\mu\dots})$ notation denotes a tensor dual to $A_{\mu\dots}$. The important point to notice is that the $b$ axion field is massless, unlike the ordinary axion field.

We notice at this stage, that for the case of the electromagnetic field, the term $bF_{\mu\nu}\widetilde{F}^{\mu\nu}$ becomes (up to total derivative terms) a Chern-Simons (CS) form in four space-time dimensions

$$\int bF_{\mu\nu}\widetilde{F}^{\mu\nu} \propto S_{\rm CS} = \int B_\mu A_\nu F_{\rho\sigma}\epsilon^{\mu\nu\rho\sigma}\,, \quad B_\mu = \epsilon_{\mu\alpha\beta\gamma}H^{\alpha\beta\gamma}\,, \; H_{\alpha\beta\gamma} = \epsilon_{\alpha\beta\gamma\delta}\partial^\delta b(x). \tag{35}$$

Notice that $B_\mu$ is nothing but our axial vector coupled to the fermions in (10), but here is not a background but a full fledged quantum field. In fact, when considering the coupling of charged fermions (*e.g.* electrons or quarks) with the electromagnetic field $A_\mu$, the presence of such CS terms may affect the quantum photon propagator. This subject is still controversial, and we postpone a detailed discussion for a forthcoming publication [34].

For the purposes of the current work, we notice that, following ref. [20], we may couple (via appropriate Yukawa interactions of strength $y_a$ ) the (right-handed) Majorana fermions to an ordinary axion field, $a(x)$, which is allowed to mix (via the corresponding kinetic terms $\gamma\int\partial_\mu b\,\partial^\mu a$, with $|\gamma| < 1$) with the KR axion $b(x)$. It is convenient to diagonalize the axion kinetic terms by redefining the KR axion field

$$b(x) \to b'(x) \equiv b(x) + \gamma a(x)$$

and canonically normalise the axion field $a$. The $b'$ field decouples, then, leaving an effective axion-fermion action [20]:

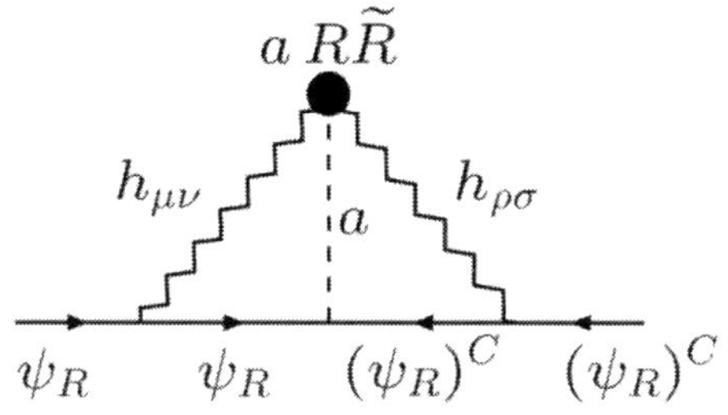


**Figure 1.** *Feynman graph giving rise to anomalous fermion mass generation. The black circle denotes the operator $a(x)\,R_{\mu\nu\lambda\rho}\widetilde{R}^{\mu\nu\lambda\rho}$ induced by torsion. Wavy lines are gravitons, dashed lines pertain to axion $a(x)$ propagators, while continuous lines denote Majorana spinors.*

$$\begin{aligned}\mathcal{S}_a \;=\; \int d^4x\sqrt{-g}\,\Big[\frac{1}{2}(\partial_\mu a)^2 - \frac{\gamma a(x)}{192\pi^2 f_b\sqrt{1-\gamma^2}}R^{\mu\nu\rho\sigma}\widetilde{R}_{\mu\nu\rho\sigma} -\\ \frac{iy_a}{\sqrt{1-\gamma^2}}\,a(x)\left(\overline{\psi}_R^{\;C}\psi_R - \overline{\psi}_R\psi_R^{\;C}\right) + \frac{1}{2f_b^2}J^5_\mu J^{5\mu}\Big]\,. \end{aligned} \tag{36}$$

The mechanism for the anomalous Majorana mass generation is shown in Fig. 1. We may now estimate the two-loop Majorana neutrino mass in quantum gravity with an effective UV energy cut-off $\Lambda$ by adopting the effective field-theory framework of [45]. This leads to a gravitationally induced Majorana mass $M_R$:

$$M_R \sim \frac{\sqrt{3}\,y_a\,\gamma\,\kappa^5\Lambda^6}{49152\sqrt{8}\,\pi^4(1-\gamma^2)}\,. \tag{37}$$

In a UV complete theory such as strings, $\Lambda$ and $M_P$ are related, since $\Lambda$ is proportional to $M_s$ and the latter is related to $M_P$ (or $\kappa$) via the strng coupling and the compactification volume. Obviously, the generation of $M_R$ is highly model dependent. Taking, for example, the quantum gravity scale to be $\Lambda = 10^{17}$ GeV, we find that $M_R$ is at the TeV scale, for $y_a = 10^{-3}$ and $\gamma = 0.1$. However, if we take the quantum gravity scale to be close to the GUT scale, i.e. $\Lambda = 10^{16}$ GeV, we obtain a right-handed neutrino mass $M_R \sim 16$ keV, for the choice $y_a = \gamma = 10^{-3}$. This is in the preferred ballpark region for the sterile neutrino $\psi_R$ to qualify as a warm dark matter [18].

In a string-theoretic framework, many axions might exist that could mix with each other. Such a mixing can give rise to reduced UV sensitivity of the two-loop graph shown in Fig. 1. To make this point explicit, let us therefore consider a scenario with a number $n$ axion fields, $a_{1,2,\dots,n}$. Of this collection of $n$ pseudoscalars, only $a_1$ has a kinetic mixing term $\gamma$ with the KR axion $b$ and only $a_n$ has a Yukawa coupling $y_a$ to right-handed neutrinos $\psi_R$. The other axions $a_{2,3,\dots,n}$ have a next-to-neighbour mixing pattern. In such a model, the anomalously generated Majorana mass may be estimated to be [20]

$$M_R \sim \frac{\sqrt{3}\,y_a\,\gamma\,\kappa^5\Lambda^{6-2n}(\delta M_a^2)^n}{49152\sqrt{8}\,\pi^4(1-\gamma^2)}\,, \tag{38}$$

for $n \leq 3$, and thus independent of $\Lambda$ for $n = 3$. Of course, beyond the two loops, $M_R$ will depend on higher powers of the energy cut-off $\Lambda$, i.e. $\Lambda^{n>6}$, but if $\kappa\Lambda \ll 1$, these higher-order effects are expected to be subdominant. In the above $n$-axion-mixing scenarios, the anomalously generated Majorana mass term will only depend on the mass-mixing parameters $\delta M_a^2$ of the axion fields and not on their masses themselves, as long as $n \leq 3$.

## 5. Conclusions and Outlook

In this note we have discussed ways of obtaining leptogenesis/baryogenesis, which do not follow the Sakharov paradigm and involve non-trivial background geometries of the early universe that violate Lorentz symmetry. As a specific example we considered a string-inspired theory involving anstisymmetric Kalb-Ramond (KR) tensor fields of spin 1, which in four space-time dimensions are equivalent to a pseudoscalar degree of freedom (the KR axion). The KR field provides the geometry with an appropriate totally antisymmetric torsion. The latter couples to all matter fermions both charged and neutral, but it is the coupling to right-handed Majorana neutrinos that plays a crucial rôle in providing microscopic processes of neutrino/antineutrino oscillations underlying the generation of matter-antimatter asymmetry in the lepton sector at high temperatures. The latter is then communicated to the baryon sector via the standard baryon-minus-lepton-number conserving sphaleron processes. The string universe is assumed to undergo a phase transition at a given temperature, at which the background KR axion field vanishes (or is diminished significantly, in agreement with the stringent bounds today on the Lorentz-symmetry-violating parameter of the standard model extension that corresponds to this background).

We have also shown how quantum fluctuations of the KR torsion can generate an effective (right-handed) Majorana neutrino mass $M_R$ at two loops by gravitational interactions that involve global anomalies. The KR axion $b$ couples to both matter and gravitation and radiation gauge fields. In perturbation theory, this axion field $b(x)$ derived from torsion has derivative couplings, leading to an axion shift symmetry: $b(x) \to b(x) + c$, where $c$ is an arbitrary constant. If another axion field $a(x)$ or fields are present in the theory, the shift symmetry may be broken, giving rise to axion masses and chirality changing Yukawa couplings to massless fermions, such as right-handed Majorana neutrinos $\psi_R$. In this latter scenarios, two potential candidates for (non-supersymmetric) dark matter, right-handed neutrinos and axions, play a non-trivial rôle in generating neutrino masses in minimal extensions of the standard model beyond the standard seesaw. The cosmology of such axion/right-handed neutrino models is still to be investigated.

As a final remark, in connection with what was also discussed in this meeting by A. Aranda [46], it could be interesting to try and embed this simplified Kalb-Ramond-string model in realistic brane theories, where the Yukawa couplings of the ordinary axion fields to the Majorana neutrinos, which from a field theory point of view appear as arbitrary parameters, may be determined by means of appropriate compactifications of the underlying microscopic brane model. This is a long shot but in my opinion worths a try.

## Acknowledgments

N.E.M. thanks the organisers of the XIV Mexican Workshop on Particles and Fields, November 25-29 2013, Oaxaca (Mexico) for the invitation to give this plenary talk and for providing an excellent and thought stimulating meeting. The work of N.E.M. was supported in part by the London Centre for Terauniverse Studies (LCTS), using funding from the European Research Council via the Advanced Investigator Grant 267352, and by STFC UK under the research grant ST/J002798/1.

## References

[1] Christenson J, Cronin J, Fitch V and Turlay R 1964 *Phys.Rev.Lett.* **13** 138–140
[2] Sakharov A 1967 *Pisma Zh.Eksp.Teor.Fiz.* **5** 32–35 reprinted in *Kolb, E.W. (ed.), Turner, M.S. (ed.): The early universe* 371-373, and in *Lindley, D. (ed.) et al.: Cosmology and particle physics* 106-109, and in Sov. Phys. Usp. 34 (1991) 392-393 [Usp. Fiz. Nauk 161 (1991) No. 5 61-64]
[3] Gamow G 1946 *Phys.Rev.* **70** 572–573
[4] Kuzmin V, Rubakov V and Shaposhnikov M 1985 *Phys.Lett.* **B155** 36
[5] Shaposhnikov M 2009 *Prog.Theor.Phys.* **122** 185–203
[6] Shaposhnikov M and Tkachev I 2006 *Phys.Lett.* **B639** 414–417 (*Preprint* hep-ph/0604236)

[7] Lindner M, Merle A and Niro V 2011 *JCAP* **1101** 034 (*Preprint* 1011.4950)
[8] Kusenko A, Takahashi F and Yanagida T T 2010 *Phys.Lett.* **B693** 144–148 (*Preprint* 1006.1731)
[9] Randall L and Sundrum R 1999 *Phys.Rev.Lett.* **83** 3370–3373 (*Preprint* Randall:1999ee)
[10] Merle A and Niro V 2011 *JCAP* **1107** 023 (*Preprint* 1105.5136)
[11] Barry J, Rodejohann W and Zhang H 2011 *JHEP* **1107** 091 (*Preprint* 1105.3911)
[12] Carmona J M, Cortes J L, Das A K, Gamboa J and Mendez F 2006 *Mod.Phys.Lett.* **A21** 883–892 (*Preprint* hep-th/0410143)
[13] Streater R and Wightman A 1989 *PCT, spin and statistics, and all that* (Princeton University Press)
[14] Bertolami O, Colladay D, Kostelecky V A and Potting R 1997 *Phys.Lett.* **B395** 178–183 (*Preprint* hep-ph/9612437)
[15] Mavromatos N E 2010 *Int.J.Mod.Phys.* **A25** 5409–5485 (*Preprint* 1010.5354 and references therein.)
[16] Arnold P and McLerran L 1987 *Phys. Rev. D* **36**(2) 581–595 URL http://link.aps.org/doi/10.1103/PhysRevD.36.581
[17] Ellis J, Mavromatos N E and Sarkar S 2013 *Phys.Lett.* **B725** 407–411 (*Preprint* 1304.5433)
[18] Boyarsky A, Ruchayskiy O and Shaposhnikov M 2009 *Ann.Rev.Nucl.Part.Sci.* **59** 191–214 (*Preprint* 0901.0011)
[19] Mitsou V A 2013 *Int.J.Mod.Phys.* **A28** 1330052 (*Preprint* 1310.1072)
[20] Mavromatos N E and Pilaftsis A 2012 *Phys.Rev.* **D86** 124038 (*Preprint* 1209.6387)
[21] Pontecorvo B 1968 *Sov.Phys.JETP* **26** 984–988
[22] Bilenky S M and Pontecorvo B 1978 *Phys.Rept.* **41** 225–261
[23] Dolgov A 2010 *Phys.Atom.Nucl.* **73** 588–592 (*Preprint* 0903.4318)
[24] Carroll S M and Shu J 2006 *Phys.Rev.* **D73** 103515 (*Preprint* hep-ph/0510081)
[25] Barenboim G, Borissov L, Lykken J D and Smirnov A 2002 *JHEP* **0210** 001 (*Preprint* hep-ph/0108199)
[26] Davoudiasl H, Kitano R, Kribs G D, Murayama H and Steinhardt P J 2004 *Phys.Rev.Lett.* **93** 201301 (*Preprint* hep-ph/0403019)
[27] Lambiase G and Mohanty S 2007 *JCAP* **0712** 008 (*Preprint* astro-ph/0611905)
[28] Lambiase G and Mohanty S 2011 *Phys.Rev.* **D84** 023509 (*Preprint* 1107.1213)
[29] Li H, Li M z and Zhang X m 2004 *Phys.Rev.* **D70** 047302 (*Preprint* hep-ph/0403281)
[30] Debnath U, Mukhopadhyay B and Dadhich N 2006 *Mod.Phys.Lett.* **A21** 399–408 (*Preprint* hep-ph/0510351)
[31] Mukhopadhyay B 2005 *Mod.Phys.Lett.* **A20** 2145–2156 (*Preprint* astro-ph/0505460)
[32] Mukhopadhyay B 2007 *Class.Quant.Grav.* **24** 1433–1442 (*Preprint* gr-qc/0702062)
[33] Sinha M and Mukhopadhyay B 2008 *Phys.Rev.* **D77** 025003 (*Preprint* 0704.2593)
[34] de Cesare M, Mavromatos N E and Sarkar S 2014 *in progress*
[35] Mavromatos N E and Sarkar S 2013 *Eur.Phys.J.* **C73** 2359 (*Preprint* 1211.0968)
[36] Gross D J and Sloan J H 1987 *Nucl.Phys.* **B291** 41
[37] Antoniadis I, Bachas C, Ellis J R and Nanopoulos D V 1989 *Nucl.Phys.* **B328** 117–139
[38] Coleman S R and Glashow S L 1999 *Phys.Rev.* **D59** 116008 (*Preprint* hep-ph/9812418)
[39] Bluhm R, Kostelecky V A and Russell N 2000 173–180 (*Preprint* hep-ph/0003223)
[40] Kharlanov O and Zhukovsky V C 2007 *J.Math.Phys.* **48** 092302 (*Preprint* 0705.3306)
[41] Bolokhov P A, Pospelov M and Romalis M 2008 *Phys.Rev.* **D78** 057702 (*Preprint* hep-ph/0609153)
[42] Pilaftsis A and Underwood T E 2004 *Nucl.Phys.* **B692** 303–345 (*Preprint* hep-ph/0309342)
[43] Pilaftsis A 2005 *Phys.Rev.Lett.* **95** 081602 (*Preprint* hep-ph/0408103)
[44] Pilaftsis A and Underwood T E 2005 *Phys.Rev.* **D72** 113001 (*Preprint* hep-ph/0506107)
[45] Bjerrum-Bohr N E J, Donoghue J F and Vanhove P 2013 (*Preprint* 1309.0804)
[46] Alvarado C, Aranda A, Corradini O, Rojas A D and Santos-Rodriguez E 2012 *Phys.Rev.* **D86** 036010 (*Preprint* 1206.5216)

XIV Mexican Workshop on Particles and Fields IOP Publishing
Journal of Physics: Conference Series **651** (2015) 012016 doi:10.1088/1742-6596/651/1/012016

# THE TEXTURIZED 2HDM (2HDM-TX) AND HIGGS SIGNATURE AT COLLIDERS

**J. Lorenzo Diaz-Cruz**[1], **Enrique Díaz**[2] **and Javier A. Orduz-Ducuara**[3]

Facultad de Ciencias Físico Matematicas, Benemérita Universidad Autónoma de Puebla, 18 sur y Av. San Claudio, C.P. 72570 Puebla, Pue., Mexico

E-mail: [1]jldiaz@fcfm.buap.mx, [2]alcmaeon@gmail.com, [3]jaorduz@alumnos.fcfm.buap.mx

**Abstract.** We explore the phenomena of flavor violation, using combinations of Yukawa matrices in order to reproduce a hermitian mass matrix. We work within the Two-Higgs doublet model with a specific four-zero texture type (2HDM-Tx). Current experimental bounds from $K-\bar{K}$ mixing, $B_s \to \mu^+\mu^-$ and $\tau \to 3\mu$ are used in order to restrict the values for the parameters model: $\tan\beta(=\frac{v_2}{v_1})$ and $\gamma$ $(0 < \gamma < 1)$. We predict the rates for the Flavor Violating decays $(h \to \tau\mu)$ and $t \to c + h$, which reach Branching ratios that could be tested at the LHC.

## 1. Introduction

Recently the Large Hadron Collider (LHC) has confirmed the existence of a Higgs-like particle with mass $m_H \approx 125$ GeV [1, 2]. This seems to confirm the linear realization of the mechanism of Electroweak symmetry breaking, which it needs to generate the masses of gauge bosons and fermions within the SM [3]. Additionally, the current experimental searches are testing the Higgs couplings, an information that can be used to discriminate between the minimal SM Higgs doublet case and other extensions of the SM, with more complicated Higgs sector [4].

For a time, a possible solution a the hierarchy problem, suffered by the SM Higgs boson, motivated many interesting proposals for extending the SM. It is expected that current and future data coming from LHC will give us the opportunity to test these scenarios. Two-Higgs Doublet Model (2HDM) is one the simplest proposals for physics Beyond the Standard Model (BSM). 2HDM was initially studied in connection with the search for the origin of CP violation [5], later on it was found that the model could be connected with the realization of new theoretical ideas, such as supersymmetry [6], extra dimensions [7] and strongly interacting models [8],[9].

Possible realizations of the general 2HDM that have been considered in the literature are known as Type I, II and III, for a review see [10]. There are also other models called X, Y, Z, but in some sense they can be considered variations of the above models. Model I has a discrete symmetry $Z_2$ , which permits a possible dark matter candidate coming from the $Z_2-$odd scalar doublet [11]. Within Type- I models, a single Higgs doublet gives mass to the up, down quarks and leptons. The type II model [12] assigns one doublet to each fermion type (this type also arises in the minimal SUSY extension of the SM [13]), which suffices to avoid Flavor Changing Neutral Currents (FCNC) mediated by the Higgs bosons[14]. Within the most general 2HDM, the mass matrix for each fermion type $f(= u, d, l)$ receives contributions from both Higgs doublets, which have vevs $v_1$ and $v_2$ after SSB, i.e.

$$M^f = \frac{1}{\sqrt{2}}\left(v_1 Y_1^f + v_2 Y_2^f\right), \tag{1}$$

where $Y_{1,2}^f$ denote the Yukawa matrices associated with each Higgs doublet. Inside this general model, the Yukawa matrices must have a structure that should reproduce the observed fermion masses and mixing angles and, simultaneously, the level of FCNC must satisfy current bounds [5, 15, 16]. One possiblity to have FCNC at acceptable levels, is the assumption that the Yukawa matrices have a certain texture form, i.e. with zeroes in different elements.

In the past this general model was called as 2HDM of type III. However, this naming scheme has become confusing, in part because it has been used to denote a different type of model [17], but also because some specific cases have acquired a relevance of their own. One can include the so-called Minimal Flavor violating 2HDM (MFV) in the relevant sub-cases of the general 2HDM [18], which provides precisely the minimal level of FCNC consistent with data; MFV could be studied from a pure phenomenological point of view[19] or as arising from flavor symmetries [20]. Although the so-called 2HDM with Alignment does not contain flavor violation, it is another possibility one can use to obtain realistic models [21]. Thus, in order to clarify the notation and to single out the use of textures within the 2HDM, from now on we shall call the two-higgs doublet model with textures as 2HDM-Tx.

Early studies of the 2HDM [22] considered the specific texture with six-zeroes, but also the so-called cyclic model. There, it was identified a specific pattern of FCNC Higgs-fermion couplings, known nowadays as the Cheng-Sher ansatz, which can be written as $\frac{\sqrt{m_i m_j}}{v}$. The extension of the 2HDM-Tx with four-zero texture was presented in [23, 24]. The phenomenological consequences of these matrix textures (Hermitian 4-textures or non-hermitian 6-textures) were considered in [25], whilst further phenomenological studies were presented in [26, 27, 28]. In all these cases, it happens that FCNC bounds are satisfied with Higgs masses lighter than O(TeV).

Nevertheless, most of the above studies assumed that both $Y_1^f$ and $Y_2^f$ have the same texture form, a pattern that can be called "Paralell textures" namely:

$$Y_1 = \begin{pmatrix} 0 & a_1 & 0 \\ a_1^* & c_1 & b_1 \\ 0 & b_1^* & a_1 \end{pmatrix}, \quad Y_2 = \begin{pmatrix} 0 & a_2 & 0 \\ a_2^* & c_2 & b_2 \\ 0 & b_2^* & a_2 \end{pmatrix}. \tag{2}$$

However, $Y_1^f$ and $Y_2^f$ could have different forms, but in such a way that the resulting mass matrix has a specific type of texture [3]. In this paper we shall discuss different combinations of Yukawa matrices that result in a mass matrix of the four-zero texture form. We shall discuss first the different types of Yukawa matrices that result in a mass matrix with four-zero textures 2. Section 3 contains the Lagrangian for the 2HDM-III, written in terms of mass eigenstates. Low energy constraints are discussed in section 4, including $K - \bar{K}$ mixing, $B_s \to \mu^+\mu^-$ and $\tau \to 3\mu$. We explored one particular case in section 5, while other cases can be found in [42]. Section 6 contains results and conclusions.

## 2. Classification of the Textures

Here, we shall assume that at most one element from each of the Yukawa matrices $Y_{1,2}^f$ contributes to each entry of the full mass matrix; namely:

Case 1:

$$Y_1 = \begin{pmatrix} 0 & d & 0 \\ d & c & b \\ 0 & b & 0 \end{pmatrix}, Y_2 = \begin{pmatrix} 0 & 0 & 0 \\ 0 & 0 & 0 \\ 0 & 0 & a \end{pmatrix};$$

Case 2:

$$Y_1 = \begin{pmatrix} 0 & 0 & 0 \\ 0 & c & 0 \\ 0 & 0 & 0 \end{pmatrix}, Y_2 = \begin{pmatrix} 0 & d & 0 \\ d & 0 & b \\ 0 & b & a \end{pmatrix}$$

Case 3:

$$Y_1 = \begin{pmatrix} 0 & d & 0 \\ d & 0 & 0 \\ 0 & 0 & 0 \end{pmatrix}, Y_2 = \begin{pmatrix} 0 & 0 & 0 \\ 0 & c & b \\ 0 & b & a \end{pmatrix};$$

Case 4:

$$Y_1 = \begin{pmatrix} 0 & 0 & 0 \\ 0 & 0 & b \\ 0 & b & 0 \end{pmatrix}, Y_2 = \begin{pmatrix} 0 & d & 0 \\ d & c & 0 \\ 0 & 0 & a \end{pmatrix}$$

Case 5:

$$Y_1 = \begin{pmatrix} 0 & d & 0 \\ d & c & 0 \\ 0 & 0 & 0 \end{pmatrix}, Y_2 = \begin{pmatrix} 0 & 0 & 0 \\ 0 & 0 & b \\ 0 & b & a \end{pmatrix};$$

Case 6:

$$Y_1 = \begin{pmatrix} 0 & d & 0 \\ d & 0 & b \\ 0 & b & 0 \end{pmatrix}, Y_2 = \begin{pmatrix} 0 & 0 & 0 \\ 0 & c & 0 \\ 0 & 0 & a \end{pmatrix}$$

Case 7:

$$Y_1 = \begin{pmatrix} 0 & d & 0 \\ d & c & b \\ 0 & b & a \end{pmatrix}, Y_2 = \begin{pmatrix} 0 & 0 & 0 \\ 0 & 0 & 0 \\ 0 & 0 & a' \end{pmatrix}$$

Then, the above combinations of Yukawa matrices produce a mass matrix with a 4-zero texture:

$$M = \begin{pmatrix} 0 & D & 0 \\ D^* & C & B \\ 0 & B^* & A \end{pmatrix}.$$

## 3. The Lagrangian for the model

The Yukawa Lagrangian in the 2HDM-III is given by,

$$\mathcal{L} = Y_1^u \overline{Q}_L^0 \tilde{\Phi}_1 u_R^0 + Y_2^u \overline{Q}_L^0 \tilde{\Phi}_2 u_R^0 + Y_1^d \overline{Q}_L^0 \Phi_1 d_R^0 + Y_2^d \overline{Q}_L^0 \Phi_2 d_R^0 + h.c.$$

where,

$$Q_L^0 = \begin{pmatrix} u_L \\ d_L \end{pmatrix}, \quad \overline{Q}_L^0 = \left(\overline{u}_L, \overline{d}_L\right), \quad \Phi_1 = \begin{pmatrix} \phi_1^{\pm} \\ \phi_1 \end{pmatrix}. \tag{3}$$

and

$$\Phi_2 = \begin{pmatrix} \phi_2^{\pm} \\ \phi_2 \end{pmatrix}, \quad \tilde{\Phi}_j = i\sigma_2 \Phi_j^* = \begin{pmatrix} \phi_j{}^* \\ -\phi_i^{\mp} \end{pmatrix}. \tag{4}$$

After introducing the mass eigenstates and diagonalizing the spinor fields ($d = Vd'$), the rotated Yukawa are $\widetilde{Y}_i^d = V^\dagger Y_i^d V$ and $Y_1^d = Y_1^{d\dagger}$, $Y_2^d = Y_2^{d\dagger}$. A more compact notation to write the Higgs-fermion interactions, is given by:

$$\mathcal{L}_N^d = \left(g\frac{\sqrt{m_i m_j}}{\sqrt{2}M_W}\right)\overline{d}_i'\left[\left(\eta^{(1)}\right)_{ij} h^o + \left(\eta^{(2)}\right)_{ij} H^0 + i\left(\eta^{(3)}\right)_{ij} A^0\gamma^5\right]d_j'$$

where the $\eta$ factors are defined as follows $\eta_{ij}^{(n)} = \chi_{ij}^{(1)} f_n(\alpha,\beta) + \chi_{ij}^{(2)} g_n(\alpha,\beta)$ and we defined: $f_1(\alpha,\beta) = -\frac{\sin\alpha}{\cos\beta}$, $g_1(\alpha,\beta) = \frac{\cos\alpha}{\sin\beta}$, $f_2(\alpha,\beta) = \frac{\cos\alpha}{\cos\beta}$, $g_2(\alpha,\beta) = \frac{\sin\alpha}{\sin\beta}$, $f_3(\alpha,\beta) = -\tan\beta$, $g_3(\alpha,\beta) = \cot\beta$.

## 4. Constraints from Low Energy

In order to find the allowed regions of parameter space we need to consider all relevant low energy and collider constraints. For the low energy constraints, this includes:

### *4.1. K-K Mixing*

The Feynman diagram is,

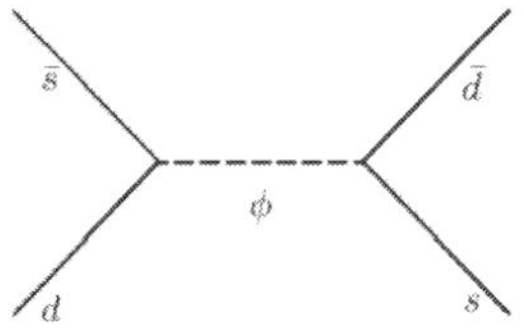


(5)

the effective Hamiltonian is given by,

$$H_{eff}^{\Delta S=2} = \frac{G_F^2 M_W^2}{16\pi^2} \sum_i C_i Q_i.$$

with,

$$M_{12}^K = \frac{4}{3} F_K^2 \eta_2 \bar{B}_K (m_d m_s) \frac{1}{v_1^2} \sum_{a=1}^{3} \left[ \bar{P}_2^{LR} \frac{U_{2a} U_{1a}}{m_{\phi_a}^2} + \bar{P}_1^{SLL} \left( \frac{U_{2a}^2}{m_{\phi_a}^2} + \frac{U_{1a}^2}{m_{\phi_a}^2} \right) \right]$$

### *4.2. Decay $\tau \to \mu\mu\mu$*

The Feynman diagram is

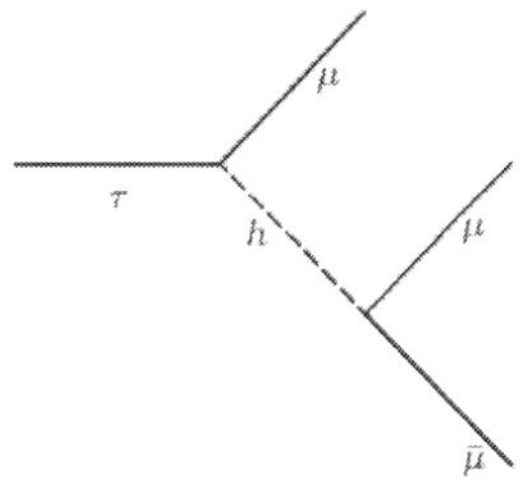


(6)

where,

$$\Gamma(\tau^- \to \mu^- \mu^+ \mu^-) = \frac{m_\tau V_{22}^2 V_{ij}^2}{192\pi^3 r_\phi^2 \left( R_\phi^2 + r_\phi^2 \right)} \left( 1 - 12 r_\mu^3 - 36 r_\mu^{1/2} + 4 r_\mu \right) \quad (7)$$

with,

$$V_{ij}^2 = \left( \frac{1}{2v} \right)^2 \left( - \frac{(y_1)_{ij} \sin(n-\beta)}{\cos\beta} + \frac{(y_2)_{ij} \cos(n-\beta)}{\sin\beta} \right)^2$$

Journal of Physics: Conference Series **651** (2015) 012016 doi:10.1088/1742-6596/651/1/012016

### 4.3. Decay $B_s \to \mu^+\mu^-$

The Feynman diagram representing is,

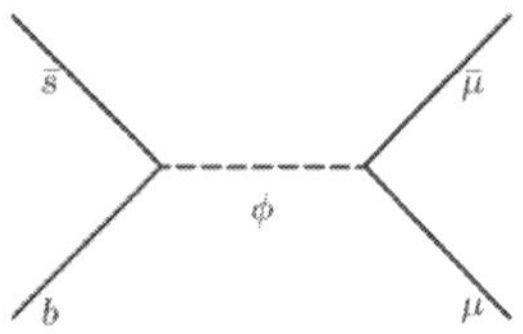


(8)

where,

$$\mathbf{Br}\left(B_s \to \mu^+\mu^-\right) = \frac{G_F^2\alpha_{em}^2}{16\pi^3} M_B\tau_B \left|V_{ts}V_{tb}^*\right|^2 \sqrt{1-\frac{4m_l^2}{M_B^2}} \times \left[\left|F_s^d\right|^2\left(1-\frac{4m_l^2}{M_B^2}\right) + \left|F_p^d + 2m_l F_A^d\right|^2\right]$$

the Wilson coefficients are,

$$C_p = \frac{2\pi\sqrt{m_2^q m_3^q}\, m_2^l}{G_F V_{tb}^* V_{ts}\alpha_{em} M_b\, v^2}\left[\frac{1}{4M_{A^0}^2}\prod_{r=q,l}\left(\left(\chi_r^{(2)}\right)_{23} g_3(\alpha,\beta) + \left(\chi_r^{(1)}\right)_{23} f_3(\alpha,\beta)\right)\right] \quad (9)$$

and

$$C_s = \frac{2\pi\sqrt{m_2^q m_3^q}\, m_2^l}{G_F V_{tb}^* V_{ts}\alpha_{em} M_b\, v^2}\left[\frac{1}{4M_{H^0}^2}\prod_{r=q,l}\left(\left(\chi_r^{(2)}\right)_{23} g_2(\alpha,\beta) + \left(\chi_r^{(1)}\right)_{23} f_2(\alpha,\beta)\right) + \frac{1}{4M_{h^0}^2}\prod_{r=q,l}\left(\left(\chi_r^{(2)}\right)_{23} g_1(\alpha,\beta) + \left(\chi_r^{(1)}\right)_{23} f_1(\alpha,\beta)\right)\right]$$

## 5. A case Exploring

For illustration we shall display here some of the elements of the $\chi$ factors for case 5, the other cases can find in [42]. In general, each component has a complicated expression relating the masses of the particles and our parameter $\gamma$. For instance, the 12-Yukawa elements are given by,

$$\left(\chi_1\right)_{12} = \sqrt{\frac{1}{1-\frac{m_2\gamma}{m_3}}}\left(1-\frac{|m_1|}{m_2}\right) - \left(\frac{m_1}{m_2}+\gamma+1\right) \qquad \left(\chi_2\right)_{12} = 2\gamma + \left(\frac{m_2}{m_3}\gamma - 1\right)\gamma$$

We obtain the figure 1

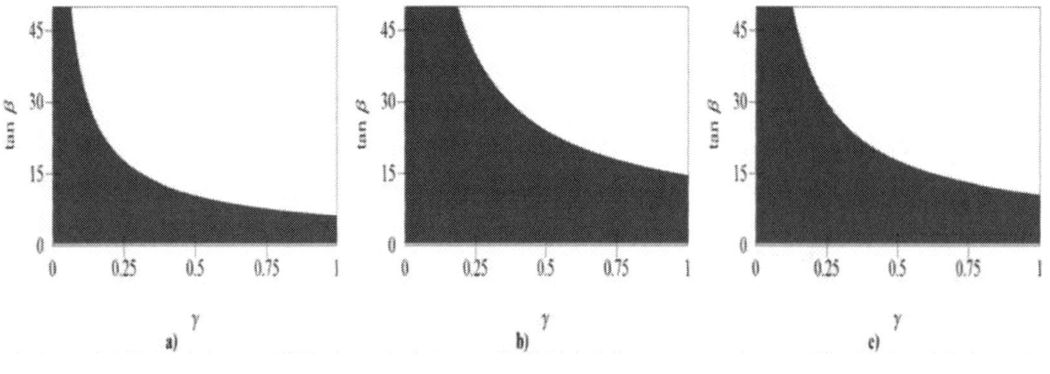


**Figure 1.** Allowed regions (blue) in the plane $\tan\beta - \gamma$ for case 6, where a) $\alpha - \beta = 0$, b) $\alpha - \beta = \pi/2$ and c)$\alpha - \beta = \pi/3$.

Figure 1 shows the allowed regions (blue) in the plane $\tan\beta-\gamma$ for case 5, where a) $\alpha-\beta=0$, b) $\alpha-\beta=\pi/2$ and c)$\alpha-\beta=\pi/3$. Thus, we find that the Cheng-Sher ansatz still holds, up to corrections, for case 5 of complementary Yukawa matrices; and this remains valid for all the cases we are considering here.

## 6. Results and Conclusions

We predict that the LFV Higgs decays ($h\to\tau\mu$), as well as the rare top decay $t\to c+h$, could reach significant levels, which may be searched at the coming phases of LHC.
Branching ratios for Case 5 where: $\alpha-\beta=\frac{\pi}{3}$ and the first row corresponds to $\tan\beta=10$, the second row $\tan\beta=15$ and the third row $\tan\beta=20$ . The second part of the table has an analogous arrangement of $\tan\beta$.

| | $\gamma=0.1$ | $\gamma=0.2$ | $\gamma=0.3$ |
|---|---|---|---|
| | 1.5 | – | – |
| $\text{Br}(h\to\tau\mu)\times10^5$ | – | 4.0 | – |
| | – | 4.2 | 7.2 |
| | 1.4 | – | – |
| $\text{Br}(t\to ch)\times10^5$ | – | 8.1 | – |
| | – | 1.5 | 2.6 |

## Acknowledgments

We also acknowledge support from CONACYT-SNI (Mexico).

## References

[1] G. Aad *et al.* [ATLAS Collaboration], Phys. Lett. B **716**, 1 (2012) [arXiv:1207.7214 [hep-ex]].
[2] S. Chatrchyan *et al.* [CMS Collaboration], Phys. Lett. B **716**, 30 (2012) [arXiv:1207.7235 [hep-ex]].
[3] J. F. Gunion, H. E. Haber, G. L. Kane and S. Dawson, Front. Phys. **80**, 1 (2000).
[4] J. L. Diaz-Cruz and D. A. Lopez-Falcon, Phys. Lett. B **568**, 245 (2003) [hep-ph/0304212].
[5] N. G. Deshpande and E. Ma, Phys. Rev. D **18**, 2574 (1978).
[6] S. P. Martin, In *Kane, G.L. (ed.): Perspectives on supersymmetry II* 1-153 [hep-ph/9709356].
[7] M. Quiros, hep-ph/0606153.
[8] A. Pomarol, CERN Yellow Report CERN-2012-001, 115-151 [arXiv:1202.1391 [hep-ph]].
[9] A. Aranda, J. L. Diaz-Cruz, J. Hernandez-Sanchez and R. Noriega-Papaqui, Phys. Lett. B **658**, 57 (2007) [arXiv:0708.3821 [hep-ph]].
[10] G. C. Branco, P. M. Ferreira, L. Lavoura, M. N. Rebelo, M. Sher and J. P. Silva, Phys. Rept. **516**, 1 (2012) [arXiv:1106.0034 [hep-ph]].
[11] E. Ma, Mod. Phys. Lett. A **21**, 1777 (2006) [hep-ph/0605180];
R. Barbieri, L. J. Hall and V. S. Rychkov, Phys. Rev. D **74**, 015007 (2006) [hep-ph/0603188];
L. Lopez Honorez, E. Nezri, J. F. Oliver and M. H. G. Tytgat, JCAP **0702**, 028 (2007) [hep-ph/0612275];
I. F. Ginzburg, K. A. Kanishev, M. Krawczyk and D. Sokolowska, Phys. Rev. D **82**, 123533 (2010) [arXiv:1009.4593 [hep-ph]].
[12] I. F. Ginzburg and M. Krawczyk, Phys. Rev. D **72**, 115013 (2005) [hep-ph/0408011].
[13] J. F. Gunion and H. E. Haber, Nucl. Phys. B **307**, 445 (1988) [Erratum-ibid. B **402**, 569 (1993)].
[14] S. L. Glashow and S. Weinberg, Phys. Rev. D **15**, 1958 (1977).
[15] G. C. Branco, D. Emmanuel-Costa and R. Gonzalez Felipe, Phys. Lett. B **477**, 147 (2000) [hep-ph/9911418].
[16] L. J. Hall and S. Weinberg, Phys. Rev. D **48**, 979 (1993) [hep-ph/9303241].
[17] W. Altmannshofer, S. Gori and G. D. Kribs, Phys. Rev. D **86**, 115009 (2012) [arXiv:1210.2465 [hep-ph]].
[18] A. J. Buras, M. V. Carlucci, S. Gori and G. Isidori, JHEP **1010**, 009 (2010) [arXiv:1005.5310 [hep-ph]].
[19] G. D'Ambrosio, G. F. Giudice, G. Isidori and A. Strumia, Nucl. Phys. B **645**, 155 (2002) [hep-ph/0207036].
[20] A. Aranda, C. Bonilla and J. L. Diaz-Cruz, Phys. Lett. B **717**, 248 (2012) [arXiv:1204.5558 [hep-ph]].
[21] A. Pich and P. Tuzon, Phys. Rev. D **80**, 091702 (2009) [arXiv:0908.1554 [hep-ph]].
[22] T. P. Cheng and M. Sher, Phys. Rev. D **35**, 3484 (1987).
[23] Y. -F. Zhou, J. Phys. G **30**, 783 (2004) [hep-ph/0307240].

[24] J. L. Diaz-Cruz, R. Noriega-Papaqui and A. Rosado, Phys. Rev. D **69**, 095002 (2004) [hep-ph/0401194].
[25] J. L. Diaz-Cruz, R. Noriega-Papaqui and A. Rosado, Phys. Rev. D **71**, 015014 (2005) [hep-ph/0410391].
[26] Y. -L. Wu and Y. -F. Zhou, Eur. Phys. J. C **36**, 89 (2004) [hep-ph/0403252].
[27] W. -j. Li, Y. -d. Yang and X. -d. Zhang, Phys. Rev. D **73**, 073005 (2006) [hep-ph/0511273].
[28] A. E. Carcamo Hernandez, R. Martinez and J. A. Rodriguez, Eur. Phys. J. C **50**, 935 (2007) [hep-ph/0606190].
[29] M. Gomez-Bock and R. Noriega-Papaqui, J. Phys. G **32**, 761 (2006) [hep-ph/0509353].
[30] A. Denner, S. Heinemeyer, I. Puljak, D. Rebuzzi and M. Spira, Eur. Phys. J. C **71**, 1753 (2011) [arXiv:1107.5909 [hep-ph]].
[31] A. Pilaftsis, Phys. Lett. B **285**, 68 (1992).
[32] J. L. Diaz-Cruz and J. J. Toscano, Phys. Rev. D **62**, 116005 (2000) [hep-ph/9910233].
[33] T. Han and D. Marfatia, Phys. Rev. Lett. **86**, 1442 (2001) [hep-ph/0008141];
K. A. Assamagan, A. Deandrea and P. -A. Delsart, Phys. Rev. D **67**, 035001 (2003) [hep-ph/0207302];
D. McKeen, M. Pospelov and A. Ritz, Phys. Rev. D **86**, 113004 (2012) [arXiv:1208.4597 [hep-ph]].
[34] J. L. Diaz-Cruz, JHEP **0305**, 036 (2003) [hep-ph/0207030];
A. Brignole and A. Rossi, Nucl. Phys. B **701**, 3 (2004) [hep-ph/0404211].
E. Arganda, A. M. Curiel, M. J. Herrero and D. Temes, Phys. Rev. D **71**, 035011 (2005) [hep-ph/0407302].
S. Kanemura, T. Ota and K. Tsumura, Phys. Rev. D **73**, 016006 (2006) [hep-ph/0505191].
[35] J. L. Diaz-Cruz, D. K. Ghosh and S. Moretti, Phys. Lett. B **679**, 376 (2009) [arXiv:0809.5158 [hep-ph]].
M. Arana-Catania, E. Arganda and M. J. Herrero, arXiv:1304.3371 [hep-ph];
A. Arhrib, Y. Cheng and O. C. W. Kong, Phys. Rev. D **87**, 015025 (2013) [arXiv:1210.8241 [hep-ph]];
R. Harnik, J. Kopp and J. Zupan, JHEP **1303**, 026 (2013) [arXiv:1209.1397 [hep-ph]].
A. Goudelis, O. Lebedev and J. -h. Park, Phys. Lett. B **707**, 369 (2012) [arXiv:1111.1715 [hep-ph]].
[36] G. Eilam, J. L. Hewett and A. Soni, Phys. Rev. D **44**, 1473 (1991) [Erratum-ibid. D **59**, 039901 (1999)].
[37] B. Mele, S. Petrarca and A. Soddu, Phys. Lett. B **435**, 401 (1998) [hep-ph/9805498].
[38] S. Bejar, J. Guasch and J. Sola, hep-ph/0101294.
[39] B. Mele, hep-ph/0003064.
[40] J. L. Diaz-Cruz, H. -J. He and C. P. Yuan, Phys. Lett. B **530**, 179 (2002) [hep-ph/0103178].
[41] J. A. Aguilar-Saavedra and G. C. Branco, Phys. Lett. B **495**, 347 (2000) [hep-ph/0004190].
[42] M. Arroyo, J. L. Diaz-Cruz, E. Diaz and J. A. Orduz-Ducuara, arXiv:1306.2343 [hep-ph].

Journal of Physics: Conference Series **651** (2015) 012017 doi:10.1088/1742-6596/651/1/012017

# Scalar dark matter and its connection with neutrino physics

**E. Peinado**

INFN, Laboratori Nazionali di Frascati,
Via Enrico Fermi 40, I-00044 Frascati, Italy

E-mail: epeinado@lnf.infn.it

**Abstract.** The existence of non-baryonic Dark Matter is well established by cosmological and astrophysical probes, however its detailed nature still remains elusive. Among the extensions of the Standard Model (SM) explaining the DM relic abundance, the simplest one is the inert dark matter, where a scalar eld is added to the Standard Model which is stabilized by a $Z_2$ symmetry. We intend to give a brief review of this scenario and its possible connection with neutrino physics. In particular the discrete dark matter mechanism will be outlined. This mechanism consists in an extended SM with a non-Abelian flavor symmetry. When the flavor symmetry is spontaneously broken by the electroweak symmetry breaking mechanism, it explains the neutrino mixing patterns and at the same time renders the dark matter stable.

## 1. Introduction

on-baryonic Dark atter (D ) is one of the most compelling problems of modern cosmology. Despite the fact that its existence is well established by cosmological and astrophysical probes, its nature remains elusive. Still, observations can constraint the properties of dark matter and give some hints about its identity. For instance, a fundamental requirement for a viable dark matter candidate is the stability over cosmological times, that is $\approx 10^{18}$ seconds, and depending on the nature of this, if it decays causes uxes of photons, anti-protons and/or positrons larger than the observed its life-time should be larger than $\approx 10^{26}$ seconds. This suggests the existence of an exact or slightly-broken symmetry protecting or suppressing its decay.

The inert dark matter model [1] is a minimal extension of the S that can account for the dark matter in the universe. This model contains an extra $SU(2)$ iggs doublet which is odd under a $Z_2$ symmetry. This symmetry render stable the lightest member of the $SU(2)$ multiplet which in order to be a good candidate should be electrically neutral.

As the S is even respect to this $Z_2$ symmetry, the only interactions of this new scalar fiend, lets call it $\eta$ will be with the gauge bosons, through the kinetic term and with the iggs field through the quartic coupling. Therefore, the quartic coupling with the iggs will give us the so-called iggs portal . This is important because this can lead to the direct detection of the D in the elastic scattering with the nucleons. This extension has been studied extensively in the past years, see for instance [2, 3, 4, 5, 6]. In this model there are basically two allowed regions of masses for the D candidate, one which goes from few GeVs to $100's$ GeV, and another region of heavy masses [7]. This is basically because in the iggs potential, there are two $\mu$ parameters (the quadratic terms in the potential), one for the S iggs and the other one for the inert iggs. The fact that the S iggs acquire vev give us the constraint that we can write this $\mu$ parameter as a combination of the quartic couplings and the vev of the iggs, therefore this $\mu$ term is related to the electroweak scale. On the other hand, as the vev of the inert is zero, the $\mu$ term for this is free and it can push the mass of this inert scalar to higher scales. We will come back to this point in the discussion of the models with avor symmetries.

## 2. The inert scalar model and its connection with neutrino masses

In this section we will discuss brie y the scotogenic model which is an extension of the inert model introduced by Ernest a [8]. In this model the inclusion of some right-handed (R ) neutrinos odd under the $Z_2$ symmetry play a role in the generation of the light neutrino mass. This of course will give us a Dirac interaction with the Left- anded leptons. As one condition for the inert D is that the neutral component has zero vev. This will not contribute to the Dirac mass of the neutrino. The neutrino mass is generated through a loop, where inside the loop enters the inert sector, namely the inert scalar and the right-handed neutrino, see figure 1.

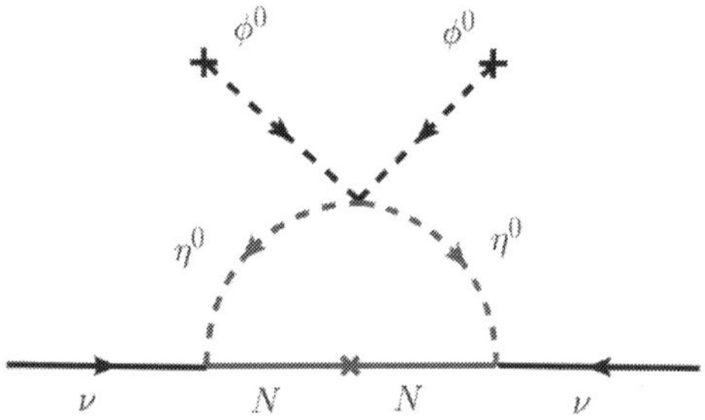


**Figure 1.** eutrino mass generation in the Scotogenic model. The fields charged under the $Z_2$ symmetry are in blue.

| | $L$ | $l^c$ | $N$ | $\Phi$ | $\eta$ |
|---|---|---|---|---|---|
| $SU(2)$ | 2 | 1 | 1 | 2 | 2 |
| $Z_2$ | + | + | − | + | − |

**Table 1.** Summary of relevant model quantum numbers

The charge assignment under the $SU(2)_L \times U(1)_Y \times Z_2$ of the relevant matter fields is as shown in Table 1. With this charge assignment, the lepton Yukawa Lagrangian is as follows [8]:

$$\mathcal{L}_Y = f_{ij} L \Phi l^c_j + h_{ij} L \eta N_j + H.c. \quad (1)$$

The a orana mass term for the Right-handed neutrino is

$$\frac{1}{2} M_i N_i N_i + H.c.$$

The iggs potential in this model is

$$\begin{aligned} V &= m_1^2 \Phi^\dagger \Phi + m_2^2 \eta^\dagger \eta + \frac{1}{2}\lambda_1 (\Phi^\dagger \Phi)^2 + \frac{1}{2}\lambda_2 (\eta^\dagger \eta)^2 + \lambda_3 (\Phi^\dagger \Phi)(\eta^\dagger \eta) \\ &+ \lambda_4 (\Phi^\dagger \eta)(\eta^\dagger \Phi) + \frac{1}{2}\lambda_5 [(\Phi^\dagger \eta)^2 + H.c.]. \quad (2) \end{aligned}$$

The relevant term in this potential for the neutrino mass is the term proportional to $\lambda_5$. The iggs field $\Phi$ acquire a vev breaking the electroweak symmetry while the field $\phi$ remains with zero vev. With this the $Z_2$ symmetry remains a symmetry of the theory stabilizing the lightest particle odd under this symmetry, which we are assuming is the scalar $\phi^0$[1]. The diagram of Fig. 1 give the following contribution to the neutrino mass

$$(\mathcal{M}_\nu)_{ij} = \sum_k \frac{h_{ik}h_{jk}M_k}{16\pi^2}\left[\frac{m_R^2}{m_R^2 - M_k^2}\ln\frac{m_R^2}{M_k^2} - \frac{m_I^2}{m_I^2 - M_k^2}\ln\frac{m_I^2}{M_k^2}\right], \tag{3}$$

where $m_R$ and $m_I$ are the masses of $\sqrt{2}Re\eta^0$ and $\sqrt{2}Im\eta^0$ respectively. If $m_R^2 - m_I^2 = 2\lambda_5 v^2$ is assumed to be small compared to $m_0^2 = (m_R^2 + m_I^2)/2$, then

$$(\mathcal{M}_\nu)_{ij} = \frac{\lambda_5 v^2}{8\pi^2}\sum_k \frac{h_{ik}h_{jk}M_k}{m_0^2 - M_k^2}\left[1 - \frac{M_k^2}{m_0^2 - M_k^2}\ln\frac{m_0^2}{M_k^2}\right]. \tag{4}$$

If $M_k^2 >> m_0^2$, then

$$(\mathcal{M}_\nu)_{ij} = \frac{\lambda_5 v^2}{8\pi^2}\sum_k \frac{h_{ik}h_{jk}}{M_k}\left[\ln\frac{M_k^2}{m_0^2} - 1\right]. \tag{5}$$

If $m_0^2 \simeq M_k^2$, then

$$(\mathcal{M}_\nu)_{ij} \simeq \frac{\lambda_5 v^2}{16\pi^2}\sum_k \frac{h_{ik}h_{jk}}{M_k}. \tag{6}$$

This is a possible scenario where the D is connected with the a orana neutrino mass generation. In the following section we will discus another possibility where the D stability is closed related with the neutrino masses and mixing pattern.

## 3. Discrete dark matter mechanism

In the past 10 years or so, avor symmetries has been very popular mainly because they can lead to the prediction of some mixing patterns, the tri-bimaximal [9], bi-maximal [10], Golden ratio [11, 12], for a review see for instance [13]. ost popular models where based in $A_4$ [14, 15] basically because this is the smallest group that contains a triplet irreducible representation, which can accommodate the three leptons.

The discrete dark matter (DD ) mechanism contains a stable particle, where its stability is obtained by means a residual $Z_N$ symmetry. This residual symmetry arise from the spontaneously broken non-Abelian discrete avor symmetry. In the first model, the group of even permutations of four ob ects was used, $A_4$, and it was spontaneously broken by the electroweak symmetry breaking mechanism into a $Z_2$ sub-group, $A_4 \to Z_2$ [16].

The model is obtained by adding to the S three iggs doublets which transform as a triplet under $A_4$, and four right handed neutrinos transforming as a singlet and a triplet of the avor group. In table 3 we present the relevant quantum numbers for the matter fields.

The group of even permutations of four ob ects, $A_4$, has two generators: S, and T, which obeys the properties $S^2 = T^3 = (ST)^3 = I$. $S$ is a $Z_2$ generator while $T$ is a $Z_3$ generator. $A_4$ has four irreducible representations, three singlets 1, 1′, and 1″ and one triplet. The generators of $A_4$ for each irreducible representation are

$$\begin{array}{lll} 1 & S=1 & T=1 \\ 1' & S=1 & T=\omega^2 \\ 1'' & S=1 & T=\omega \end{array} \quad 3: \; S = \begin{pmatrix} 1 & 0 & 0 \\ 0 & -1 & 0 \\ 0 & 0 & -1 \end{pmatrix} \quad T = \begin{pmatrix} 0 & 1 & 0 \\ 0 & 0 & 1 \\ 1 & 0 & 0 \end{pmatrix} \tag{7}$$

[1] There is the possibility of the $N$ to be the DM candidate when it is the lightest particle charged under the $Z_2$ symmetry [8]

| | $L_e$ | $L_\mu$ | $L_\tau$ | $l_e^c$ | $l_\mu^c$ | $l_\tau^c$ | $N_T$ | $N_4$ | $H$ | $\eta$ |
|---|---|---|---|---|---|---|---|---|---|---|
| $SU(2)$ | 2 | 2 | 2 | 1 | 1 | 1 | 1 | 1 | 2 | 2 |
| $A_4$ | 1 | $1'$ | $1''$ | 1 | $1''$ | $1'$ | 3 | 1 | 1 | 3 |

**Table 2.** Summary of relevant model quantum numbers

where $\omega^3 = 1$. The Yukawa Lagrangian of the model is

$$\begin{aligned} \mathcal{L} &= y_e L_e l_e^c H + y_\mu L_\mu l_\mu^c H + y_\tau L_\tau l_\tau^c H + +y_1^\nu L_e (N_T\eta)_1 + y_2^\nu L_\mu (N_T\eta)_{1''} + \\ &\quad +y_3^\nu L_\tau (N_T\eta)_{1'} + y_4^\nu L_e N_4 H + M_1 N_T N_T + M_2 N_4 N_4 + \text{h.c.} \end{aligned} \quad (8)$$

This way $H$ is responsible for quark and charged lepton masses, the latter automatically diagonal[2]. eutrino masses arise from $H$ and $\eta$. One solution for the minimum of the iggs potential [16] is

$$\langle H^0 \rangle = v_h \neq 0, \quad \langle \eta_1^0 \rangle = v_\eta \neq 0 \quad \langle \eta_{2,3}^0 \rangle = 0\,, \quad (9)$$

which means the vev alignment for the $A_4$ triplet of the form $\langle\eta\rangle \sim (1,0,0)$. This alignment is invariant under the $S$ generator[3], see eq. (7), which means that the minimum of the potential breaks spontaneously $A_4$ into a $Z_2$ subgroup generated by $S$. All the fields in the model singlets under $A_4$ are even under the residual $Z_2$, the triplets transform as:

$$\begin{array}{lll} N_1 \to +N_1\,, & \eta_1 \to +\eta_1 \\ N_2 \to -N_2\,, & \eta_2 \to -\eta_2 \\ N_3 \to -N_3\,, & \eta_3 \to -\eta_3. \end{array} \quad (10)$$

The D candidate is the lightest particle charged under $Z_2$ i. e. the lightest combination of the scalars $\eta_2$ and $\eta_3$, which we will denote generically by $\eta_{DM}$. We list below all interactions of $\eta_{DM}$:

(i) Yukawa interactions

$$\eta_{DM}\, \overline{\nu}_i N_{2,3}\,, \quad (11)$$

where $i = e,\ \mu,\ \tau$.

(ii) iggs-Vector boson couplings

$$\begin{array}{ll} \eta_{DM}\eta_{DM} ZZ\,, & \eta_{DM}\eta_{DM} WW\,, \\ \eta_{DM}\eta_{2,3}^{\pm} W^{\pm} Z\,, & \eta_{DM}\eta_{2,3}^{\pm} W^{\pm}\,, \\ \eta_{DM} A_{2,3} Z\,. & \end{array} \quad (12)$$

(iii) Scalar interactions from the iggs potential:

$$\begin{array}{ll} \eta_{DM}\, A_1 A_2 h\,, & \eta_{DM}\, A_1 A_3 h_1\,, \\ \eta_{DM}\, A_1 A_2 h_1\,, & \eta_{DM}\, A_1 A_3 h\,, \\ \eta_{DM}\, A_2 A_3 h_3\,, & \eta_{DM}\, h_1 h_3 h \\ \eta_{DM}\eta_{DM} hh\,, & \eta_{DM}\eta_{DM} h_1 h_1\,. \end{array} \quad (13)$$

After electroweak symmetry breaking, the vevs the iggs fields acquire vacuum expectation values, $v_h$ and $v_\eta$ for the singlet and the first component of the triplet respectively, additional terms are obtained from those in Eq. (13) by replacing $h \to v_h$ and $h_1 \to v_\eta$. These vertices

[2] For quark mixing angles generated through higher dimension operators see reference [1 ].
[3] $H_s$ is in the 1 representation of $A_4$ and its vev also respect the generator $S$.

Journal of Physics: Conference Series **651** (2015) 012017 doi:10.1088/1742-6596/651/1/012017

are relevant for direct detection [18]. The phenomenology of dark matter of this model has been studied in detail in [18]. The model accommodates W AP and collider constraits and is consistent with Xenon100 and CD S on one hand and CoGent or DA A on the other hand [18].

The allowed region with this constraints for the D mass vs the mass of the iggs[4] is presented in figure 2

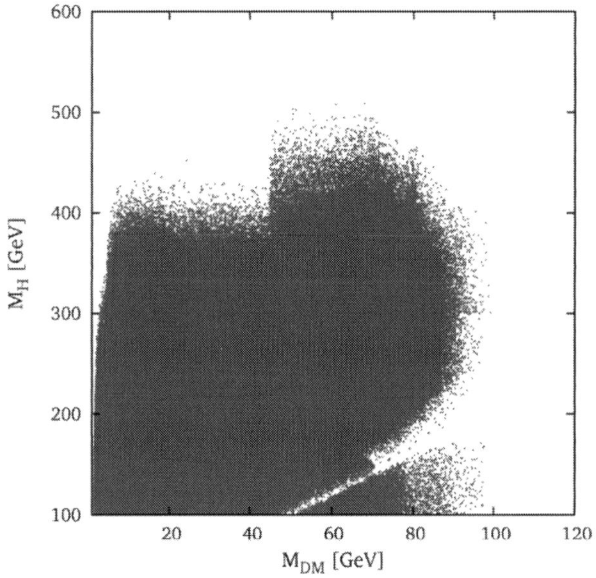


**Figure 2.** Regions in the plane D mass ($M_{DM}$) - lightest iggs boson $M_H$ allowed by collider constraints and leading to a D relic abundance compatible with W AP measurements.

The plot for direct D detection is presented in Figure [?]

## 4. Neutrino masses phenomenology

The model contains four heavy right-handed neutrinos it is a special case, called (3,4), of the general type-I seesaw mechanism. After electroweak symmetry breaking, it is characterized by Dirac and a orana mass-matrix:

$$m_D = \begin{pmatrix} x_1 & 0 & 0 & x_4 \\ x_2 & 0 & 0 & 0 \\ x_3 & 0 & 0 & 0 \end{pmatrix}, \quad M_R = \begin{pmatrix} M_1 & 0 & 0 & 0 \\ 0 & M_1 & 0 & 0 \\ 0 & 0 & M_1 & 0 \\ 0 & 0 & 0 & M_2 \end{pmatrix}. \tag{14}$$

where $x_1, x_2, x_3$ and $x_4$ are respectively proportional to $y_1^\nu$, $y_2^\nu$, $y_3^\nu$ and $y_4^\nu$ of eq. (16) and are of the order of the electroweak scale, while $M_{1,2}$ are assumed to be close to the unification scale. Light neutrinos get a orana masses by means of the type-I seesaw relation and the light-neutrinos mass matrix has the form:

$$m_\nu = -m_{D_{3\times4}} M^{-1}_{R_{4\times4}} m^T_{D_{3\times4}} \equiv \begin{pmatrix} y^2 & ab & ac \\ ab & b^2 & bc \\ ac & bc & c^2 \end{pmatrix}. \tag{15}$$

[4] When this paper was published the mass of the Higgs was unknown, but as we can see from the gure the range of the DM mass remains more or less the same

Journal of Physics: Conference Series **651** (2015) 012017 doi:10.1088/1742-6596/651/1/012017

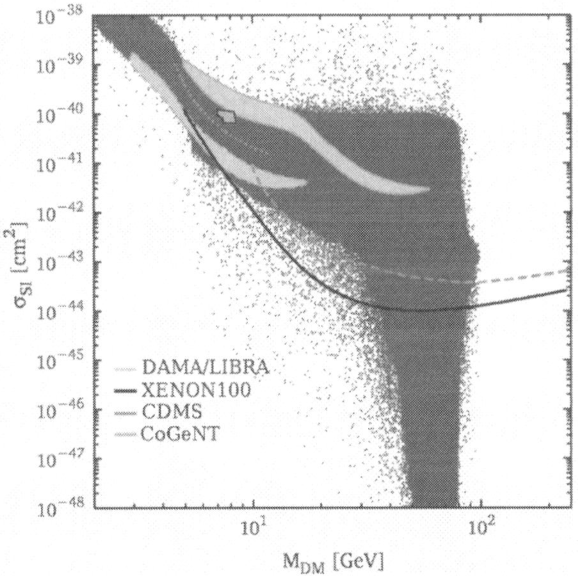


**Figure 3.** Spin-independent D scattering cross section off-protons as a function of the dark matter mass. The orange regions delimited by the dashed (solid) line show the DA A/LIBRA annual modulation regions including (neglecting) the channeling effect [19]. The green region corresponds to the COGE T data [20]. Dashed and dotted red lines correspond to the upper bound from CD S (respectively from [21] and [22]). XE O 100 bounds [23] are shown as a solid red line.

This texture of the light neutrino mass matrix has a null eigenvalue $m_3 = 0$ corresponding to the eigenvector $(0, -b/c, 1)^T$, [5] implying a vanishing reactor mixing angle $\theta_{13} = 0$ and inverse hierarchy. The atmospheric angle, the solar angle and the two square mass differences can be fitted. The model implies a neutrinoless double beta decay effective mass parameter in the range 0.03 to 0.05 eV at 3 $\sigma$, within reach of upcoming experiments. The fact that only two R neutrinos are $Z_2$ even tell us that basically in the even sector we have a two R neutrino see-saw and this is the reason why we have a massless state.

In a modification of this model [24] we found that as is expected, by adding an extra R neutrino we can generate a mass for the lightest state. In Tab.(3) we summarized the model quantum numbers. In contrast to [16], the right-handed neutrino $N_4$ is assigned to $1'$ instead of 1 and we introduced one more right-handed neutrino $N_5$ assigned to $1''$.

| | $L_e$ | $L_\mu$ | $L_\tau$ | $l_e^c$ | $l_\mu^c$ | $l_\tau^c$ | $N_T$ | $N_4$ | $N_5$ | $H$ | $\eta$ |
|---|---|---|---|---|---|---|---|---|---|---|---|
| $SU(2)$ | 2 | 2 | 2 | 1 | 1 | 1 | 1 | 1 | 1 | 2 | 2 |
| $A_4$ | 1 | $1'$ | $1''$ | 1 | $1''$ | $1'$ | 3 | $1'$ | $1''$ | 1 | 3 |

**Table 3.** *Summary of relevant model quantum numbers*

[5] Note that if we were to stick to the minimal (3,3)-type-I seesaw scheme, with ust 3 SU(2) singlet states, one would nd a pro ective nature of the effective tree-level light neutrino mass matrix with two zero eigenvalues, hence phenomenologically inconsistent. That is why we adopted the (3,4) scheme.

The operators needed to generate neutrinos masses are the following:

$$\begin{aligned}\mathcal{L} &= y_1^\nu L_e(N_T\eta)_1 + y_2^\nu L_\mu(N_T\eta)_{1''} + y_3^\nu L_\tau(N_T\eta)_{1'} + \\ &+ y_4^\nu L_\tau N_4 H + y_5^\nu L_\mu N_5 H + M_1 N_T N_T + M_2 N_4 N_5 + \text{h.c.}\end{aligned}$$

The scalar and the lepton charged current sectors of our model are the same as in [16] and we refer to that paper for all the details. After electroweak symmetry breaking, two of the iggs doublets contained in the $A_4$ triplet $\eta$ do not take vev and we have:

$$\langle H^0\rangle = v_h \neq 0, \qquad \langle \eta_1^0\rangle = v_\eta \neq 0 \qquad \langle \eta_{2,3}^0\rangle = 0\,. \tag{16}$$

As a consequence, the Dirac and a orana neutrino mass matrices are:

$$m_D = \begin{pmatrix} y_1^\nu v_\eta & 0 & 0 & 0 & 0 \\ y_2^\nu v_\eta & 0 & 0 & 0 & y_5^\nu v_h \\ y_3^\nu v_\eta & 0 & 0 & y_4^\nu v_h & 0 \end{pmatrix} \qquad m_M = \begin{pmatrix} M_1 & 0 & 0 & 0 & 0 \\ 0 & M_1 & 0 & 0 & 0 \\ 0 & 0 & M_1 & 0 & 0 \\ 0 & 0 & 0 & 0 & M_2 \\ 0 & 0 & 0 & M_2 & 0 \end{pmatrix}. \tag{17}$$

sing the See-Saw type-I formula we get the following light neutrino mass matrix:

$$m_\nu = \begin{pmatrix} a^2 & ab & ac \\ ab & b^2 & bc+k \\ ac & bc+k & c^2 \end{pmatrix}, \tag{18}$$

where we defined:

$$a = \frac{y_1^\nu v_\eta}{\sqrt{M_1}} \qquad b = \frac{y_2^\nu v_\eta}{\sqrt{M_1}} \tag{19}$$

$$c = \frac{y_3^\nu v_\eta}{\sqrt{M_1}} \qquad k = \frac{y_4^\nu y_5^\nu v_h^2}{M_2}. \tag{20}$$

This mass matrix is only compatible with a normal spectrum. Among all possible correlations, we found that the $(\theta_{13}-\theta_{23})$ one is quite interesting, as shown in fig.(4). As we can see, there is a strong correlation between the atmospheric and the reactor mixing angles. The other interesting neutrino observable that we want to discuss is the effective mass $|m_{ee}|$ entering the neutrinoless double beta decay rate. Since this is only compatible with the normal hierarchy, it is expected to be quite small for small $m_{\nu_1}$. The result of our simulation can be found in fig.(5), where we can see that a lower bound $(|m_{\nu_1}|, |m_{ee}|) \sim (2\cdot 10^{-3}, 4\cdot 10^{-4})\, eV$ can be set.

The extension of this mechanism in such a way we can include the quarks is not trivial, we did some attempts, first with the group $D_4$ where the quarks transform as different singlets representation [30]. In that model some avor changing neutral currents are induced but under control. Again in that model a massless neutrino state is predicted as well as a zero reactor mixing angle. All this models are ruled out by the results of Daya-Bay on the reactor neutrino mixing angle.

## 5. A Discrete Dark Matter model for quarks and leptons

We where looking for a group $G$ that contains at least two irreducible representations of dimension larger than one, namely $r_a$ and $r_b$. We also require that all the components of the irreducible representation $r_a$ transform trivially under an abelian subgroup of $G \supset Z_N$ (with $N = 2, 3$) while at least one component of the irreducible representation $r_b$ is charged with under

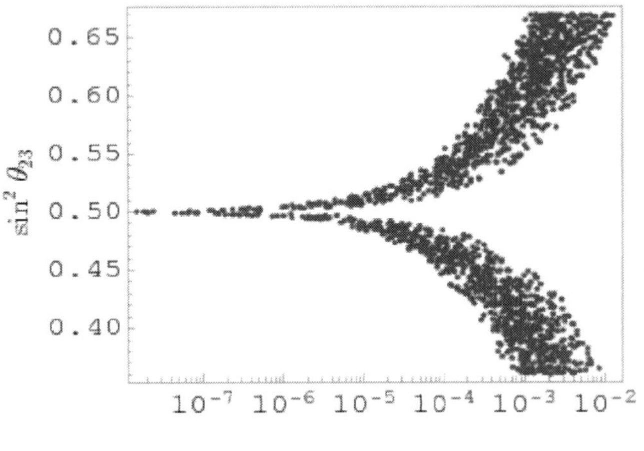


**Figure 4.** *Correlation among the $\theta_{13}$ and $\theta_{23}$ angles as predicted by the model.*

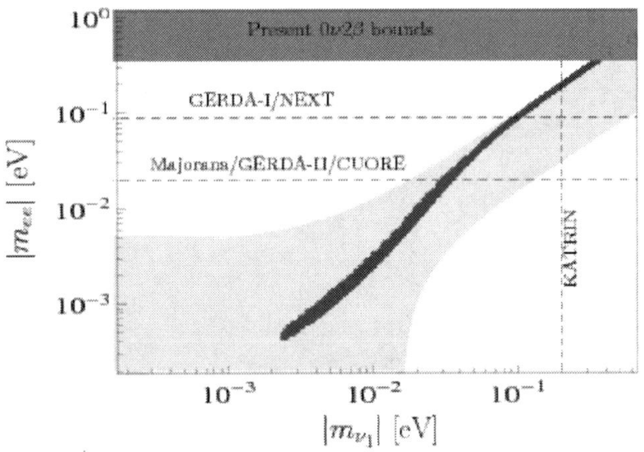


**Figure 5.** *Values of the effective mass $|m_{ee}|$ (in eV) allowed in our model. The two dashed horizontal lines represent the experimental sensitivities of some of the forthcoming experiments while the dashed vertical line is the upper limit from tritium $\beta$-decay experiment. For references to experiments see [25, 26, 27, 28, 29].*

$Z_N$. The stability of the lightest component of the matter fields transforming as $r_b$ is guaranteed by $Z_N$ giving a potential [6] D candidate.

The smallest group with this property we found is (54), isomorphic to $(Z_3 \times Z_3) \rtimes S_3$. This group contains four triplet irreducible representations, $\mathbf{3_{1,2,3,4}}$, in addition, (54) contains four different doublets $\mathbf{2_{1,2,3,4}}$ and two singlet irreducible representations, $\mathbf{1_{\pm}}$. The product rules for the doublets are as follows:

[6] Other requirements must ful lled in order to have a viable DM candidate, such as neutrality, correct relic abundance, and consistency with constraints from DM search experiments.

- The product of two equal doublets

$$\mathbf{2_k} \times \mathbf{2_k} = \mathbf{1_+} + \mathbf{1_-} + \mathbf{2_k} \tag{21}$$

- The product of two different doublets give us the other two doublets, for instance:

$$\mathbf{2_1} \times \mathbf{2_2} = \mathbf{2_3} + \mathbf{2_4} \tag{22}$$

Of the four doublets $\mathbf{2_1}$ is invariant under the $P \equiv (Z_3 \times Z_3)$ subgroup of (54), while the others transform non-trivially, for example $\mathbf{2_3} \sim (\chi_1, \chi_2)$, which transforms as $\chi_1\,(\omega^2, \omega)$ and $\chi_2\,(\omega, \omega^2)$ respectively, where $\omega^3 = 1$. We can see that by taking $r_a = \mathbf{2_1}$ and $r_b = \mathbf{2_3}$ that (54) is a perfect choice for our purpose [31].

| | $\overline{L}_e$ | $\overline{L}_D$ | $e_R$ | $l_D$ | $H$ | $\chi$ | $\eta$ | |
|---|---|---|---|---|---|---|---|---|
| $SU(2)$ | 2 | 2 | 1 | 1 | 2 | 2 | 2 | 3 |
| (54) | $\mathbf{1_+}$ | $\mathbf{2_1}$ | $\mathbf{1_+}$ | $\mathbf{2_1}$ | $\mathbf{1_+}$ | $\mathbf{2_1}$ | $\mathbf{2_3}$ | $\mathbf{2_1}$ |

**Table 4.** Lepton and higgs boson assignments of the model.

| | $Q_{1,2}$ | $Q_3$ | $(u_R, c_R)$ | $t_R$ | $d_R$ | $s_R$ | $b_R$ |
|---|---|---|---|---|---|---|---|
| $SU(2)$ | 2 | 2 | 1 | 1 | 1 | 1 | 1 |
| (54) | $\mathbf{2_1}$ | $\mathbf{1_+}$ | $\mathbf{2_1}$ | $\mathbf{1_+}$ | $\mathbf{1_-}$ | $\mathbf{1_+}$ | $\mathbf{1_+}$ |

**Table 5.** Quark gauge and avor representation assignments.

In this way we can choose the scalars in the model to belong to the $\mathbf{2_3}$ for instance while the active scalars to belong to the $\mathbf{2_1}$ [31]. The relevant quantum numbers of the model are in Tables 4 and 5. In table 4 $L_D \equiv (L_\mu, L_\tau)$ and $l_D \equiv (\mu_R, \tau_R)$. There are five $SU(2)_L$ doublets of iggs scalars: the $H$ is a singlet of (54), while $\eta = (\eta_1, \eta_2) \sim \mathbf{2_3}$ and $\chi = (\chi_1, \chi_2) \sim \mathbf{2_1}$ are doublets. In order to preserve a remnant $P$ symmetry, the doublet $\eta$ is not allowed to take vacuum expectation value (vev).

Lets start discussing the quark sector. In Ref. [16, 18, 24] quarks were considered blind under the avor symmetry to guarantee the stability of the D . Consequently the generation of quark mixing angles was di cult, see [17].

A nice feature of our current model is that with (54) we can assign quarks to the singlet and doublet representations as shown in table 4, in such a way that we can fit the CK mixing parameters.

The resulting up- and down-type quark mass matrices in our model are given by

$$M_d = \begin{pmatrix} ra_d & rb_d & rd_d \\ -a_d & b_d & d_d \\ 0 & c_d & e_d \end{pmatrix}, \quad M_u = \begin{pmatrix} ra_u & b_u & d_u \\ b_u & a_u & rd_u \\ c_u & rc_u & e_u \end{pmatrix}. \tag{23}$$

where $r = \langle\chi_2\rangle/\langle\chi_1\rangle$. ote that the iggs fields $H$ and $\chi$ are common to the lepton and the quark sectors and in particular the parameter $r$. In order to fit of all quark masses and mixings provided $r$ lies in the range of about $0.1 < r < 0.2$ [31]. We turn to the leptonic sector, the charged leptons are basically equal to the up quarks under the avor symmetry, so the charged lepton mass matrix is of the same form of the up quark mass matrix in Eq. (23). The neutrino masses are generated through the type-II see-saw mechanism [32]. For that we include in the model an $SU_L(2)$ iggs triplet scalar field $\sim \mathbf{2_1}$. Regarding dark matter, note that the lightest $P$-charged particle in $\eta_{1,2}$ can play the role of inert D [2], as it has no direct couplings to matter. The conceptual link between dark matter and neutrino phenomenology arises from the fact that the D stabilizing symmetry is a remnant of the underlying avor symmetry which accounts for the observed pattern of oscillations. Choosing the vev alignment $\langle \quad \rangle \sim (1,1)$ and $\langle \chi_1 \rangle \neq \langle \chi_2 \rangle$, consistent with the minimization of the scalar potential one finds that

$$M_\nu \propto \begin{pmatrix} 0 & \delta & \delta \\ \delta & \alpha & 0 \\ \delta & 0 & \alpha \end{pmatrix}, \tag{24}$$

where $\delta = y_a \langle \quad \rangle$, $\alpha = y_b \langle \quad \rangle$ [31]. ote that this matrix has two free parameters and gives us a neutrino mass sum rule of the form $m_1^\nu + m_3^\nu = m_2^\nu$ in the complex plane, which has implications for neutrinoless double beta decay [33], as seen in Fig. (6).

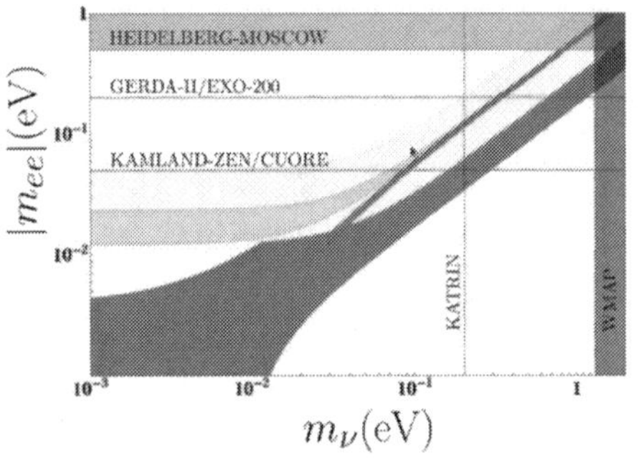


**Figure 6.** Effective neutrinoless double beta decay parameter $m_{ee}$ versus the lightest neutrino mass. The thick upper and lower branches correspond the avor-generic inverse (yellow) and normal (gray) hierarchy neutrino spectra, respectively. The model predictions are indicated by the green and red (darker-shaded) regions, respectively. They were obtained by taking the $3\sigma$ band on the mass squared differences. Only these sub-bands are allowed by the (54) model.

We now turn to the second prediction. For simplicity, we consider in what follows only real parameters and we fix the intrinsic neutrino CP signs [34] as $\eta = diag(-,+,+)$, where $\eta$ is defined so that the CP conservation condition in the charged current weak interaction reads $U^\star = U\eta$, $U$ being the lepton mixing matrix. The correlations we have are presented in Fig. 7.

## 6. Conclusions

We have present here some extensions of the S where the generation of the neutrino mass and the D are connected. First we present the scotogenic model where the neutrino mass is generated at one loop involving the dark sector. We have present tthe DD mechanism, in particular we have discussed the stability of the dark matter particle arises from a avor symmetry. The $A_4$ non-abelian discrete group accounts for both the observed pattern of neutrino mixing and for D stability. We have analyzed the constraints that follow from electroweak precision tests, collider searches and perturbativity. We have also analyzed the prospects for direct and indirect dark matter detection and found that, although the former already excludes a large region in parameter space, we cannot constrain the mass of the D candidate.

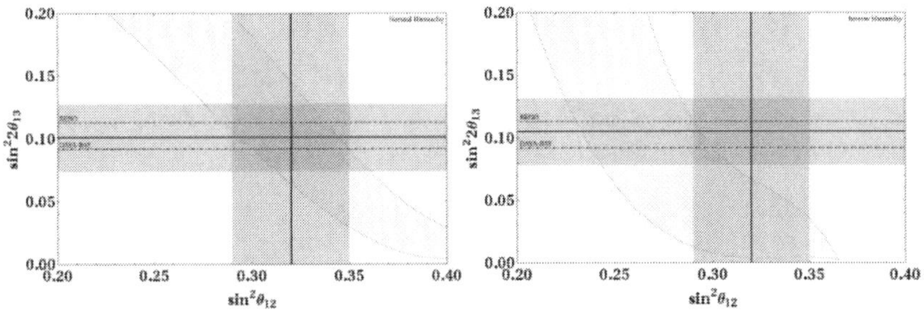


**Figure 7.** The left figure (right figure) is the correlation for normal hierarchy (inverse hierarchy). The shaded (yellow) curved band gives the predicted correlation between solar and reactor angles when the solar and atmospheric squared mass splittings are varied within $2\sigma$ for the normal hierarchy spectrum. The solid (black) line gives the global best fit values for $\theta_{12}$ and $\theta_{13}$, along with the corresponding two-sigma bands, from Ref. [35]. The dashed lines correspond to the central values of the recent published reactor measurements [36, 37]. ote that $\theta_{23}$ is also within $2\sigma$

All of the above relies mainly on the properties of the scalar sector responsible for the breaking of the gauge and avor symmetries. The motivation of our approach is to link the origin of dark matter to the origin of neutrino mass and the understanding of the pattern of neutrino mixing, two of the most outstanding challenges in particle physics today. At this level one may ask what are the possible tests of this idea in the neutrino sector. We have studied other possibilities based on other avor groups, giving as predictions mainly one massless neutrino and a very tiny neutrino reactor mixing angle. When we move to a bigger group where the quarks and charged leptons also transform as non-trivial representation of the avor group, it is possible to generate the correct neutrino mixing parameters and at the same time, the quark mixing angles can be accomodated.

## Acknowledgments

I thank the organizers of the XIV exican Workshop on Particles and Fields for the invitation and the finantial support. I also thank the EPLA ET program for the support during my visit in mexico and the hospitality of CI VESTAV.

## References

[1] N. G. Deshpande and E. Ma, Phys. Rev. D **18**, 25 4 (1 8).
[2] R. Barbieri, L. . Hall and V. S. Rychkov, Phys. Rev. D **74**, 01500 (2006) [hep-ph/0603188].
[3] D. Ma umdar and A. Ghosal, Mod. Phys. Lett. A **23**, 2011 (2008) [hep-ph/060 06 ].
[4] T. Hambye and M. H. G. Tytgat, Phys. Lett. B **659**, 651 (2008) [ar iv:0 0 .0633 [hep-ph]].
[5] L. Lopez Honorez, E. Nezri, . F. Oliver and M. H. G. Tytgat, CAP **0702**, 028 (200 ) [hep-ph/06122 5].
[6] L. Lopez Honorez and C. E. aguna, HEP **1009**, 046 (2010) [ar iv:1003.3125 [hep-ph]].
[ ] M. Cirelli, N. Fornengo and A. Strumia, Nucl. Phys. B **753**, 1 8 (2006) [hep-ph/05120 0].
[8] E. Ma, Phys. Rev. D **73**, 0 301 (2006) [hep-ph/0601225].
[ ] P. F. Harrison, D. H. Perkins and W. G. Scott, Phys. Lett. B **530**, 16 (2002) [hep-ph/02020 4].
[10] G. Altarelli, F. Feruglio and L. Merlo, HEP **0905**, 020 (200 ) [ar iv:0 03.1 40 [hep-ph]].
[11] . a iyama, M. Raidal and A. Strumia, Phys. Rev. D **76** (200 ) 11 301 [ar iv:0 05.455 [hep-ph]].
[12] L. L. Everett and A. . Stuart, Phys. Rev. D **79**, 085005 (200 ) [ar iv:0812.105 [hep-ph]].
[13] C. H. Albright, A. Dueck and W. Rode ohann, Eur. Phys. . C **70**, 10 (2010) [ar iv:1004.2 8 [hep-ph]].
[14] . S. Babu, E. Ma and . W. F. Valle, Phys. Lett. B **552**, 20 (2003) [hep-ph/02062 2].
[15] G. Altarelli and F. Feruglio, Nucl. Phys. B **720**, 64 (2005) [hep-ph/0504165].
[16] M. Hirsch, S. Morisi, E. Peinado and . W. F. Valle, Phys. Rev. D **82**, 116003 (2010) [ar iv:100 .08 1 [hep-ph]].

Journal of Physics: Conference Series **651** (2015) 012017 doi:10.1088/1742-6596/651/1/012017

[1 ] R. de Adelhart Toorop, F. Bazzocchi and S. Morisi, Nucl. Phys. B **856** (2012) 6 0 [ar iv:1104.56 6 [hep-ph]].
[18] M. S. Boucenna, M. Hirsch, S. Morisi, E. Peinado, M. Taoso and . W. F. Valle, HEP **1105** (2011) 03 [ar iv:1101.28 4 [hep-ph]].
[1 ] N. Fornengo, S. Scopel, and A. Bottino, "Discussing direct search of dark matter particles in the Minimal Supersymmetric extension of the Standard Model with light neutralinos, [ar iv:1011.4 43 [hep-ph]].
[20] C. Savage, G. Gelmini, P. Gondolo, and . Freese, " ENON10/100 dark matter constraints in comparison with CoGeNT and DAMA: examining the Leff dependence, [ar iv:1006.0 2 [astro-ph.CO]].
[21] . Ahmed *et al.*, "Dark Matter Search Results from the CDMS II Experiment, *Science*, vol. 32 , pp. 161 1621, 2010.
[22] . Ahmed *et al.*, "Results from a Low-Energy Analysis of the CDMS II Germanium Data, *Phys.Rev.Lett.*, 2010.
[23] E. Aprile *et al.*, "First Dark Matter Results from the ENON100 Experiment, *Phys.Rev.Lett.*, vol. 105, p. 131302, 2010.
[24] D. Meloni, S. Morisi and E. Peinado, Phys. Lett. B **697** (2011) 33 [ar iv:1011.13 1 [hep-ph]].
[25] A. Osipowicz *et al.* [ ATRIN Collaboration], ar iv:hep-ex/010 033.
[26] V. E. Guiseppe *et al.* [Ma orana Collaboration], ar iv:0811.2446 [nucl-ex].
[2 ] A. A. Smolnikov [GERDA Collaboration], ar iv:0812.41 4 [nucl-ex].
[28] A. Giuliani [CUORE Collaboration], . Phys. Conf. Ser. **120** (2008) 052051.
[2 ] . . Gomez-Cadenas, . Martin-Albo, M. Sorel, P. Ferrario, F. Monrabal, . Munoz and P. Novella, ar iv:1010.5112 [hep-ex].
[30] D. Meloni, S. Morisi and E. Peinado, Phys. Lett. B **703** (2011) 281 [ar iv:1104.01 8 [hep-ph]].
[31] M. S. Boucenna, S. Morisi, E. Peinado, . Shimizu and . W. F. Valle, Phys. Rev. D **86** (2012) 0 3008 [ar iv:1204.4 33 [hep-ph]].
[32] . Schechter and . W. F. Valle, Phys. Rev. D **22** (1 80) 222 .
[33] L. Dorame, D. Meloni, S. Morisi, E. Peinado and . W. F. Valle, Nucl. Phys. B **861** (2012) 25 [ar iv:1111.5614 [hep-ph]].
[34] . Schechter and . W. F. Valle, Phys. Rev. D **24** (1 81) 1883 [Erratum-ibid. D **25** (1 82) 283].
[35] M. Tortola, . W. F. Valle, and D. Vanegas, ar iv:1205.4018 which updates previous analysis before the recent reactor results, e. g. T. Schwetz, M. Tortola and . W. F. Valle, New . Phys. **13**, 063004 (2011) New .Phys. **13**, 10 401 (2011).
[36] Daya-Bay Collaboration, F. An *et al.*, 1203.166 , 5 gures.
[3 ] Soo-Bong im for RENO collaboration, 1204.0626.

# QCD Phase Diagram and the Constant Mass Approximation

**A. Ahmad[1,2], A. Ayala[3], A. Bashir[1], E. Gutiérrez[1], A. Raya[1,4].**
[1]Instituto de Física y Matemáticas, Universidad Michoacana de San Nicolás de Hidalgo. Edi cio C-3, Ciudad Universitaria, Morelia 58040, Michoacán, México.
[2]Department of Physics, Gomal University, 2 220, D.I. han, .P. ., Pakistan.
[3]Instituto de Ciencias Nucleares, Universidad Nacional Autónoma de México. Circuito Exterior s/n. Apartado Postal 0-543, C.P. 04510, México, D.F., México.
[4] Facultad De Ciencias, Ponti cia Universidad Católica de Chile, Casilla 306, Santiago 22, Chile.

E-mail: (aftabahmad@ifm.umich.mx, ayala@nucleares.unam.mx, adnan@ifm.umich.mx, enif@ifm.umich.mx, raya@ifm.umich.mx)

**Abstract.** Dynamical generation of quark masses in the infrared region of QCD plays an important role to understand the peculiar nature of the physics of hadrons. As it is known, the solution of QCD gap equation for the quark mass function is flat for low momentum, but smoothly evolves to the perturbative behavior at high momentum. In this work, we use an effective truncation of QCD gap equation valid up to 1 GeV, and implement it at nite temperature and chemical potential to understand the QCD phase diagram for chiral symmetry breaking-chiral symmetry restoration, and con nement-decon nement phase transitions from the Schwinger-Dysin equations point of view. Our effective kernel contains a gluon dressing function with two light quark flavors $N_f = 2$, with current quark mass 0.0035 GeV. An effective coupling, ad usted to reproduce the behavior of the chiral condensate at nite $T$ complements our truncation. We nd the critical end point of the phase diagram located at the temperature $T^E = 0.1245$ GeV and the baryonic chemical potential $\mu_B^E = 0.211$ GeV.

## 1. Introduction

QCD at finite temperature and chemical potential plays an important role to understand the different transitions that took place in the early universe, after a few micro seconds from the Big Bang. As it is known, the observable degrees of freedom of quantum chromodynamics (QCD) at low temperature are the color-singlet hadrons, while at high temperature, the interaction between quark and gluons becomes weaker, causing hadrons to split up in a new phase where the dominant degrees of freedom are the quarks and gluons. This type of phase transition is referred to as confinement-deconfinement transition.

The vanishing of dynamically generated quark mass at high temperature $T$ and/or chemical potential $\mu$ corresponds to another type of transition i.e., chiral symmetry restoration, while at zero $T$ and $\mu$, chiral symmetry is broken. Thus, when the strength of the QCD interaction diminishes with increasing $T$ and $\mu$, only the current quark masses survive when $T$ and $\mu$ exceed a set of critical values. This is the chiral symmetry breaking-restoration phase transition. As for experiment is concerned, implications of chiral symmetry breaking for the elastic and transition form factors of mesons and baryons form an integral part of the planned program at the 12

GeV upgrade of the Thomas Jefferson ational Accelerator Facility in Virginia [1]. There are other experiments around the world that might help to understand the confinement and chiral transitions like R IC in Brookhaven, L C at CER , and future experiment proposals like CB at FAIR, in Germany.

In this work we use the Schwinger-Dyson Equations (SDE) approach to study the above mentioned phase transitions from first principles. We exploit the idea that dynamically generated masses for the quark are constants in the infrared region for momentum lower than 1 GeV, and promote this to finite temperature and density to understand the QCD phase transitions. For our effective kernel, we use a model for the gluon propagator which is consistent with the refined Gribov-Zwanziger scenario for confinement and is in agreement with lattice QCD. We consider the case in which the propagator has quark avor dependence, Ref. [2], and dress our kernel with a momentum dependent effective coupling as in Refs. [3, 4]. We calculate the chiral quark condensate for different values of temperature and match with lattice data, Ref. [5], at finite temperature. We then extend this parametriazation of the condensate at finite density. For the confinement-deconfinement transition, we use the spatial average of the scalar part of the quark propagator at zero spatial and explore violation of re ection positivity. The contribution is organized as follows: In. Sect. 2 we present the SDE for the quark propagator at finite temperature and density. Section 3 is devoted to introduce the Constant ass Approximation and our effective kernel. Discussion and conclusions of our findings are presented in Sect. 4.

## 2. QCD Gap Equation at $T \neq 0$ and $\mu \neq 0$

The quark propagator is a basic ob ect to analyze dynamical chiral symmetry breaking and confinement. At finite $T$ and $\mu$, we start from the general form of this two-point function

$$S^{-1}(\vec{p},\tilde{\omega}_n) = (i\vec{\gamma}.\vec{p})A(\vec{p}^2,\tilde{\omega}_n^2) + i\gamma_0\omega_n C(\vec{p}^2,\tilde{\omega}_n^2) + B(\vec{p}^2,\tilde{\omega}_n^2), \tag{1}$$

where $\tilde{w}_n = w_n + i\mu$, with $\mu$ representing the chemical potential and $w_n = (2n+1)\pi T$ are the fermionic atsubara frequencies. $A(\vec{p}^2, w_n^2)$, $B(\vec{p}^2, w_n^2)$ and $C(\vec{p}^2, w_n^2)$ are the scalar functions to be self consistently determined through solving the corresponding SDE or gap equation

$$S^{-1}(\vec{p},\tilde{\omega}_n) = (i\vec{\gamma}\cdot\vec{p}) + i\gamma_0\tilde{\omega}_n + m + \quad(\vec{p},\tilde{\omega}_n), \tag{2}$$

in which $m$ is the bare mass for the quark, the self energy has the form

$$(\vec{p},\tilde{\omega}_n) \;=\; T\sum_{l=-\infty}^{l=+\infty}\int\frac{d^3\vec{k}}{(2\pi)^3}g^2 D_{\mu\nu}(\vec{p}-\vec{k},\Omega_{nl};T,\mu)\frac{\lambda^a}{2}\gamma_\mu S(\vec{k},\omega_l)\frac{\lambda^a}{2}\Gamma_\nu(\vec{k},\vec{p},\omega_l,\omega_n), \tag{3}$$

$\Omega_{nl} = \omega_n - \omega_l$ and $\Gamma_\nu(\vec{k},\vec{p},\omega_l,\omega_n)$ is the full quark-gluon vertex. Due to the heat bath, the gluon propagator splits into longitudinal and transverse parts, namely

$$D_{\mu\nu}(\vec{q}^2,\Omega_{nl}^2) = P_{\mu\nu}^T D^T(\vec{q}^2,\Omega_{nl}^2) + P_{\mu\nu}^L D^L(\vec{q}^2,\Omega_{nl}^2,m_D^2). \tag{4}$$

The longitudinal part is modified by an extra term, the Debye screening mass $m_D^2 = (4/3)T^2 + \mu^2/\pi^2$ [6]. In Eq. (4), $P_{\mu\nu}^T$ is the transverse and $P_{\mu\nu}^L$ is the longitudinal pro ection operators, given by $P_{\mu\nu}^L = \delta_{\mu\nu} - Q_\mu Q_\nu/Q^2 - P_{\mu\nu}^T$, $P_{44}^T = P_{00}^T = 0$, $P_{ij}^T = \delta_{ij} - q_i q_j/\vec{q}^2$. We complete our gap equation kernel by considering

$$g^2 D_{\mu\nu}(\vec{q}^2,\Omega_{nl}^2)\Gamma_\nu(\vec{k},\vec{p},\omega_l,\omega_n) \to g_{eff}^2(\vec{q}^2,\Omega_{nl}^2)D_{\mu\nu}(\vec{q}^2,\Omega_{nl}^2)\zeta(T)\gamma_\nu, \tag{5}$$

where we introduce $\zeta(T)$ to match with the lattice results.

## 3. Constant Mass Approximation

The constant mass approximation approach (C A) has been motivated in the field of magnetic catalysis of dynamical chiral symmetry breaking [7]. It is based on the observation that the kernel of gap equation yields only contribution from the small momentum region, where the mass function is roughly constant. From Eqs.(1) and (2), and using our effective kernel, Eq. (5), after taking the trace we reach to the following expression for our $B(\vec{p}^2, w_n^2)$ function

$$\begin{aligned} B(\vec{p}^2, \tilde{\omega}_n^2) &= m \\ &-\frac{4}{3}T\zeta(T)\sum_{l=-\infty}^{l=+\infty}\int\frac{d^3\vec{k}}{(2\pi)^3}\sigma_B(\vec{k}^2,\tilde{\omega}_l^2)g_{eff}^2(\vec{q}^2,\Omega_{nl}^2)\left[D^L(\vec{q}^2,\Omega_{nl}^2,m_D^2)+2D^T(\vec{q}^2,\Omega_{nl}^2)\right].(6) \end{aligned}$$

ere $\sigma_B(\vec{p}^2,\tilde{\omega}_n^2) = -B(\vec{p}^2,\tilde{\omega}_n^2)/(\vec{p}^2A^2(\vec{p}^2,\tilde{\omega}_n^2)+\omega_n^2C^2(\vec{p}^2,\tilde{\omega}_n^2)+B^2(\vec{p}^2,\tilde{\omega}_n^2))$. The mass function for the quark is $M(\vec{p}^2,\omega_n^2) = B(\vec{p}^2,\omega_n^2)/A(\vec{p}^2,\omega_n^2)$. We set $A(\vec{k}^2,\omega_l^2) = C(\vec{k}^2,\omega_l^2) = 1$. We further take $\vec{p} = 0$ and take all the $B$ functions to be independent of momentum. Replacing $B(\vec{p}^2 = 0, w_n^2) = M_0(w_n^2)$, we have the following tower of relations of constant masses for every atsubara frequency

$$\begin{aligned} M_0(\omega_n^2) &= m \\ &+\frac{2T\zeta(T)}{\pi^2}\sum_l\int_0^{\Lambda^2}dk\vec{k}^2\frac{M_0(\omega_l^2)g_{eff}^2(\vec{k}^2,\Omega_{nl}^2)}{\vec{k}^2+\omega_l^2+M_0^2(\omega_l^2)}\left[D^L(\vec{k}^2,\Omega_{nl}^2,m_D^2)+2D^T(\vec{k}^2,\Omega_{nl}^2)\right]. \quad (7) \end{aligned}$$

For the gluon propagator, we take $D^L(K, m_D^2) = D^T(K, 0) \equiv D(K)$, where we have selected the following form of the avor dependent dressing for the gluon in Landau gauge

$$D(K^2) = \frac{K^2 + M_1^2}{K^4 + K^2(M_1^2 - \frac{13g_1^2<A_1^2>}{24}) + M_1^2m_0^2}, \quad (8)$$

where $K^2 = \vec{k}^2 + \Omega_{nl}^2$. The parameters in this dressing function are the following: $M_1 = 4.85$ [GeV$^2$], which is related to the condensate of auxiliary fields emerging when incorporating the horizon condition to the action, $g_1^2 < A_1^2 >= 0.474(16.406 - N_f)$ [GeV$^2$] is related to the dimension-two gluon condensate and $m_0 = 1.011(9.161 - N_f)^{-1/2}$ [GeV$^2$]. We complete our kernel with the effective interaction $g_{eff}^2(\vec{k}^2, \Omega_{nl}^2) = 4\pi a_2\nu(\vec{k}^2, \Omega_{nl}^2)$ where $\nu(\vec{k}^2, \Omega_{nl}^2)$ is given by as in Refs. [3, 4] by

$$\nu(K^2) = \left[\frac{a + b\left(\frac{K}{\Lambda}\right)^2}{1 + c\left(\frac{K}{\Lambda}\right)^2 + d\left(\frac{K}{\Lambda}\right)^4 + o\left(\frac{K}{\Lambda}\right)^6 + \frac{\pi\gamma}{\log\left[e+\left(\frac{K}{\Lambda}\right)^2\right]}}\right], \quad (9)$$

with $a = 1.47$; $b = 0.881$; $c = 0.314$; $d = 0.00986$; $o = 0.00168$; $\gamma = 12/25$; $\Lambda = 0.234$; $a_2 = 0.7$. The parameter $\zeta(T)$ is fitted with lattice data is in the form (see discussion below)

$$\zeta(T) = \frac{a_1 + b_1T + c_1T^2}{1 + d_1T + e_1T^2 + f_1T^3} \quad (10)$$

where $a_1 = 0.355$; $b_1 = -2.83$; $c_1 = 6.627$; $d_1 = -6.74$; $e_1 = 4.45$; $f_1 = 43.35$. The chiral condensate, in our case is

$$- < \psi\psi >= \frac{2N_cT}{\pi^2}\sum_n\int_0^{\Lambda^2}dk\vec{k}^2\frac{M_0(\omega_n^2)}{\vec{k}^2 + \omega_n^2 + M_0^2(\omega_n^2)}. \quad (11)$$

and we use the ultraviolet cut-off $\Lambda^2 = 1$ GeV$^2$. The confinement order parameter is defined through the function [8]

$$(\tau) = T\sum_n e^{i\omega_n\tau}\frac{M_0(\omega_n^2)}{\omega_n^2 + M_0^2(\omega_n^2)}. \quad (12)$$

According to the axiom of re ection positivity, if $(\tau)$ is positive for all $\tau$, then the particle is stable, otherwise is confined. The inverse of the value of $\tau$ at which the first crossing occurs $(\tau > 0)$ is taken as the order parameter for confinement.

## 4. Discussion and Conclusion

We solve the gap equation Eq.(2) for different atsubara frequencies. We then calculate the chiral quark condensate at values of the temperature and fit with Lattice data [5] at finite temperature to determine the parameters of $\zeta(T)$ in eq. (10). The comparison of our fitted condensate versus the lattice data is shown in Fig. 1. Our gap equation solution provide us the full plateau for evolution of the chiral quark condensate from lower to higher temperature domain.

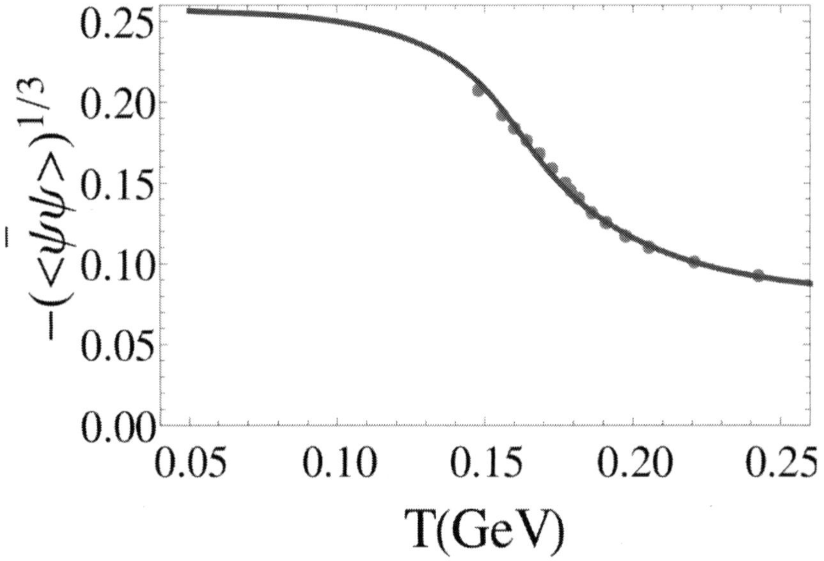


**Figure 1.** Chiral condensate as a function of temperature and at zero chemical potential: the blue curve represent the C A result, while the red dots are the Lattice QCD data for the chiral condensate.

Working with finite quark masses, the chiral condensate is not a true order parameter for the chiral transitions. evertheless, we still can extract the pseudo-critical temperature by taking the derivative of the condensate, with respect to the temperature i.e., the temperature gradient of the chiral condensate as shown in Fig. 2. The maximum lies at $T_c = 0.161$ GeV (pseudo-critical temperature), while for Lattice it is $T_c^{lat} = 0.154$ GeV. According to the Lattice prediction, the mentioned pseudo-critical temperature at zero chemical potential is the origin of the cross over phase transition in the QCD phase diagram.

ext, we include the quark chemical potential $\mu_q$ (or baryonic chemical potential $\mu_B = 3\mu_q$) and calculate the chiral quark condensate. It is plotted as a function of the temperature for different values of the chemical potential in Fig. 3. The plot shows smooth pointwise

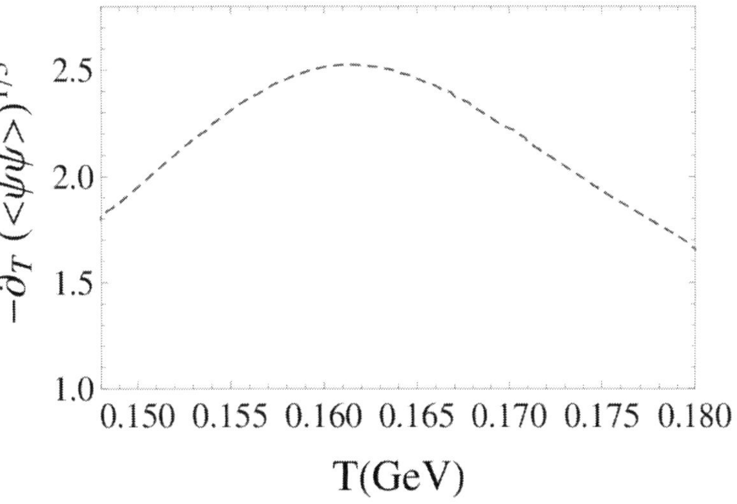


**Figure 2.** The thermal gradient of the chiral condensate, peak at $T = 0.161$ GeV and at zero chemical potential.

evolution of the condensate upto $\mu_q = 0.07$ ($\mu_B = 0.210$) GeV, while for higher than $\mu_q = 0.07$ ($\mu_B = 0.210$)GeV, a discontinuity occurs.

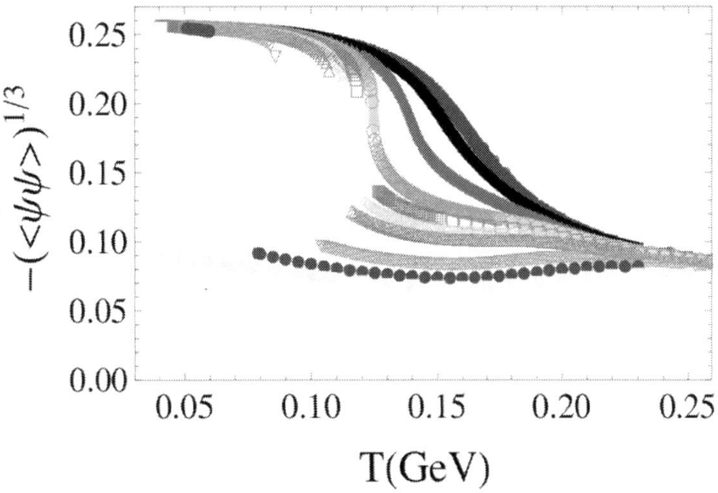


**Figure 3.** Chiral quark condensate as function of temperature and chemical potential: red dots represent Lattice QCD data at $\mu_q = 0$ GeV, C Result: blue squares represent, $\mu_q = 0$ GeV , black diamonds,$\mu_q = 0.03$ GeV, brown triangles $\mu_q = 0.05$ GeV, Green inverted triangles $\mu_q = 0.07$ GeV, orange open circles $\mu_q = 0.0705$ GeV, gray open squares$\mu_q = 0.08$ GeV, cyan open diamonds, $\mu_q = 0.09$ GeV, pink open triangles $\mu_q = 0.1$ GeV, magenta inverted open triangles $\mu_q = 0.15$ GeV, purple circles $\mu_q = 0.2$ GeV, yellow squares $\mu_q = 0.25$ GeV.

In Fig. 4 we take the temperature gradient of the chiral quark condensate for different values of chemical potential. The plot shows the highest change occurs near $T_E = 0.1245$ GeV and chemical potential $\mu_{q_E} = 0.07044$ ($\mu_{B_E} = 0.21132$) GeV.

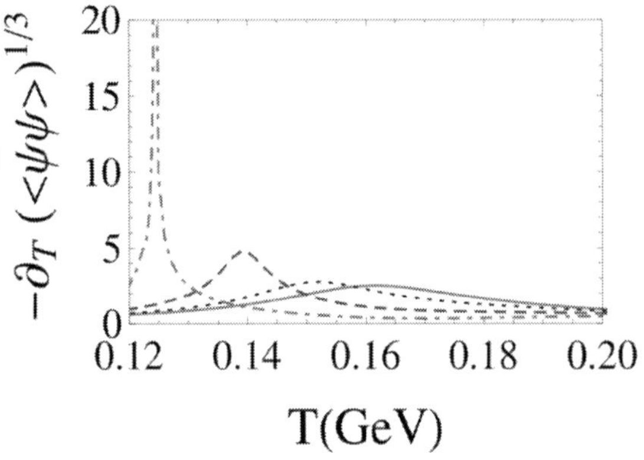


**Figure 4.** The thermal derivative of the chiral condensate at different chemical potential: red line represent $\mu_q = 0$ GeV, black dots represent $\mu_q = 0.03$ GeV, blue dashes represent $\mu_q = 0.05$ GeV, brown dashed dots $\mu_q = 0.07$ GeV.

In Fig. 5 we plot $\nu_T(\mu_B)$ i.e, the height of the thermal gradient $-\partial_T \langle \bar{\psi}\psi \rangle_{\mu_B}^{1/3}$ as a function of temperature for different value of the chemical potential. It shows the domains where different transitions occur. The $\nu_T(\mu_B)$ drawn on the vertical line, becomes singular as it approaches the dashed line, suggesting a change in the nature of phase transition from a simple cross-over to a first order phase transition.

We draw QCD phase diagram in Fig. 6, where temperature is along the vertical axis and baryonic potential $\mu_B$ along the horizontal axis. We locate the critical end point of the cross over and start of a first order phase transitions at ($T_E = 0.1245$ GeV, $\mu_E = 0.21132$ GeV). Thus we find that C A approach describes that, there is cross-over in small region of chemical potential, while at higher chemical potential the transition becomes of first order. There are chiral symmetry breaking and restoration and confinement-deconfinement transition. Other approaches, which include chiral models and JL-inspired models, place this point in the region $(\mu_E/T_c, T_E/T_c) \sim (1.0 - 2.0, 0.4 - 0.8)$ [9], whereas some mathematical extensions of lattice techniques [10] yield $(\mu_E/T_c, T_E/T_c) \sim (1.0 - 1.4, 0.9)$. We observe that the location $(\mu_E/T_c, T_E/T_c) \sim (1.3, 0.7)$ established by the C A gives a smaller value of $\mu_E$ than the one derived from a truncation of the gap equation considering the full momentum dependence of the quark propagator [6], but in fair agreement with other approaches. It could be due to the lack of control of the confinement scale with the chemical potential in the effective kernel of the gap equation. This issue and details of the confinement test will be addressed elsewhere [11].

## Acknowledgments

We acknowledge CO ACYT and IF - S for support. AA acknowledges support to the organizers of the XIV WPF, Oaxaca 2013. We also acknowledge Axel Weber for valuable discussions.

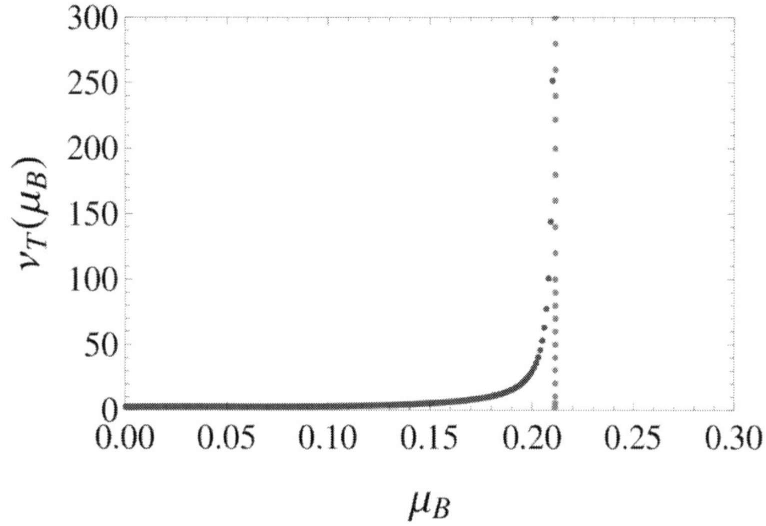


**Figure 5.** The height of the thermal derivative $\nu_T(\mu_B)$ is plotted as a function of the baryonic chemical potential ($\mu_B$), that shows the domains where different transitions occur.

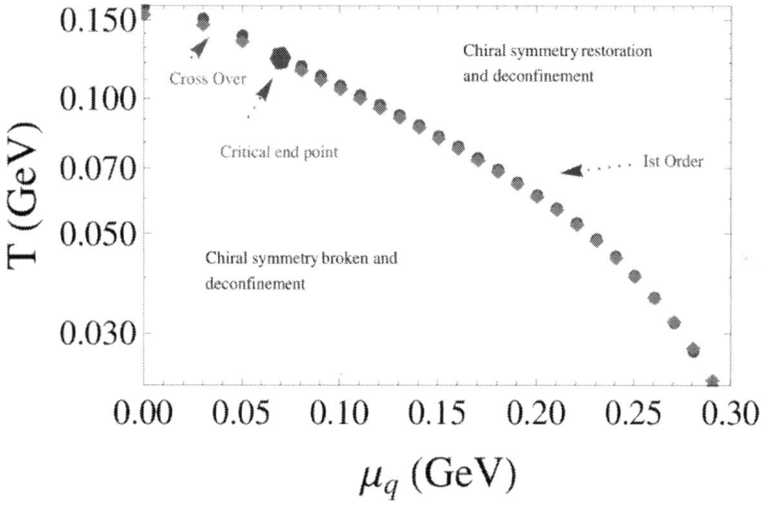


**Figure 6.** QCD phase diagram: Temperature vs baryonic chemical potential, the red diamonds represent the confinement-deconfinement transition, while the blue circles represent the chiral symmetry breaking-restoration. The purple dot represent the critical end point at ($T_E = 0.1245$ GeV, $\mu_E = 0.21132$ GeV).

## References

[1] Aznauryan I G *et. al.* 2013, *Int. J. Mod. Phys. E* **22** 1330015.
[2] Bashir A, Raya A and Rodriguez-Quintero 2013, *Phys. Rev. D* **88** 054003.
[3] Ahmad A and Raya A 2013, *Jour. Phys. Conf. Ser.* **418** 01200 .
[4] Alkofer R, Fischer C S, Smekal L V 2002, *Phys. Rev. D* **52** 1 1
Maris P, Raya A, Roberts C D and Schmidt S M, 2003,*Eur. Phys. J. A***18** 231 .
[5] Bazavov A, Bhattacharya T, Cheng M, DeTar C, Ding H T, Gottlieb S, Gupta R, Hegde P, Heller U M,

arsch F 2012,*Phys. Rev. D* **85** 054503.
[6] Gutierrez E, Ahmad A, Ayala A, Bashir A, Raya A 2013, ar iv: 1304.8065v1 [hep-ph] .
[ ] limenko G 1 2, *Z. Phys. C* **54** 323
Gusynin V P, Miransky V A, Shovkovy I A 1 5, *Phys. Rev. D* **52** 4 4
Gusynin V P, Miransky V A, Shovkovy I A 1 6, Nucl. *Phys. B* **464** 24 .
[8] Roberts C D 1 8, ar iv: 806088v1 [nucl-th]
Roberts C D (1 ), *Phys. Part.Nucl* **30** 223.
[ ] Schaefer B , Pawlowski M and Wambach 200 , *Phys. Rev. D* **76** 0 4023
ovács P and Szép 2008, *Phys. Rev. D* **77** 065016
Sasaki C, Friman B and Redlich 2008, *Phys. Rev. D* **77**, 034024
Costa P, Ruivo M C and de Sousa C A 2008, *Phys. Rev. D* **77** 0 6001
Fu W- , hang and Liu -x 2008, *Phys. Rev. D* **77** 014006
Abuki H, Anglani R, Gatto R, Nardulli G and Ruggieri M 2008, *Phys. Rev. D* **78** 034034
Schaefer B and Wagner M 200 , *Phys. Rev. D* **79** 014018
Costa P, Hansen H, Puivo M C and de Sousa C A 2010, *Phys. Rev. D* **81** 01600
Loewe M, Marquez F and Villavicencio C 2013, *Phys. Rev. D* **88** 056004
[10] Fodor and atz S D 2002, *J. High Energy Phys.* **03** 014
Gavai R V and Gupta S 2005, *Phys. Rev. D* **71** 114014
de Forcrand P and ratochvila S 2006, *Nucl. Phys. B, Proc. Suppl.* **153** 62
Li A, Alexandru A, Meng , and Liu F 200 , *Nucl. Phys. A* **830** 633C.
[11] Ahmad A *et. al.* 2014, work is in progress .

# Failure of Breit-Wigner and success of dispersive descriptions of the $\tau^- \to K^-\eta\nu_\tau$ decays

**Pablo Roig**
Instituto de Física, Universidad Nacional Autónoma de México, Apartado Postal 20-364, 01000 México D.F., México

E-mail: pabloroig@fisica.unam.mx

**Abstract.** The $\tau^- \to K^-\eta\nu_\tau$ decays have been studied using Chiral Perturbation Theory extended by including resonances as active fields. We have found that the treatment of final state interactions is crucial to provide a good description of the data. The Breit-Wigner approximation does not resum them and neglects the real part of the corresponding chiral loop functions, which violates analyticity and leads to a failure in the confrontation with the data. On the contrary, its resummation by means of an Omnès-like exponentiation of through a dispersive representation provides a successful explanation of the measurements. These results illustrate the fact that Breit-Wigner parametrizations of hadronic data, although simple and easy to handle, lack a link with the underlying strong interaction theory and should be avoided.

As a result of our analysis we determine the properties of the $K^\star(1410)$ resonance with a precision competitive to its traditional extraction using $\tau^- \to (K\pi)^-\nu_\tau$ decays, albeit the much limited statistics accumulated for the $\tau^- \to K^-\eta\nu_\tau$ channel. We also predict the soon discovery of the $\tau^- \to K^-\eta'\nu_\tau$ decays.

## 1. Introduction

Hadronic tau decays provide a clean laboratory to test the non-perturbative strong interaction involving mesons in rather clean conditions [1]. At the inclusive level this allows to extract fundamental parameters of the Standard Theory, like the strong coupling at the tau mass scale, the CKM matrix element $V_{us}$ or the mass of the strange quark [2]. On the exclusive side, the non-trivial strangeness-changing processes studied more accurately are the $\tau^- \to (K\pi)^-\nu_\tau$ decays, even more with the advent of the B-factory measurements performed by the BaBar and Belle Collaborations [3]. These very precise data have triggered a number of dedicated theoretical studies [4] which have taken advantage of the $K^\star(892)$-exchange dominance in these decays to determine very precisely the mass, width and couplings of this resonance. Although subleading, the contribution of its first excitation, the $K^\star(1410)$ meson, can also be probed through the $K\pi$ tau decays. As a result, its parameters were also evaluated in the quoted studies, even though with much less precision than in the case of its lightest copy.

Phase space considerations suggest that the related $\tau^- \to K^-\eta\nu_\tau$ decays may be specially sensitive to the properties of the $K^\star(1410)$ resonance and to its interplay with the $K^\star(892)$ meson. To test this possibility was one of the motivations of our study [5] aiming to evaluate the hadronic matrix element and give sensible estimates of invariant mass spectrum and decay

Journal of Physics: Conference Series **651** (2015) 012019 doi:10.1088/1742-6596/651/1/012019

width that could be confronted to the BaBar and Belle measurements [6]. We also wanted to improve the description of the strange spectral function with a better understanding of this decay channel and to provide TAUOLA [7], the standard Monte Carlo generator for tau lepton decays, with theory-based hadronic currents [8] for this process that, up to know, was modeled very crudely using a constant form factor. In this sense, Belle's paper requested more elaborated theoretical analyses of tau decays including $\eta$ mesons beyond the classic works [9] to match the precision of current measurements. This improvement is also requested from the Monte Carlo point of view [10].

BaBar and Belle measurements [6] of the $\tau^- \to K^-\eta\nu_\tau$ decays yield the respective branching fractions $(1.58 \pm 0.05 \pm 0.09) \cdot 10^{-4}$ and $(1.42 \pm 0.11 \pm 0.07) \cdot 10^{-4}$. The partner $\tau^- \to K^-\eta'\nu_\tau$ decays have not been detected yet, but an upper branching ratio of $2.4 \cdot 10^{-6}$ at the 90% confidence level was placed by BaBar [11]. Branching ratios at the level of $10^{-6}$ should be within reach at the forthcoming Belle-II experiment. Therefore, another target of our analysis was to determine if a soon discovery of this decay mode could be expected.

## 2. Matrix element and decay width

The differential decay width for the considered processes reads

$$\frac{d\Gamma\left(\tau^- \to K^-\eta^{(\prime)}\nu_\tau\right)}{d\sqrt{s}} = \frac{G_F^2 M_\tau^3}{32\pi^3 s} S_{EW} \left|V_{us} f_+^{K^-\eta^{(\prime)}}(0)\right|^2 \left(1-\frac{s}{M_\tau^2}\right)^2 \tag{1}$$
$$\left\{\left(1+\frac{2s}{M_\tau^2}\right) q^3_{K\eta^{(\prime)}}(s)\left|\tilde{f}_+^{K^-\eta^{(\prime)}}(s)\right|^2 + \frac{3\Delta^2_{K\eta^{(\prime)}}}{4s} q_{K\eta^{(\prime)}}(s)\left|\tilde{f}_0^{K^-\eta^{(\prime)}}(s)\right|^2\right\},$$

where

$$s = \left(p_{\eta^{(\prime)}} + p_K\right)^2, \quad q_{PQ}(s) = \frac{\sqrt{s^2 - 2s\Sigma_{PQ} + \Delta^2_{PQ}}}{2\sqrt{s}},$$
$$\Sigma_{PQ} = m_P^2 + m_Q^2, \quad \Delta_{PQ} = m_P^2 - m_Q^2, \quad \tilde{f}_{+,0}^{K^-\eta^{(\prime)}}(s) = \frac{f_{+,0}^{K^-\eta^{(\prime)}}(s)}{f_{+,0}^{K^-\eta^{(\prime)}}(0)}, \tag{2}$$

and $S_{EW} = 1.0201$ [12] represents an electroweak correction factor. $\left|V_{us} f_+^{K^-\eta}(0)\right| = \left|V_{us} f_+^{K^-\pi^0}(0)\cos\theta_P\right|$ and $\left|V_{us} f_+^{K^-\eta'}(0)\right| = \left|V_{us} f_+^{K^-\pi^0}(0)\sin\theta_P\right|$, where $\theta_P = (-13.3 \pm 1.0)^\circ$ [13]. From these values we can already guess the suppression of the $\tau^- \to K^-\eta'\nu_\tau$ decays. Unitarization effects will increase the scalar form factor contribution to this decay channel making it more important than the vector form factor effect, but it will still be roughly two orders of magnitude less frequent than the $\tau^- \to K^-\eta\nu_\tau$ decays. We will use the value $\left|V_{us} f_+^{K^-\pi^0}(0)\right| = 0.21664 \pm 0.00048$, obtained analyzing semi-leptonic Kaon decay data [14].

In eq. (1) the strong interaction dynamics has been encoded in the $f_{+,0}^{K^-\eta^{(\prime)}}(0)$ form factors defined by

$$\left\langle K^-\eta^{(\prime)}\right|\bar{s}\gamma^\mu u\left|0\right\rangle = \left[\left(p_{\eta^{(\prime)}} - p_K\right)^\mu + \frac{\Delta_{K\eta^{(\prime)}}}{s} q^\mu\right] c^V_{K^-\eta^{(\prime)}} f_+^{K^-\eta^{(\prime)}}(s) + \frac{\Delta_{K\pi}}{s} q^\mu c^S_{K^-\eta^{(\prime)}} f_0^{K^-\eta^{(\prime)}}(s), \tag{3}$$

with the normalization condition

$$f_+^{K^-\eta^{(\prime)}}(0) = -\frac{c^S_{K^-\eta^{(\prime)}}}{c^V_{K^-\eta^{(\prime)}}}\frac{\Delta_{K\pi}}{\Delta_{K\eta^{(\prime)}}} f_0^{K^-\eta^{(\prime)}}(0) \tag{4}$$

and the values $c^V_{K\eta^{(\prime)}} = -\sqrt{\frac{3}{2}}$, $c^S_{K^-\eta} = -\frac{1}{\sqrt{6}}$, $c^S_{K^-\eta'} = \frac{2}{\sqrt{3}}$. The use of the tilded $\widetilde{f}_{+,0}^{K^-\eta^{(\prime)}}(s)$ form factors in eq. (1) yields more compact expressions than previously used, which are explicitly symmetric under the exchange $\eta \leftrightarrow \eta'$.

## 3. Scalar and vector form factors in Resonance Chiral Theory

Nowadays, there is not any analytic way known of obtaining the relevant $\widetilde{f}_{+,0}^{K^-\eta^{(\prime)}}(s)$ form factors employing the quantum field theory of strong interactions, QCD [1]. This, however, does not mean that its Lagrangian is useless for obtaining them. In particular, at very low energies, QCD exhibits and approximate (chiral) symmetry in the limit of massless light quarks. This property allows to build an effective field theory dual to it in this regime, Chiral Perturbation Theory ($\chi PT$) [15].

However, as the energy increases, it does not suffice to include higher and higher order computations in $\chi PT$ [16]. On the contrary, it is necessary to incorporate the next lighter states, the lowest-lying light-flavoured resonances, as active degrees of freedom into the action. In the context of tau decays this feature was studied in Ref. [17].

A complementary and equivalent view comes from considering the expansion parameter of $\chi PT$, the ratio between momenta and masses of the lightest pseudoscalar ($\pi$, $K$ and $\eta$) mesons over the chiral symmetry breaking scale, of order 1 GeV. It is obvious that when its value starts to be comparable to one the chiral expansion will no longer converge. This scale is (parametrically) of the same order of the mass of the lightest mesons, i.e. the $\rho(770)$ resonance. Therefore, the need of finding an alternative expansion parameter valid in the GeV-region also leads to an extension of $\chi PT$. A successful candidate was provided by the $1/N_C$ expansion of QCD [18], whose predictions for a theory including resonances are corroborated experimentally [19].

A realization of these ideas is provided by the Resonance Chiral Theory ($R\chi T$) [20], which includes the meson resonances in the antisymmetric tensor field representation. This formalism brings in the advantage that there is no need to include the local contact interactions at subleading orders in $\chi PT$ because they are reproduced upon integrating the resonances out. The values of the $R\chi T$ couplings are not restricted by symmetry requirements. However, the resonance Physics is supposed to interpolate between the known chiral and perturbative regimes. Consequently, the matching between the operator product expansion and the $R\chi T$ results for Green functions is performed [20, 21], rendering relations among and predictions of the Lagrangian couplings which increase the predictability of the theory. Remarkably, within $R\chi T$ the successful notion of vector meson dominance is not an *a priori* assumption but a dynamical result. The application of $R\chi T$ to the study of hadronic tau decays has proved successful in a variety of decay channels [4, 5, 8, 22] and related processes [23].

The relevant effective Lagrangian for the lightest resonance nonets is

$$\mathcal{L}_{\mathrm{R}\chi\mathrm{T}} \doteq \mathcal{L}_{\mathrm{kin}}^{\mathrm{V,S}} + \frac{F^2}{4}\langle u_\mu u^\mu + \chi_+\rangle + \frac{F_V}{2\sqrt{2}}\langle V_{\mu\nu} f_+^{\mu\nu}\rangle + i\frac{G_V}{\sqrt{2}}\langle V_{\mu\nu}u^\mu u^\nu\rangle + c_d\langle S u_\mu u^\mu\rangle + c_m\langle S\chi_+\rangle\,, \tag{5}$$

[1] Lattice simulations do not provide these form factors at the moment.

where all coupling constants are real: $F$ is the pion decay constant, $F_V$ ($c_m$) gives the coupling of the vector (scalar) resonances to the $W$ current (scalar source) and $G_V$ ($c_d$) provides the coupling of the vector (scalar) mesons to pairs of pseudoscalars. The definition of the chiral tensors entering eq. (5) can be found in Refs. [20].

The resulting vector form factors are

$$\tilde{f}_+^{K^-\eta}(s) = \frac{f_+^{K^-\eta}(s)}{f_+^{K^-\eta}(0)} = 1 + \frac{F_V G_V}{F^2}\frac{s}{M_{K^\star}^2 - s} = \frac{f_+^{K^-\eta'}(s)}{f_+^{K^-\eta'}(0)} = \tilde{f}_+^{K^-\eta'}(s)\,, \tag{6}$$

where the values of $f_+^{K^-\eta^{(\prime)}}(0)$ were used.

The strangeness changing scalar form factors and associated S-wave scattering within $R\chi T$ have been investigated in a series of papers by Jamin, Oller and Pich [24]. Our expressions for the scalar form factors can be written in terms of the $f_0^{K^-\eta_8}(s)$, $f_0^{K^-\eta_1}(s)$ form factors given in Refs. [24]. It should be noted that using our conventions, the tilded scalar form factors become simply

$$\tilde{f}_0^{K^-\eta}(s) = \frac{f_0^{K^-\eta}(s)}{f_0^{K^-\eta}(0)} = 1 + \frac{c_d c_m}{4F^2}\frac{s}{M_S^2 - s} = \frac{f_+^{K^-\eta'}(s)}{f_0^{K^-\eta'}(0)} = \tilde{f}_0^{K^-\eta'}(s)\,, \tag{7}$$

which are more compact than the corresponding results in Refs. [24] and display the same symmetry $\eta \leftrightarrow \eta'$ than the vector form factors in eq. (6).

The contribution of heavier resonances can be included within $R\chi T$ in the spirit of the $N_C \to \infty$ limit. However, within this infinite tower of states only the $K^\star(1410)$ resonance will play a relevant role in the considered processes (in the vector form factor). In the scalar one, it should be pointed out that the resonance labeled $S$ in eq. (7) corresponds to the $K_0^\star(1430)$, since the $\kappa(800)$ meson is dynamically generated through $K\pi$ rescattering [25].
Then, the vector form factor in eq. (6) shall be replaced by

$$\tilde{f}_+^{K^-\eta^{(\prime)}}(s) = 1 + \frac{F_V G_V}{F^2}\frac{s}{M_{K^\star}^2 - s} + \frac{F_V' G_V'}{F^2}\frac{s}{M_{K^{\star\prime}}^2 - s}\,, \tag{8}$$

where the operators with couplings $F_V'$ and $G_V'$ are defined in analogy with the corresponding unprimed couplings in eq. (5).

The vanishing of the $f_+^{K^-\eta^{(\prime)}}(s)$ and $f_0^{K^-\eta^{(\prime)}}(s)$ form factors for $s \to \infty$ at least as $1/s$ [26] determines the short-distance constraints

$$F_V G_V + F_V' G_V' = F^2\,, \quad 4 c_d c_m = F^2\,, \quad c_d - c_m = 0\,, \tag{9}$$

which yield the form factors

$$\tilde{f}_+^{K^-\eta}(s) = \frac{M_{K^\star}^2 + \gamma s}{M_{K^\star}^2 - s} - \frac{\gamma s}{M_{K^{\star\prime}}^2 - s} = \tilde{f}_+^{K^-\eta'}(s)\,, \tag{10}$$
$$\tilde{f}_0^{K^-\eta}(s) = \frac{M_S^2}{M_S^2 - s} = \tilde{f}_0^{K^-\eta'}(s)\,,$$

where $\gamma = -\frac{F_V' G_V'}{F^2} = \frac{F_V G_V}{F^2} - 1$ [4]. The modifications introduced by heavier resonances to this relation and to the definition of $\gamma$ are negligible.

Journal of Physics: Conference Series **651** (2015) 012019 doi:10.1088/1742-6596/651/1/012019

## 4. Treatment of final state interactions

The form factors in eq. (10) diverge for $s = M^2$, while the hadronic observables are peaked (but not divergent) at these energies. The solution comes by including a subleading effect in the large-$N_C$ expansion as given by the finite width of the resonances. In fact, some of them are wide enough so that the energy dependence of their widths becomes an issue. This topic has been studied within $R\chi T$ in Ref. [27] were a precise and consistent definition for the spin-one meson off-shell widths was given. Its application to the $K^\star(892)$ case yields an expression that does quite a good job in the comparisons with data. However, at the present level of precision it is preferable to include the corrections to the corresponding on-shell width value to write

$$\Gamma_{K^*}(s) \;=\; \Gamma_{K^*}\left(M_{K^*}^2\right)\frac{s}{M_{K^*}^2}\frac{\sigma_{K\pi}^3(s)+\cos^2\theta_P\sigma_{K\eta}^3(s)+\sin^2\theta_P\sigma_{K\eta'}^3(s)}{\sigma_{K\pi}^3(M_{K^*}^2)}\,, \quad (11)$$

where it was used that the only absorptive cut at the $M_{K^\star}$-scale is given by the elastic contribution and $\sigma_{PQ}(s) = \frac{2q_{PQ}(s)}{\sqrt{s}}\theta\left(s-(m_P+m_Q)^2\right)$.

Contrary to the $K^\star(892)$ case, there is no restriction from the chiral limit that applies to the width of the $K^\star(1410)$. Assuming that the lightest $K\pi$ absorptive contribution dominates one has

$$\Gamma_{K^{*\prime}}(s) = \Gamma_{K^{*\prime}}\left(M_{K^{*\prime}}^2\right)\frac{s}{M_{K^{*\prime}}^2}\frac{\sigma_{K\pi}^3(s)}{\sigma_{K\pi}^3(M_{K^{*\prime}}^2)}\,. \quad (12)$$

The scalar resonance width can also be computed in $R\chi T$ similarly yielding, for the $K_0^\star(1430)$

$$\Gamma_S(s) = \Gamma_S\left(M_S^2\right)\left(\frac{s}{M_S^2}\right)^{3/2}\frac{g(s)}{g\left(M_S^2\right)}\,, \quad (13)$$

with

$$\begin{aligned} g(s) \;=\; & \frac{3}{2}\sigma_{K\pi}(s)+\frac{1}{6}\sigma_{K\eta}(s)\left[\cos\theta_P\left(1+\frac{3\Delta_{K\pi}+\Delta_{K\eta}}{s}\right)+2\sqrt{2}\sin\theta_P\left(1+\frac{\Delta_{K\eta}}{s}\right)\right]^2 \\ & +\frac{4}{3}\sigma_{K\eta'}(s)\left[\cos\theta_P\left(1+\frac{\Delta_{K\eta'}}{s}\right)-\frac{\sin\theta_P}{2\sqrt{2}}\left(1+\frac{3\Delta_{K\pi}+\Delta_{K\eta'}}{s}\right)\right]^2\,. \end{aligned} \quad (14)$$

The final state interactions (FSI) encoded in the chiral loop functions are not small and the real parts of these functions are not negligible. Moreover, even if they were numerically small, disregarding them while keeping the corresponding imaginary part (giving the meson widths) would violate analyticity. The key point is how to handle the resummation of these FSI. We will show that a Breit-Wigner approximation (which neglects the real part of the chiral loop functions and does not resum FSI) fails dramatically in the $\tau^- \to K^-\eta\nu_\tau$ decays. On the contrary, two kinds of resummations of FSI (Omnès-like and dispersive representation) do provide a good agreement with the data. The latter, which also gives the best results, is preferable because it is analytic by construction, while in the former analyticity only holds perturbatively. We emphasize that despite Breit-Wigner parametrizations of hadronic data are very simple and manageable, they violate known properties of the underlying strong interaction (analyticity, unitarity, asymptotic behaviour, ...). Therefore very little can be learned about QCD by using them. If that is our purpose (as it is supposed to be), they should not be employed.

Specifically, three different options will be considered, in increasing degree of soundness, for the treatment of FSI. The relevant form factors will be obtained from eqs.(10) in each case by:

- Dipole model (Breit-Wigner): $M_R^2 - s$ will be replaced by $M_R^2 - s - iM_R\Gamma_R(s)$ with $\Gamma_{K^\star}(s)$ and $\Gamma_S(s)$ given by eqs. (11) and (13). Since the scalar form factors will not be unitarized, the $\kappa(800)$ resonance contribution is lost (or at best badly modeled) in this approach. Analyticity is violated already at the lowest order.
- Exponential parametrization (JPP): The Breit-Wigner vector form factor described above is multiplied by the exponential of the real part of the chiral loop function. In this way, this part of the loop function is resummed, while the imaginary one (the width) is kept unresummed in the denominator, $M_R^2 - s - iM_R\Gamma_R(s)$. This violates analyticity at the next-to-next-to-leading order in the chiral expansion, which is a small effect numerically. The unitarized scalar form factor [24], which is a solution of the Muskelishivili-Omnès problem for three channel case ($K\pi$, $K\eta$, $K\eta'$) will be employed.
- Dispersive representation (BEJ): A three-times subtracted dispersion relation will be used for the vector form factor. The input phaseshift will be defined using the vector form factor in eq. (10)

$$\tan\delta(s) = \frac{Im\left[\tilde{f}_+^{K^-\eta}(s)\right]}{Re\left[\tilde{f}_+^{K^-\eta}(s)\right]}, \tag{15}$$

so that the output form factor is

$$\tilde{f}_+(s) = \exp\left[\lambda'_+\frac{s}{m_\pi^2} + \frac{1}{2}\lambda''_+\frac{s^2}{m_\pi^4} + \frac{s^3}{\pi}\int_{s_{K\pi}}^{s_{cut}} ds'\frac{\delta(s')}{(s')^3(s'-s-i0)}\right], \tag{16}$$

where $s_{K\pi} = (m_K + m_\pi)^2$. $\Gamma_{K^\star}(s)$ includes only the $K\pi$ cut and the whole complex loop function is resummed in the denominator, which keeps analyticity exactly. This vector form factor neglects inelastic coupled channel effects, which is in principle a questionable approximation. We anticipate that the agreement with data does not call for including these effects at the moment. The slope parameters, $\lambda_+^{\prime(\prime)}$, encode the very low energy dynamics. The unitarized scalar form factor will also be used [24].

## 5. Predicting the $\tau^- \to K^-\eta\nu_\tau$ decays

Eqs. (10) also hold for the $\tilde{f}_{+,0}^{(K\pi)^-}(s)$ form factors. Therefore, the detailed knowledge of the $K\pi$ form factors [4] could in principle be used to predict the the $\tau^- \to K^-\eta^{(\prime)}\nu_\tau$ decays. While this is true for the vector form factors, the unitarization procedure of the scalar form factors breaks this universality and different $\tilde{f}_0^{K^-\pi^0/\eta/\eta'}(s)$ are obtained. Then, the unitarization procedure can be tested through the effect of the scalar form factors in the $\tau^- \to K^-\eta^{(\prime)}\nu_\tau$ decays. While the suppression of the scalar contribution in the $\eta$ case makes it difficult to check finely the unitarized results, its leading role in the $\eta'$ case would give us un opportunity to probe the $\tilde{f}_0^{K^-\eta'}(s)$ form factor.

Taking this discussion into account, we have predicted the $\tau^- \to K^-\eta\nu_\tau$ decays (and later on the $\tau^- \to K^-\eta'\nu_\tau$ decays in section 7) as explained below:

- In the dipole model, we have taken the $K^\star(892)$, $K^\star(1410)$ and $K_0^\star(1430)$ mass and width from the PDG [28] because Breit-Wigner parametrizations are employed in this reference. We estimated the relative weight of them using $\gamma = \frac{F_V G_V}{F^2} - 1 = -0.021 \pm 0.031$.
- In the JPP parametrization, we have used the best fit results of the second reference in [4] for the vector form factor. The scalar form factor was obtained from the solutions (6.10)

Journal of Physics: Conference Series **651** (2015) 012019 doi:10.1088/1742-6596/651/1/012019

and (6.11) of the second reference in [24]. The scalar form factors have also been treated alike in the BEJ approach.

- In the BEJ representation, one would use the best fit results of the last reference in [4] to obtain our vector form factor. However, the slope form factor parameters, $\lambda'_+$ and $\lambda''_+$, are very sensitive to isospin breaking corrections on the particle masses. For this reason we estimated the corresponding parameters for the $K^-\pi^0$ case of interest for the $\tau^- \to K^-\eta^{(\prime)}\nu_\tau$ decays. We have therefore used the results in the middle column of table 1 to predict the $\tau^- \to K^-\eta^{(\prime)}\nu_\tau$ decays. More details on this procedure can be found in Ref. [5].

**Table 1.** Results for the fit to Belle $\tau^- \to K_S\pi^-\nu_\tau$ data [3] with a three-times subtracted dispersion relation including two vector resonances in $\widetilde{f}_+^{K\pi}(s)$, according to eq. (10), and resumming the loop function in the denominator; as well as the unitarized scalar form factor [24]. The middle column is obtained using the masses of the $K^-$ and $\pi^0$ mesons and the last column using the $K_S$ and $\pi^-$ masses actually corresponding to the data.

| Parameter | Best fit with $K^-\pi^0$ masses | Best fit with $K_S\pi^-$ masses |
|---|---|---|
| $\lambda'_+ \times 10^3$ | $22.2 \pm 0.9$ | $24.7 \pm 0.8$ |
| $\lambda''_+ \times 10^4$ | $10.3 \pm 0.2$ | $12.0 \pm 0.2$ |
| $M_{K^\star}$ (MeV) | $892.1 \pm 0.6$ | $892.0 \pm 0.9$ |
| $\Gamma_{K^\star}$ (MeV) | $46.2 \pm 0.5$ | $46.2 \pm 0.4$ |
| $M_{K^{\star\prime}}$ (GeV) | $1.28 \pm 0.07$ | $1.28 \pm 0.07$ |
| $\Gamma_{K^{\star\prime}}$ (GeV) | $0.16^{+0.10}_{-0.07}$ | $0.20^{+0.06}_{-0.09}$ |
| $\gamma$ | $-0.03 \pm 0.02$ | $-0.04 \pm 0.02$ |

Using these inputs we have found the differential decay distributions for the three different approaches considered using eq. (1). This one is, in turn, related to the experimental data by using

$$\frac{dN_{events}}{dE} = \frac{d\Gamma}{dE}\frac{N_{events}}{\Gamma_\tau BR(\tau^- \to K^-\eta\nu_\tau)}\Delta E_{bin}\,. \tag{17}$$

In Fig.1 we show our predictions based on the $K\pi$ system according to BW, JPP and BEJ. In this figure we have normalized the BaBar data to Belle's using eq. (17). We notice some tension between the BaBar and Belle data sets and strange oscillations of some Belle points that may hint to a systematic issue or an underestimation of the errors. In this figure, the one-sigma contours for the three approaches are also shown. The corresponding branching ratios are displayed in table 2, where the $\chi^2/dof$ is also given. It is seen that the BW approximation fails both in the decay width and the differential decay distribution shape. On the contrary, the JPP and BEJ treatments give already a good agreement with both of them. These results point to the BW modelization being a too rough approximation to the underlying dynamics. We understand this fact since, as we have discussed, this approach does not resum FSI and violates analyticity and unitarity severely. We will check this conclusion in the next section by fitting BaBar and Belle data to discard the possibility that the input parameters for the BW prediction were inappropriate.

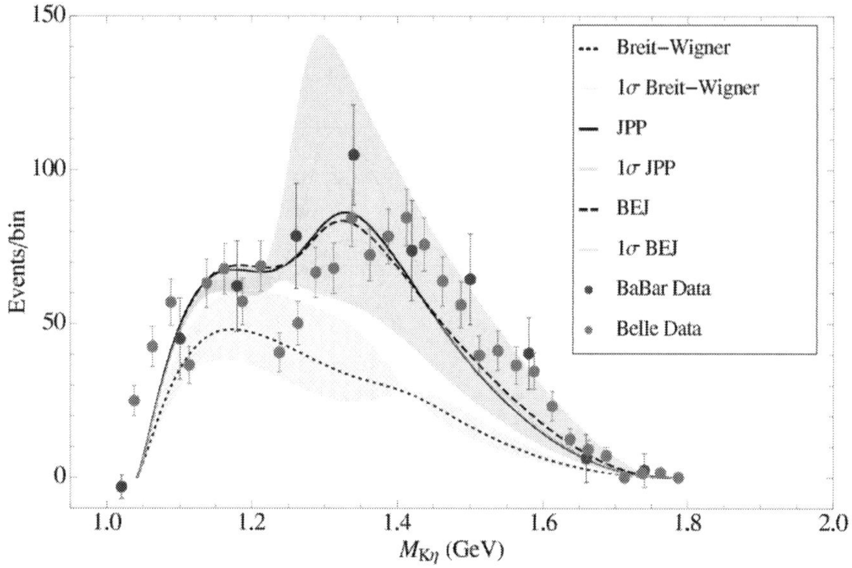


**Figure 1.** BaBar (blue) and Belle (red) [6] data for the $\tau^- \to K^-\eta\nu_\tau$ decays are confronted to the predictions obtained in the BW (dotted), JPP (solid) and BEJ (dashed) approaches (see the main text for details) which are shown together with the corresponding one-sigma error bands in yellow, light blue and light green, respectively.

**Table 2.** Predicted branching ratio of the $\tau^- \to K^-\eta\nu_\tau$ decays according to the different approaches used. The corresponding $\chi^2/dof$ values are also shown and the PDG branching fraction is given for reference.

| Source | Branching ratio | $\chi^2/dof$ |
|---|---|---|
| Dipole Model (BW) | $\left(0.78^{+0.17}_{-0.10}\right) \cdot 10^{-4}$ | 8.3 |
| JPP | $\left(1.47^{+0.14}_{-0.08}\right) \cdot 10^{-4}$ | 1.9 |
| BEJ | $(1.49 \pm 0.05) \cdot 10^{-4}$ | 1.5 |
| Experimental value | $(1.52 \pm 0.08) \cdot 10^{-4}$ | - |

## 6. Fitting the $\tau^- \to K^-\eta\nu_\tau$ data

We have considered different kinds of fits to BaBar and Belle $\tau^- \to K^-\eta\nu_\tau$ data. We have first assessed, in full generality, that these decays are insensitive to the mass and width of the $K^\star(892)$ resonance, as it could be expected since phase space for the $K\eta$ channel opens above the region of $K^\star(892)$ dominance. This is even more the case for the slope parameters of BEJ, which are determined by the Physics immediately above the $K\pi$ threshold. For this reason, we have borrowed these parameters from the corresponding predictions used in the previous section. We have thus fitted only the $K^\star(1410)$ parameters in all three approaches [2].

[2] The dependence on the $K_0^\star(1430)$ mass and width in the dipole (BW) model is very mild and can be neglected.

Our best fit results for the branching ratios are given in table 3, including the corresponding $\chi^2/dof$. These results are obtained with the best fit parameter values shown in table 4, which can be compared to the reference values used for the predictions in the previous section (these are recalled in table 5). The corresponding decay distributions including one-sigma error bands are plotted in Fig. 2.

These results show that the BW model does not approximate the underlying physics for any value of its parameters and should be discarded. On the contrary, JPP and BEJ yield good fits to data with $\chi^2/dof$ values close to unity. This confirms that the simplified treatment of FSI in BW, which misses the real part of the two-meson rescatterings, violates analyticity by construction and does not resum FSI, is responsible for the failure.

**Table 3.** The branching ratios and $\chi^2/dof$ obtained in BW, JPP and BEJ fitting $\gamma$ only and also the $K^\star(1410)$ parameters are displayed. Other parameters were fixed to the reference values used in section 5. The PDG branching fraction is also given for reference.

| Source | Branching ratio | $\chi^2/dof$ |
|---|---|---|
| Dipole Model (BW) (Fit $\gamma$) | $\left(0.96^{+0.21}_{-0.15}\right)\cdot 10^{-4}$ | 5.0 |
| Dipole Model (BW) (Fit $\gamma$, $M_{K^{\star\prime}}$, $\Gamma_{K^{\star\prime}}$) | Unphysical result | - |
| JPP (Fit $\gamma$) | $\left(1.50^{+0.19}_{-0.11}\right)\cdot 10^{-4}$ | 1.6 |
| JPP (Fit $\gamma$, $M_{K^{\star\prime}}$, $\Gamma_{K^{\star\prime}}$) | $(1.42\pm 0.04)\cdot 10^{-4}$ | 1.4 |
| BEJ (Fit $\gamma$) | $\left(1.59^{+0.22}_{-0.16}\right)\cdot 10^{-4}$ | 1.2 |
| BEJ (Fit $\gamma$, $M_{K^{\star\prime}}$, $\Gamma_{K^{\star\prime}}$) | $(1.55\pm 0.08)\cdot 10^{-4}$ | 0.8 |
| Experimental value | $(1.52\pm 0.08)\cdot 10^{-4}$ | - |

**Table 4.** The best fit parameter values corresponding to the different alternatives considered in table 3 are given. These can be compared to the reference values, which are given in table 5. BEJ results for the mass and width of the $K^\star(1410)$ correspond to pole values, while JPP figures are given for the model parameter as in the original literature.

| Fitted value \ Approach | Dipole Model (BW) | JPP | BEJ |
|---|---|---|---|
| $\gamma$ | $-0.174\pm 0.007$ | $-0.063\pm 0.007$ | $-0.041\pm 0.021$ |
| $\gamma$ | Unphysical | $-0.078^{+0.012}_{-0.014}$ | $-0.051^{+0.012}_{-0.036}$ |
| $M_{K^{\star\prime}}$ (MeV) | best fit | $1356\pm 11$ | $1327^{+30}_{-38}$ |
| $\Gamma_{K^{\star\prime}}$ (MeV) | parameters | $232^{+30}_{-28}$ | $213^{+72}_{-118}$ |

These results are plotted in Fig. 2. Looking at the JPP and BEJ results in more detail one can notice that:

- The $\chi^2/dof$ of both approaches improves by $15 \leftrightarrow 20\%$ fitting only $\gamma$. Fitted and reference values are consistent (see table 5). Both the $\tau^- \to (K\pi)^-\nu_\tau$ and the $\tau^- \to K^-\eta\nu_\tau$ decays are sensitive to the interplay between the first two vector resonances and this agreement is a good autoconsistency test.

**Table 5.** Reference values (used in section 5) corresponding to the best fit parameters appearing in table 4. Again BEJ results are pole values and JPP ones are model parameters. The latter are converted to resonance pole values at the end of this section and the results are discussed in section 8.

| Reference value \ Approach | Dipole Model (BW) | JPP | BEJ |
|---|---|---|---|
| $\gamma$ | $-0.021 \pm 0.031$ | $-0.043 \pm 0.010$ | $-0.029 \pm 0.017$ |
| $M_{K^{\star\prime}}$ (MeV) | $1414 \pm 15$ | $1307 \pm 17$ | $1283 \pm 65$ |
| $\Gamma_{K^{\star\prime}}$ (MeV) | $232 \pm 21$ | $206 \pm 49$ | $163 \pm 68$ |

- Fitting also the $K^\star(1410)$ parameters improves the results by $\sim 13\%$ in JPP and by $\sim 33\%$ in BEJ. The three-parameter fits do not yield physical results in BW. Specifically, $K^\star(1410)$ mass and width approach to the $K^\star(892)$ values and $|\gamma|$ is one order of magnitude larger than the reference values, which makes us discard this result. Although the branching ratios of both JPP and BEJ agree with the PDG value, the JPP branching ratios are closer to its lower limit while BEJ is nearer the upper one. Deviations of the three-parameter best fit values with respect to the input values lie within errors in BEJ, as it so happens with $\Gamma_{K^{\star\prime}}$ in JPP. However, there are small tensions between the reference and best fit values of $M_{K^{\star\prime}}$ and $\gamma$ in JPP.

Although the BW curve in Fig. 2 has improved with respect to Fig. 1 and seems to agree well with the data in the higher-energy half of the spectrum, it fails completely at lower energies. On the contrary, JPP and BEJ provide good quality fits to data which are satisfactory along the whole phase space. Even though the vector form factor gives the dominant contribution to the decay width, the scalar form factor is not negligible and gives $\sim (3 \leftrightarrow 4)\%$ of the branching fraction in the JPP and BEJ cases. In the BW model this contribution is $\sim 7\%$.

We have then translated the JPP model values appearing in tables 4 and 5 to the physical pole values [30], yielding $M_{K^{\star\prime}} = 1332^{+16}_{-18}$, $\Gamma_{K^{\star\prime}} = 220^{+26}_{-24}$ (best fit values) and $M_{K^{\star\prime}} = 1286^{+26}_{-28}$, $\Gamma_{K^{\star\prime}} = 197^{+41}_{-45}$ (reference values), with energy units given in MeV. Remarkable agreement is found between our best fit values in the JPP and BEJ cases, since the latter yields $M_{K^{\star\prime}} = 1327^{+30}_{-38}$, $\Gamma_{K^{\star\prime}} = 213^{+72}_{-118}$.

## 7. Predicting the $\tau^- \to K^-\eta'\nu_\tau$ decays

We can finally predict the $\tau^- \to K^-\eta'\nu_\tau$ decays. In this case, the good understanding of the tau decays into $K\pi$ and $K\eta$ processes can only be exploited to a limited extent in the $K\eta'$ decays. This is because while the vector form factor dominates the former decays, the scalar one essentially saturates the contribution to the latter. Therefore, this prediction will be more a test of the unitarization results obtained for the corresponding scalar form factor. The only information that one has for this decay channel is the upper limit fixed at ninety percent confidence level by the BaBar Collaboration [11], $BR < 4.2 \cdot 10^{-6}$. We will see that our predictions respect this bound and hint to the soon discovery of this decay channel at Belle-II.

We have ellaborated these predictions using our best fit results in the BW (one-parameter fit) JPP and BEJ (three-parameter fits) cases and the unitarized scalar form factors in the last two approaches. Our results are plotted in Fig. 3 and the branching ratios shown in table 6. It is seen that the decay width is indeed dominated by the scalar form factor, with the vector one

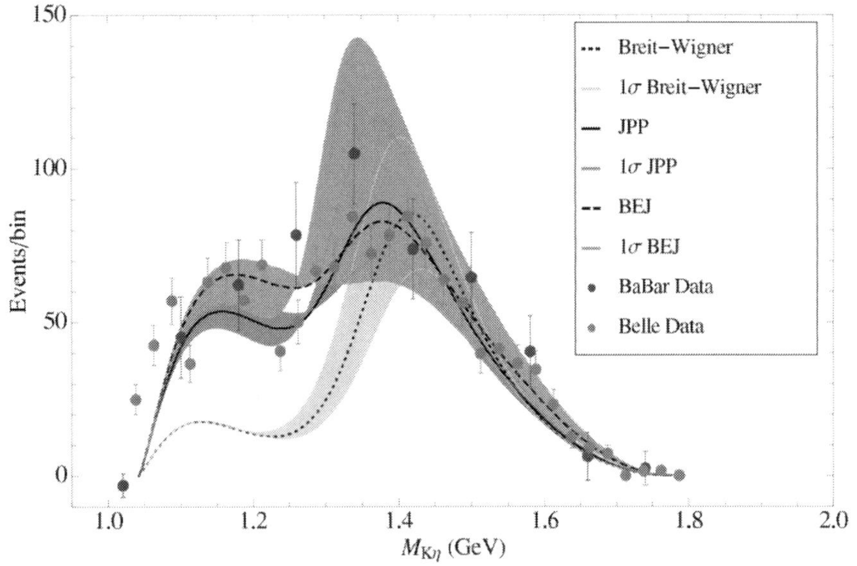


**Figure 2.** BaBar (blue) and Belle (red) [6] data for the $\tau^- \to K^-\eta\nu_\tau$ decays are confronted to the best fit results obtained in the BW (dotted), JPP (solid) and BEJ (dashed) approaches (see the main text for details) which are shown together with the corresponding one-sigma error bands in light green, pink and orange, respectively. The BW curve corresponds to the one-parameter fit while the JPP and BEJ ones correspond to three-parameter fits.

contributing in the interval $(9 \leftrightarrow 15)\%$ to the corresponding decay width. The BW prediction is only shown for reference, but its associated (large) error bands are not shown for clarity. Since the same scalar form factor is used JPP and BEJ and it basically saturates the decay width, the differences between them are tiny.

**Table 6.** Predicted branching ratios for the $\tau^- \to K^-\eta'\nu_\tau$ decays. The BaBar upper limit is also shown [11].

| Source | Branching ratio |
|---|---|
| Dipole Model (BW) (Fit) | $(1.45^{+3.80}_{-0.87}) \cdot 10^{-6}$ |
| JPP (Fit) | $(1.00^{+0.37}_{-0.29}) \cdot 10^{-6}$ |
| BEJ (Fit) | $(1.03^{+0.37}_{-0.29}) \cdot 10^{-6}$ |
| Experimental bound | $< 4.2 \cdot 10^{-6}$ at 90% C.L. |

## 8. Conclusions and outlook

Hadronic tau decays constitute and ideal tool to study the hadronization of QCD currents in a clean environment. Here we have reported our study of the $\tau^- \to K^-\eta^{(\prime)}\nu_\tau$ decays which was triggered by the recent measurements and searches performed by the BaBar and Belle Collaborations [6, 11]. The $K\eta$ channel is dominated by the vector form factor contribution, which can

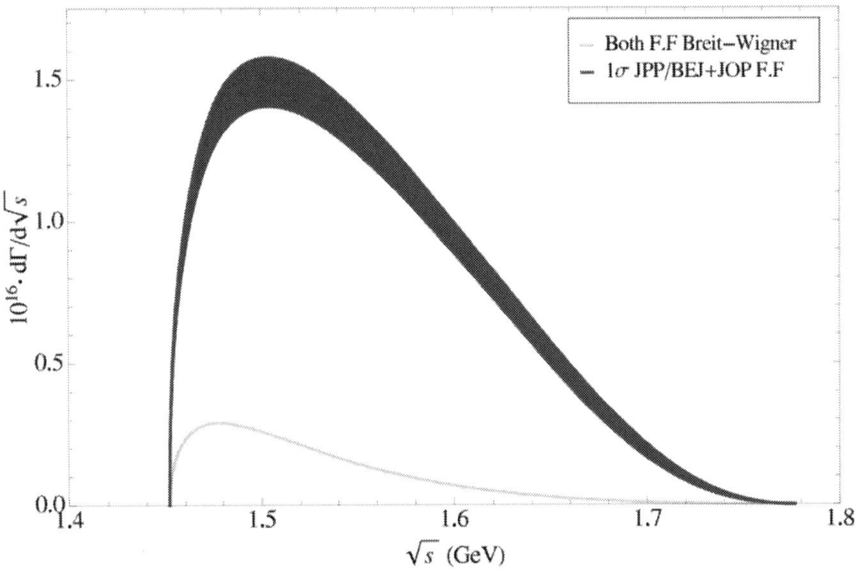


**Figure 3.** The predicted $\tau^- \to K^-\eta'\nu_\tau$ decay width according to BW (green, its big uncertainty is not shown for clarity of the figure), JPP (blue with lower band in red) and BEJ (blue with upper part in pink) is shown. In these last two the scalar form factor corresponds to Ref. [24], which is represented by the author's initials, JOP, in the figure's legend. The corresponding vector form factor contributions, which are subleading are plotted in orange (solid), blue (dashed) and purple (dotted).

be predicted accurately on the basis of previous studies of the $K\pi$ system. In fact, although the information on the $K^\star(892)$ resonance needs to be borrowed from the $K\pi$ studies, the $K\eta$ decays are extremely sensitive to the characteristics of the $K^\star(1410)$ resonance, whose properties can therefore be determined with precission in this analysis.

We have proposed a description of these processes encoding the hadronization properties in the tilded form factors, which yield more compact and symmetric expressions than used previously. After deriving the participating vector and scalar form factors within $R\chi T$ we have discussed the treatment of FSI, which turns out to be crucial in the considered decays. The popular Breit-Wigner (dipole) parametrizations do not resum FSI and break analyticity at leading order by neglecting the real part of the chiral loop functions. Omnès-like resummations resum only the real part of these functions through their exponentiation which violates analyticity (slightly at the numerical level). Finally, a dispersive form factor resums the whole loop function and respects analyticity by construction.

We have found that the Breit-Wigner model fails dramatically in accounting for the data, while the Omnès-like resummation and the dispersive representation do provide good fits to data. Although on theory grounds the simple Breit-Wigner models are always poor approximations its eventual agreement with data in decays with an easy dynamics still motivates its wide use in the analysis of hadronic data. These results for the $\tau^- \to K^-\eta\nu_\tau$ decays show neatly that also phenomenology suggests not to employ them.

Our results for the $K^\star(1410)$ pole mass and width (in MeV) and interaction strength are

$$M_{K^{\star\prime}} = 1327^{+30}_{-38}, \quad \Gamma_{K^{\star\prime}} = 213^{+72}_{-118}, \quad \gamma = -0.051^{+0.012}_{-0.036}, \tag{18}$$

in the dispersive representation (BEJ) and

$$M_{K^{\star\prime}} = 1332^{+16}_{-18}, \quad \Gamma_{K^{\star\prime}} = 220^{+26}_{-24}, \quad \gamma = -0.078^{+0.012}_{-0.014}, \tag{19}$$

in the exponential parametrization (JPP). Our determination of these parameters happens to be competitive with its traditional extraction from $\tau^- \to (K\pi)^-\nu_\tau$ decays. In order to illustrate this, we average the JPP and BEJ determinations from the $K\pi$ [4] and $K\eta$ systems, respectively, to find

$$M_{K^{\star\prime}} = 1277^{+35}_{-41}, \quad \Gamma_{K^{\star\prime}} = 218^{+95}_{-66}, \quad \gamma = -0.049^{+0.019}_{-0.016}, \tag{20}$$

from $K\pi$ and

$$M_{K^{\star\prime}} = 1330^{+27}_{-41}, \quad \Gamma_{K^{\star\prime}} = 217^{+68}_{-122}, \quad \gamma = -0.065^{+0.025}_{-0.050}, \tag{21}$$

from $K\eta$, which opens an alternative way of determining these parameters. New, more precise data on the $\tau^- \to (K\pi)^-\nu_\tau$ and $\tau^- \to K^-\eta\nu_\tau$ decays will make possible a more accurate determination of these parameters. We are currently working [29] on a combined study of Belle's data on the $\tau^- \to K_S\pi^-\nu_\tau$ and $\tau^- \to K^-\eta\nu_\tau$ decays with the purpose of obtaining more accurate isospin averaged values for the slope, $K^\star(892)$ and $K^\star(1410)$ parameters. When BaBar's data for the $\tau^- \to K^-\pi^0\nu_\tau$ and $\tau^- \to K^-\eta\nu_\tau$ data become available the effect of isospin breaking corrections on these quantities could be studied too.

We thus provide TAUOLA with QCD-motivated currents for these processes and improve the understanding of the strange spectral function thanks to our more detailed knowledge of the $\tau^- \to K^-\eta^{(\prime)}\nu_\tau$ decays.

*Acknowledgments*
The author benefited from a grant of the XIV Mexican Workshop on Particles and Fields covering his accommodation expenses and registration fee. This work has been partially funded by Conacyt and DGAPA. The support of project PAPIIT IN106913 is also acknowledged. The research reported here was supported in part by the FPI scholarship BES-2012-055371 (Sergi González-Solís), the Ministerio de Ciencia e Innovación under grants FPA2011-25948 and AIC-D-2011-0818, the European Commission under the 7thFramework Programme through the Research Infrastructures action of the Capacities Programme Call: FP7-INFRA-STRUCTURES-2008-1 (Grant Agreement N. 227431), the Spanish Consolider-Ingenio 2010 Programme CPAN (CSD2007-00042), and the Generalitat de Catalunya under grant SGR2009-00894 (Rafel Escribano, S. G.-S. and P. R.).

## References

[1] E. Braaten, S. Narison and A. Pich, Nucl. Phys. B **373** (1992) 581. E. Braaten, Phys. Rev. Lett. **60** (1988) 1606, Phys. Rev. D **39** (1989) 1458. E. Braaten and C. -S. Li, Phys. Rev. D **42** (1990) 3888. S. Narison and A. Pich, Phys. Lett. B **211** (1988) 183. A. Pich, Conf. Proc. C **890523** (1989) 416. M. Davier, A. Hocker and Z. Zhang, Rev. Mod. Phys. **78** (2006) 1043. A. Pich, "Precision Tau Physics", arXiv:1310.7922 [hep-ph], to be published in "Progress in Particle and Nuclear Physics".

[2] M. Davier, S. Descotes-Genon, A. Hocker, B. Malaescu and Z. Zhang, Eur. Phys. J. C **56** (2008) 305. M. Beneke and M. Jamin, JHEP **0809** (2008) 044. A. Pich, arXiv:1107.1123 [hep-ph]. Published in the Proc. of the Workshop on Precision Measurements of $\alpha_S$ 9-11 Feb 2011. Munich, Germany. D. Boito, M. Golterman, M. Jamin, A. Mahdavi, K. Maltman, J. Osborne and S. Peris, Phys. Rev. D **85** (2012)

Journal of Physics: Conference Series **651** (2015) 012019 doi:10.1088/1742-6596/651/1/012019

093015. R. Barate *et al.* [ALEPH Collaboration], Eur. Phys. J. C **11** (1999) 599. G. Abbiendi *et al.* [OPAL Collaboration], Eur. Phys. J. C **35** (2004) 437. K. Maltman, C. E. Wolfe, S. Banerjee, J. M. Roney and I. Nugent, Int. J. Mod. Phys. A **23** (2008) 3191. M. Antonelli, V. Cirigliano, A. Lusiani and E. Passemar, arXiv:1304.8134 [hep-ph]. K. G. Chetyrkin, J. H. Kuhn and A. A. Pivovarov, Nucl. Phys. B **533** (1998) 473 A. Pich and J. Prades, JHEP **9910** (1999) 004. J. G. Korner, F. Krajewski and A. A. Pivovarov, Eur. Phys. J. C **20** (2001) 259. J. Kambor and K. Maltman, Phys. Rev. D **62** (2000) 093023. S. Chen, M. Davier, E. Gamiz, A. Hocker, A. Pich and J. Prades, Eur. Phys. J. C **22** (2001) 31. E. Gámiz, M. Jamin, A. Pich, J. Prades and F. Schwab, JHEP **0301** (2003) 060. E. Gámiz, M. Jamin, A. Pich, J. Prades and F. Schwab, Phys. Rev. Lett. **94** (2005) 011803. P. A. Baikov, K. G. Chetyrkin and J. H. Kuhn, Phys. Rev. Lett. **95** (2005) 012003. E. Gámiz, M. Jamin, A. Pich, J. Prades and F. Schwab, PoS KAON (2008) 008.

[3] B. Aubert *et al.* [BaBar Collaboration], Phys. Rev. D **76** (2007) 051104. D. Epifanov *et al.* [Belle Collaboration], Phys. Lett. B **654** (2007) 65.

[4] M. Jamin, A. Pich and J. Portolés, Phys. Lett. B **640** (2006) 176, **664** (2008) 78. B. Moussallam, Eur. Phys. J. C **53** (2008) 401. D. R. Boito, R. Escribano and M. Jamin, Eur. Phys. J. C **59** (2009) 821, JHEP **1009** (2010) 031.

[5] R. Escribano, S. González-Solís and P. Roig, JHEP **1310** (2013) 039.

[6] P. del Amo Sanchez *et al.* [BaBar Collaboration], Phys. Rev. D **83** (2011) 032002. K. Inami *et al.* [Belle Collaboration], Phys. Lett. B **672** (2009) 209.

[7] S. Jadach, J. H. Kühn and Z. Was, Comput. Phys. Commun. **64** (1990) 275. S. Jadach, Z. Was, R. Decker and J. H. Kühn, Comput. Phys. Commun. **76** (1993) 361. N. Davidson, G. Nanava, T. Przedzinski, E. Richter-Was and Z. Was, Comput. Phys. Commun. **183** (2012) 821.

[8] O. Shekhovtsova, T. Przedzinski, P. Roig and Z. Was, Phys. Rev. D **86** (2012) 113008. I. M. Nugent, T. Przedzinski, P. Roig, O. Shekhovtsova and Z. Was, Phys. Rev. D **88** (2013) 093012.

[9] A. Pich, Phys. Lett. B **196** (1987) 561. E. Braaten, R. J. Oakes and S. -M. Tse, Int. J. Mod. Phys. A **5** (1990) 2737. B. A. Li, Phys. Rev. D **55** (1997) 1436. G. J. Aubrecht, II, N. Chahrouri and K. Slanec, Phys. Rev. D **24** (1981) 1318.

[10] S. Actis *et al.*, Eur. Phys. J. C **66** (2010) 585.

[11] J. P. Lees *et al.* [BaBar Collaboration], Phys. Rev. D **86** (2012) 092010.

[12] J. Erler, Rev. Mex. Fis. **50** (2004) 200.

[13] F. Ambrosino *et al.* [KLOE Collaboration], Phys. Lett. B **648** (2007) 267.

[14] M. Antonelli, V. Cirigliano, G. Isidori, F. Mescia, *et al.*, Eur. Phys. J. C **69** (2010) 399.

[15] S. Weinberg, Physica A **96** (1979) 327. J. Gasser and H. Leutwyler, Annals Phys. **158** (1984) 142. J. Gasser and H. Leutwyler, Nucl. Phys. B **250** (1985) 465.

[16] J. Bijnens, G. Colangelo and G. Ecker, JHEP **9902** (1999) 020, Annals Phys. **280** (2000) 100. J. Bijnens, L. Girlanda and P. Talavera, Eur. Phys. J. C **23** (2002) 539.

[17] G. Colangelo, M. Finkemeier and R. Urech, Phys. Rev. D **54** (1996) 4403.

[18] G. 't Hooft, Nucl. Phys. B **72** (1974) 461, **75** (1974) 461. E. Witten, Nucl. Phys. B **160** (1979) 57.

[19] A. V. Manohar, Published in 'Les Houches 1997, Probing the standard model of particle interactions, Pt. 2' 1091-1169. A. Pich, Published in 'Tempe 2002, Phenomenology of large $N_C$ $QCD$' 239-258. V. Cirigliano, G. Ecker, H. Neufeld and A. Pich, JHEP **0306** (2003) 012.

[20] G. Ecker, J. Gasser, A. Pich and E. de Rafael, Nucl. Phys. B **321** (1989) 311. G. Ecker, J. Gasser, H. Leutwyler, A. Pich and E. de Rafael, Phys. Lett. B **223** (1989) 425. V. Cirigliano, G. Ecker, M. Eidemüller, R. Kaiser, A. Pich and J. Portolés, Nucl. Phys. B **753** (2006) 139. K. Kampf and J. Novotny, Phys. Rev. D **84** (2011) 014036.

[21] P. D. Ruiz-Femenía, A. Pich and J. Portolés, JHEP **0307** (2003) 003. V. Cirigliano, G. Ecker, M. Eidemüller, A. Pich and J. Portolés, Phys. Lett. B **596** (2004) 96. V. Cirigliano, G. Ecker, M. Eidemüller, R. Kaiser, A. Pich and J. Portolés, JHEP **0504** (2005) 006. P. Roig and J. J. Sanz-Cillero, arXiv:1312.6206 [hep-ph].

[22] F. Guerrero and A. Pich, Phys. Lett. B **412** (1997) 382. A. Pich, J. Portolés, Phys. Rev. **D63** (2001) 093005. J. J. Sanz-Cillero and A. Pich, Eur. Phys. J. C **27** (2003) 587. D. Gomez Dumm, A. Pich and J. Portolés, Phys. Rev. D **69** (2004) 073002. Z. -H. Guo, Phys. Rev. D **78** (2008) 033004. D. G. Dumm, P. Roig, A. Pich and J. Portolés, Phys. Rev. D **81** (2010) 034031, Phys. Lett. B **685** (2010) 158. Z. -H. Guo and P. Roig, Phys. Rev. D **82** (2010) 113016. D. G. Dumm and P. Roig, Phys. Rev. D **86** (2012) 076009, Eur. Phys. J. C **73** (2013) 2528. P. Roig, arXiv:1301.7626 [hep-ph], Ph. D. Thesis, Universitat de València, Servei de Publicacions, November 2010. P. Roig, A. Guevara and G. López Castro, Phys. Rev. D **88** (2013) 033007, arXiv:1401.4099 [hep-ph]. P. Roig, arXiv:1307.8277 [hep-ph]. A. Celis, V. Cirigliano and E. Passemar, arXiv:1309.3564 [hep-ph]. Y. -H. Chen, Z. -H. Guo and H. -Q. Zheng, arXiv:1311.3366 [hep-ph].

[23] V. Mateu and J. Portolés, Eur. Phys. J. C **52** (2007) 325. R. Escribano, P. Masjuan and J. J. Sanz-Cillero, JHEP **1105** (2011) 094. Y. -H. Chen, Z. -H. Guo and H. -Q. Zheng, Phys. Rev. D **85** (2012) 054018. L. Y. Dai, J. Portolés and O. Shekhovtsova, Phys. Rev. D **88** (2013) 056001.

Journal of Physics: Conference Series **651** (2015) 012019 doi:10.1088/1742-6596/651/1/012019

[24] M. Jamin, J. A. Oller and A. Pich, Nucl. Phys. B **587** (2000) 331, **622** (2002) 279, Eur. Phys. J. C **24** (2002) 237, Phys. Rev. D **74** (2006) 074009.
[25] S. Descotes-Genon and B. Moussallam, Eur. Phys. J. C **48** (2006) 553. J. Nebreda and J. R. Peláez, Phys. Rev. D **81** (2010) 054035. S. Prelovsek, T. Draper, C. B. Lang, M. Limmer, K. -F. Liu, N. Mathur and D. Mohler, Phys. Rev. D **82** (2010) 094507. Z. -H. Guo and J. A. Oller, Phys. Rev. D **84** (2011) 034005. M. Doring and U. G. Meissner, JHEP **1201** (2012) 009.
[26] G. P. Lepage and S. J. Brodsky, Phys. Lett. B **87** (1979) 359, Phys. Rev. D **22** (1980) 2157.
[27] D. Gómez Dumm, A. Pich and J. Portolés, Phys. Rev. D **62** (2000) 054014.
[28] J. Beringer *et al.* [Particle Data Group Collaboration], Phys. Rev. D **86** (2012) 010001.
[29] R. Escribano, S. González-Solís, M. Jamin and P. Roig, work in progress.
[30] R. Escribano, A. Gallegos, J. L. Lucio M, G. Moreno and J. Pestieau, Eur. Phys. J. C **28** (2003) 107.

# Monte Carlo simulation of Tetraquark and Meson mixing in a dynamical model of the strong interaction

**I. A. Toledano Juárez**

Instituto de Física, Universidad Nacional Autónoma de México, AP 20-364, México D.F. 01000, México

E-mail: toledano.ivan@fisica.unam.mx

**G. Toledo Sánchez**

Instituto de Física, Universidad Nacional Autónoma de México, AP 20-364, México D.F. 01000, México

E-mail: toledo@fisica.unam.mx

**Abstract.**
We study the formation of a tetraquark state ($q - q - \bar{q} - \bar{q}$) from the presence of two initially uncorrelated $q - \bar{q}$ states, using a Monte Carlo simulation based on a QCD-inspired model (String-Flip), incorporating an effective many-body potential between particles. In a first step we study a two body system linked by a linear potential and elaborate on how to built a variational wave function that can mimic the exact result and from general requirements we propose an anzats for a four body system that can be bound either as two mesons or as a tetraquark. We present preliminary results of the implementation of these ideas and the procedure to evaluate the 4-body potential.

## 1. Introduction

The current understanding of the elementary particle physics phenomena is encoded in the Standard Model of Particle Physics (SM). Hadrons are described as composed of quarks; although the quarks carry a color charge, all known bound states are found to be colorless, they are also referred as color singlets. This phenomenological observation is one of the strongest guidelines to model hadrons.

Hadrons that cannot be matched by any quark-antiquark or three-quark configuration, like the so-called hybrid (quark($q$)-antiquark($\bar{q}$)-gluon), multiquark (quark-antiquark-quark-antiquark-...) or Multimeson ($M1$,$M2$,...) states, may have been already observed. The multiquark ($q - q - \bar{q} - \bar{q}$) tetraquark state is the simplest of its kind and is under study theoretically and experimentally [1, 2, 4, 3]. The main theoretical difficulty faced to describe theses states is that it is required to model the strong interaction in a low energy regime. Thus, an effective potential including the gross general properties of the strong interaction is usually invoked.

In the present work an effective model (String-Flip Model) is adopted to mimic the interaction of quarks as a many-body potential able to confine quarks within color-singlets, to study a system composed of 2 quarks and 2 anti-quarks. All the quarks (antiquarks) carry the same color (anticolor) charge.

Journal of Physics: Conference Series **651** (2015) 012020 doi:10.1088/1742-6596/651/1/012020

**Table 1.** Optimal values of the variational parameter $\lambda_0$ for a given value of $a$ and $b$.

| a\b | 1.2 | 1.1 | 1.0 | 0.9 | 0.8 |
|---|---|---|---|---|---|
| 0.7 | 1.023 | 1.186 | 1.382 | 1.621 | 1.919 |
| 0.6 | 0.982 | 1.136 | 1.322 | 1.548 | 1.829 |
| 0.5 | 0.939 | 1.085 | 1.259 | 1.473 | 1.738 |
| 0.4 | 0.894 | 1.031 | 1.195 | 1.395 | 1.644 |
| 0.3 | 0.848 | 0.975 | 1.129 | 1.315 | 1.547 |

We perform a 3-D Monte Carlo simulation of the system including quantum correlations between particles to determine wether it is energetically more favorable to form a 4-body state (tetraquark or molecular) or 2 mesons, as a function of particle density.

## 2. Variational Wave Function

### *2.1. Two body system*

Let us consider a system composed of quark and an antiquark of mass $m_1$ and $m_2$ respectively. The strong interaction between this pair can be represented by an effective linear potential $V[\vec{r}_i, \vec{r}_j] = k|\vec{r}_i - \vec{r}_j| = kr_{ij}$ where $k$ is an interaction constant. For a three-dimensional two-particle system (meson), one can analytically solve the corresponding Schrödinger equation for the radial part. The corresponding eigen-energies are given by

$$E_n = -|\xi_n|\left(\frac{k^2}{2\mu}\right)^{\frac{1}{3}}; \quad n = 1, 2, \ldots \tag{1}$$

where $\mu = \frac{m_1 m_2}{m_1 + m_2}$ is the reduced mass, and $\xi_n$ are the roots of the Airy function.
If we consider units in which $m_1 = m_2 = m = k = 1$ the corresponding energy for the ground state takes the following form: $E_0 = \xi_0 \sim 2.338$.
Previous studies have explored how to approach a variational wave function that can account for this energy [5] in the 1-Dim case. Following that result, we have explored a family of 3-D variational wave functions that can also approach this value. The general wave function take the following

$$F_\lambda(r) = r^a e^{-\lambda r^b} \tag{2}$$

where $a$, $b$ and $\lambda$ are variational parameter and $r$ is the relative coordinate of the two-body system.
We have sampled $F_\lambda$ analytically, for different values of $a$ and $b$, and found the value of $\lambda$ that minimizes the energy; this particular value is labeled as $\lambda_0$. Table 1 exhibits the corresponding sets of optimal values. In Table 2 we show the corresponding minimal energy of the system.
Our analysis shows that using $a = 1/2$ and $b = 1$, the wave function reproduces the exact minimum energy value within a 3% error, corresponding to an optimal variational parameter $\lambda_0 = 2^{1/3}$. Thus, these values will be taken as a reference when extending the discussion to larger systems.

### *2.2. Four-body system*

Using the previous results we can now extend our study to a four-body system. To study such system, we need a model able to capture the properties of the strong interaction and of the

Journal of Physics: Conference Series **651** (2015) 012020 doi:10.1088/1742-6596/651/1/012020

**Table 2.** Minimum energy $E(\lambda_0)$ for a given value of $a$ and $b$.

| a\b | 1.2 | 1.1 | 1.0 | 0.9 | 0.8 |
|---|---|---|---|---|---|
| 0.7 | 2.393 | 2.387 | 2.387 | 2.395 | 2.416 |
| 0.6 | 2.381 | 2.378 | 2.382 | 2.396 | 2.422 |
| 0.5 | 2.370 | 2.372 | 2.380 | 2.400 | 2.433 |
| 0.4 | 2.362 | 2.369 | 2.383 | 2.408 | 2.450 |
| 0.3 | 2.359 | 2.370 | 2.391 | 2.423 | 2.475 |

constituent quarks. In particular, we are interested in capturing the expected behavior as a function of the quark density. At low particle-densities, where the distance between quarks is large, the model has to reproduce a system of hadrons, clustering the particles into groups of 2 or 4 bodies. In addition, at high densities the quarks must behave like a free Fermi gas, and the transition between these two limits must be dynamically generated. The String-Flip Model [6, 7] has these characteristics and was used to made this study.

Generalizing the previous result to include different configurations, other than the meson one, at different densities, we propose the following variational wave function, which reproduces the exact energy at low densities and becomes a Fermi Gas at high ones.

$$\Psi_\lambda = \Phi_{fg}\left((r_A r_B)^{\frac{1}{2\lambda_0}}\, e^{-V}\right)^{\lambda} \tag{3}$$

where $\lambda$ is the single variational parameter, $V$ is a many-body potential, $r_A r_B$ is a product of relative distances for the interaction (see below), and $\Phi_{fg}$ is the Fermi Gas wave function given by a product of Slater determinants, one for each color-flavor combination of quarks, using the wave function of a particle in a box.

In order to recover the two-body result, we propose the following form for $r_A r_B$:

$$r_A r_B = \left(\frac{V^2 - \sum_{i=1}^{4} r_{ij}^2}{2}\right) \tag{4}$$

For a two meson system, the total potential V is the sum of two linear potentials and the $r_{ij}$ distances are just the corresponding relative distances among the particles in a single meson. For a four-body system, the corresponding potential takes a more elaborated form and the $r_{ij}$ distances become relative distances among the links required to get the lowest potential energy. We expect this form of the variational wave function to be able to describe the departure from the two-mesons case into the four-body case.

Using such variational wave function, the expectation value of the Hamiltonian operator can be separated into three parts, a purely free kinetic part ($T_{fg}$) a kinetic part afected by the interaction ($W$) and the potential energy ($V$). This structure is useful to find the variational parameter that minimizes it, in the following form:

$$\langle H\rangle_\lambda = T_{fg} + \langle W\rangle_\lambda + \langle V\rangle_\lambda \qquad \frac{\partial\langle H\rangle_\lambda}{\partial\lambda} = \frac{E(\lambda)}{\partial\lambda} = 0 \tag{5}$$

To numerically determine the expectation value we make use of Monte Carlo techniques, in particular we employ a Metropolis Algorithm. In Figure 1 we have plotted the energy of the two body system as a function of the variational parameter. The exact solution is shown as a solid

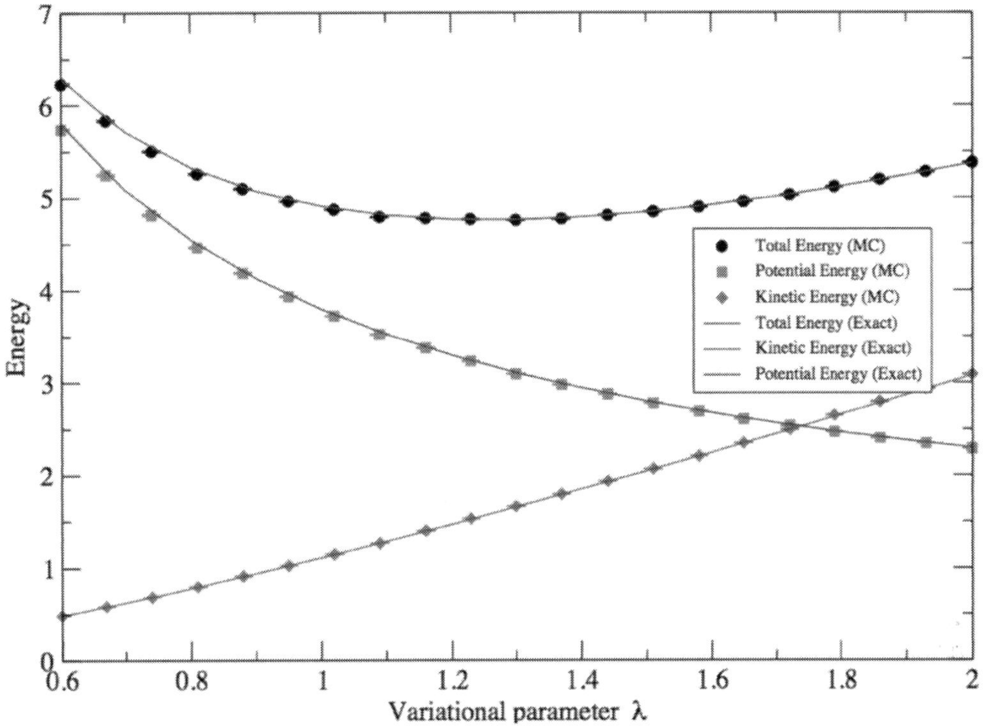


**Figure 1.** Energy comparisons between analytical and numerical results in the case of two isolated mesons.

line and the symbols correspond to the result of computing the energy using the Monte Carlo program. We can appreciate the accuracy that can be achieved to reproduce the exact result.

## 3. Many-Body Potential

The many-body potential is defined as the clustering of quarks into color-singlets objects that minimizes the potential energy. An optimal value for the potential energy of a 4-body configuration will be important to differentiate it from a meson system. In the case of four bodies, we can distinguish five possible configurations for the system: meson (two configurations), molecular meson (two configurations) and tetraquark (one configuration), as shown in Figure 3. Every line of interaction, or gluon flux tube, is simulated with an effective linear potential. In particular, a 4-body configuration requires two additional vector points $(\vec{k},\vec{l})$ for an optimal potential energy, which need to be determined in a non trivial way, this is the case for both the tetraquark and molecular configurations.

## 4. Preliminary Results

The determination of the optimal potential is at the hart of the description of the system. We present here preliminary results on the extensive study made to improve the numerical estimation and the range of validity.
Let us take for definiteness the effective tetraquark potential (this can be easily generalized to the molecular case), which is given by

$$V_{4Q-tetra} = \sum_{i=1}^{2} V\left[\vec{r}_{q_i}, \vec{k}\right] + \sum_{i=1}^{2} V\left[\vec{r}_{\bar{q}_i}, \vec{l}\right] + V\left[\vec{k}, \vec{l}\right] \tag{6}$$

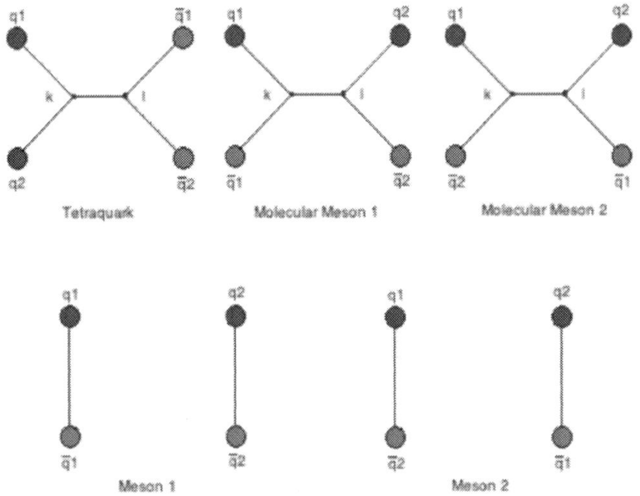


**Figure 2.** Possible configurations of the 4-particle system: meson, molecular meson, and tetraquark

We compute these additional vector points using an additional algorithm that samples discrete values for the dimensional components of these vectors, within the confined space of the 4-body system. We chose the best possible configuration of $\vec{k}$ and $\vec{l}$ and then a second sampling can be made around these new values to optimize the potential energy.
The Nelder-Mead method were also considered to compute $\vec{k}$ and $\vec{l}$. This method uses de concept of a simplex (polytope of $N+1$ vertices in $N$ dimensions) to approximate a local optimum of a problem with $N$ variables, such as the many-body potential. This method was discarded because it was too sensitive to initial conditions and approached mostly a local minimum.

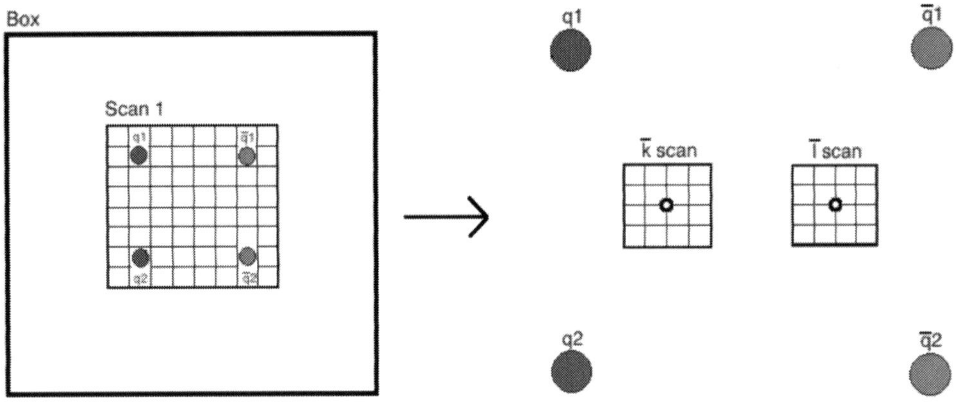


**Figure 3.** General scan for vectors $\vec{k}$ and $\vec{l}$ for a 4-body potential. Starting with a scan within the system confinement, an individual scan for the vector points is made.

## 5. Perspectives

The following steps in our study is to compute the evolution of the variational parameter $\lambda$ as a function of the particle density $\rho$, considering several cases: i) same mass for all quarks, ii) one and two different masses.

This procedure is implemented in each case by giving a value for $\rho$, which, for a fixed number of particles, means changing the size of the box where the quarks are sampled, we compute the energy of the system as a function of $\lambda$; the minimal energy determines the optimal value of $\lambda$ for the corresponding value of $\rho$.

Given these optimal values we can characterize the tetraquark and molecular state and perform a simultaneous simulation of all possible configurations to determine favorable conditions for each one.

## Acknowledgments

We acknowledge the partial support from DGPAPA-UNAM under the project PAPIITIN106913 and the Organizing committe.

## References

[1] M. Ablikim et al. (BES III Collaboration), Phys. Rev. Lett. 110, 252001 (2013).
[2] Z.Q. Liu et al. (Belle Collaboration), Phys. Rev. Lett. 110, 252002 (2013).
[3] Chengrong Deng, Jialun Ping and Fan Wang, arXiv:1402.0777 [hep-ph]
[4] J. Vijande, A. Valcarce, J. M. Richard, Phys. Rev. D 76, 114013 (2007)
[5] J. I. Casaubon and J. P. Cosentino Turk J Phys. 31 (2007), 117-121
[6] G. Toledo Sánchez and J. Piekarewicz Phys. Rev. C 65, 045208 (2001)
[7] G. Toledo Sánchez and J. Piekarewicz Phys. Rev. C 70, 035206 (2004)

Journal of Physics: Conference Series **651** (2015) 012021 doi:10.1088/1742-6596/651/1/012021

# Determining the magnetic dipole moment of the rho meson

**G. Toledo Sánchez**

Instituto de Física, Universidad Nacional Autónoma de México, AP 20-364, México D.F. 01000, México

E-mail: `toledo@fisica.unam.mx`

**D. García Gudiño**

Instituto de Física, Universidad Nacional Autónoma de México, AP 20-364, México D.F. 01000, México

E-mail: `david.garcia@fisica.unam.mx`

**Abstract.** We elaborate on the procedure followed to determine the magnetic dipole moment of the rho meson, $\mu_\rho = 2.1 \pm 0.5\,[e/2m_\rho]$, from the $e^+e^- \to \pi^+\pi^-2\pi^0$ process. Using preliminary data from the BaBar Collaboration and a vector meson dominance approach.

## 1. Introduction

The magnetic dipole moment ( D ) of the elementary particles has been a key tool to explore underlying properties of the fundamental interactions. For states made up of quarks, the properties of the strong interaction in the low energy regime can be explored. For long lived states, like the proton and neutron, the measurement of their corresponding D has helped to understand its relationship as a consequence of their quark structure. owever, to date, there is no measurement of the D of any vector meson[1], a state made up of quark an antiquark in a spin 1 configuration. This fact is related to their extremely short lifetimes ($\approx 10^{-23}$ s), which prevents experimentalists to apply standard D measurement techniques. Alternatives to determine the D , by indirect means, invoke the fact that the radiation emitted from the vector meson carries information of its electromagnetic structure[2] and thus, provided the dominant electric radiation is known, the subleading D effect can be identified. This is the same approach used to determine the D of the $W$ gauge boson, a particle with the same characteristics of a vector meson but of fundamental nature.

In this proceedings we elaborate on the procedure we followed to perform the first determination of the D of a vector meson, namely the D of the $\rho$ meson [3], obtained from the $e^+e^- \to \pi^+\pi^-2\pi^0$ process, using preliminary data from the BaBar Collaboration[4]. The $\gamma^* \to 4\pi$ vertex is modeled in the vector meson dominance (V D) approach.

## 2. The electromagnetic vertex

The $e^+e^- \to \pi^+\pi^-2\pi^0$ process has been measured for several experiments in the low energy regime. The leading channel contribution to the total cross section was found to be that involving

Journal of Physics: Conference Series **651** (2015) 012021 doi:10.1088/1742-6596/651/1/012021

the $\omega - \pi$ mesons intermediate state. easurements at increasing energies made clear that this process was also sensitive to the channel where the pions are produced in pairs through the $\rho - \rho$ meson resonant states. This particularity opens the door to explore the $\rho$ electromagnetic vertex and by consequence the corresponding electromagnetic multipoles. The study we have performed in [3] uses these facts to determine the D by fitting the experimental cross section data, modeled in the V D approach including all the relevant intermediate states, while using other observables to fix all the remaining parameters.
Let us begin with a brief discussion of the general features of such electromagnetic vertex. For a vector particle (V), the vertex for the process $V(q_1, \epsilon) \to V(q_2, \eta)\gamma(q)$ is defined from the electromagnetic current

$$< V(q_2)|J^{\mu}_{EM}(0)|V(q_1) > \equiv \eta^{\dagger}_{\nu}\epsilon_{\lambda}\Gamma^{\mu\nu\lambda}, \qquad (1)$$

where $q_i$ are the momenta and $\epsilon$ and $\eta$ are the corresponding polarization tensors. The CP conserving electromagnetic vertex $\Gamma^{\mu\nu\lambda}$ can be decomposed into the following Lorentz structures

$$\begin{aligned}\Gamma^{\mu\nu\lambda} &= \alpha(q^2)g^{\nu\lambda}(q_1+q_2)^{\mu} + \beta(q^2)(g^{\mu\nu}q^{\lambda} - g^{\mu\lambda}q^{\nu}) \\ &- \gamma(q^2)(q_1+q_2)^{\mu}q^{\nu}q^{\lambda}. \qquad (2)\end{aligned}$$

The electromagnetic form factors $\alpha(q^2)$, $\beta(q^2)$ and $\gamma(q^2)$ dependence on $q^2$ is modeled in V D by including the hadronic degrees of freedom of relevance in the corresponding energy regime. In the static limit, such form factors are related to the electromagnetic multipoles as follows: $|Q| = \alpha(0)$ is the electric charge (in $e$ units), $|\vec{\mu}| = \beta(0) \equiv 1 + \kappa + \lambda$ is the magnetic dipole moment (in $e/2M_V$ units) and the electric quadrupole is $|X_E| = \kappa - \gamma(0)M_V^2 \equiv \kappa - \lambda$ (in $e/M_V^2$ units), where we have included the relation to the parameters $\kappa$ and $\lambda$ which are also of common use in the literature to refer to the electromagnetic multipoles of spin-1 particles[5, 6]. For instance, at tree level, the standard model gauge structure predicts for the $W$ gauge boson to have $\alpha(0) = 1$, $\beta(0) = 2$ and $\gamma(0) = 0$ ($\kappa = 1$ and $\lambda = 0$), corresponding to $|Q| = 1$, $|\vec{\mu}| = 2$ and $|X_E| = 1$. These values are usually taken as a reference for vector mesons.
The predictions for the D $\rho$ meson computed using approaches to the strong interaction, have been found to lay in the region from 1.9 to 3 in $e/2M_\rho$ units.
The $e^+e^- \to \pi^+\pi^-2\pi^0$ process has been measured by several experiments in a direct way at low energies[7], and preliminary data is available from the BaBar collaboration[4] using initial radiation technique in a wider range.

## 3. Modeling the $e^+e^- \to \pi^+\pi^-2\pi^0$ process

This process involves a standard electromagnetic vertex for the electron-positron annihilation, which is properly accounted by a QED description. The description we followed for the hadronic part is based on V D. The energy range to be described goes from threshold up to 2.2 GeV. pon the pair annihilation, the photon is considered as becoming a $\rho$ and $\rho'$ neutral vector mesons, under the V D rules, with a relative phase of $180^0$.
We have considered the channels contributing to the four pions production, including the exchange of the $\pi$, $\omega$, $a_1$, $\sigma$, $f(980)$, $\rho$ and $\rho'(1450)$ mesons. Therefore, we are left with seven generic channels, each one accounting for several specific diagrams, corresponding to the allowed permutations of the momenta due to Bose-Einstein symmetry and charge con ugation. ote that not all the diagrams are independent, gauge invariance condition requires some of them to behave in such a way that the conditions is fulfilled. Of particular in the description is the case when the $\rho'$ is included. So far there is a very poor information on this particle and its decay modes. The structure of the triple vector meson vertex is expected to follow the same form as the electromagnetic vertex previously discussed but for a set of different couplings. We have made the assumption that for such a vertex involving the $\rho$ and $\rho'$ it takes exactly the same value than for the vertex involving only the $\rho$ meson. This assumption has been found to

Journal of Physics: Conference Series **651** (2015) 012021 doi:10.1088/1742-6596/651/1/012021

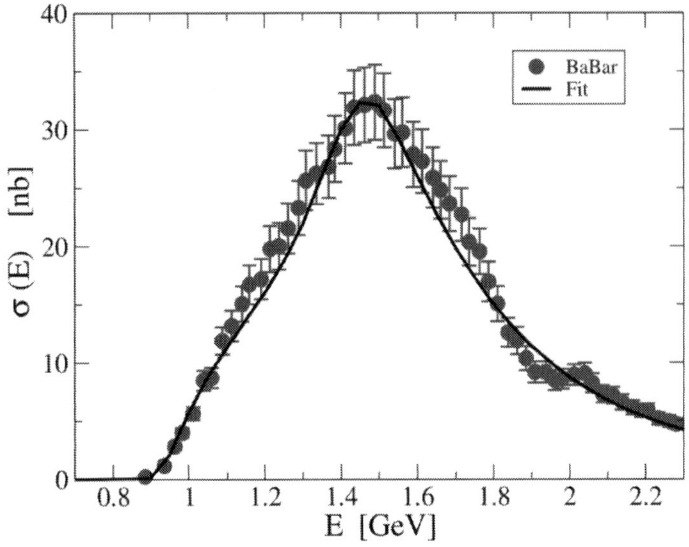


**Figure 1.** Fit to the BaBar data for the total cross section $e^+e^- \rightarrow \pi^+\pi^-2\pi^0$. We use the D as the only free parameter, while the other involved parameters are fixed from different observables.

be appealing[8] in the description of this process.
As we mentioned in the introduction, vector mesons are extremely short lived particles. In order to account for this fact a proper scheme must be used to include its unstable nature without breaking the electromagnetic gauge invariance. For this purpose, the vector meson propagator is considered to take the complex mass form.
Provided the couplings involved in the included channels are determined from different observables, the ones involving the electromagnetic vertex $\Gamma^{\nu\beta\gamma}$, namely, the $\beta$ and $\gamma$ parameters can be determined by fitting the total cross section of the $e^+e^- \rightarrow \pi^+\pi^-2\pi^0$ process.

In Figure 1, we show the total cross section data with a 10 systematic error bars [4], and the fit corresponding to $\beta$ as a free parameter and $\gamma = 0$. The fit considering $\beta$ and $\gamma$ as free parameters favors the same $\beta$ and restricts $\gamma$ to be in the range $(-1.1, 0.1)$. That is, the $\beta$ parameter accounts for the global description, while the $\gamma$ contribution enters at the end region. To determine the cross section error bars, we have taken into account the combined uncertainties coming from the couplings of the different channels, assumed as no correlated. We also explored the role of the model assumption regarding the $\rho'$ triple boson vertex, the global combination of couplings and mass was found to be consistent with data for upto a 10 deviation from the combination for the $\rho$. We determine the $\beta$ parameter error bar considering it as the responsible of the total uncertainties. In addition, to account for the model dependence, we have added a 20 error (added in quadrature). Thus, corresponding to a D

$$\mu_\rho = 2.1 \pm 0.5 \, [\frac{e}{2m_\rho}]. \tag{3}$$

## 4. Conclusions

In summary, we have determined the magnetic dipole moment of the $\rho$ meson, by studying its effect in the $e^+e^- \rightarrow \pi^+\pi^-2\pi^0$ cross section, in the V D approach. The channel that contains the electromagnetic vector meson vertex becomes relevant for energies between 1.5 and 2.2 GeV, while the remaining channels are always subdominant. We have found that the best fit to the BaBar data implies a value for the D of the $\rho$ meson of $\mu_\rho = 2.1 \pm 0.5\,[\frac{e}{2m_\rho}]$. The quoted error bar takes into account the uncertainties coming from the couplings of the different channels and model assumptions, definite data on this process and detailed information on the $\rho'$ meson will be very useful for a more refined analysis.

**Acknowledgements** We acknowledge the partial support from DGPAPA-PAPIIT, A under grants IB101012 and I 106913. We thank to the organizing committee for a very pleasant conference.

## References

[1] . Beringer et al. (Particle Data Group), Phys. Rev. D **86**, 010001(2012).
[2] V. I. akharov, L. A. ondratyuk and L. A. Ponomarev, Sov. . of Nucl. Phys. **8** 456(1 6 ).
[3] D. Garcia Gudino and G. Toledo Sanchez, ar iv: 1305.6345 [hep-ph]
[4] V. P. Druzhinin, ar iv: 0 10.3455 [hep-ex]
[5] . Hagiwara, R. D. Peccei, D. eppenfeld and . Hikasa, Nucl. Phys. B282, 253 (1 8 ).
[6] . F. Nieves and P. B. Pal, Phys. Rev. D **55** 3118 (1 ).
[ ] M. N. Achasov et al. (SND Collaboration) . Exp. Theor. Phys. **96** 8 (2003) . Exp. Theor. Phys. **109** 3 (200 ).
[8] H. Czyz, . H. uhn and A. Wapienik, Phys. Rev. D **77**, 114005 (2008)

Journal of Physics: Conference Series **651** (2015) 012022 doi:10.1088/1742-6596/651/1/012022

# Towards the low energy frontier

**J. Barranco,D. Delepine**

Departamento de Física, División de Ciencias e Ingeniería, Campus León, Universidad de Guanajuato

**Vannia Gonzalez Macias**

Department of Physics & Astronomy, University of California Riverside
Departamento de Física, División de Ciencias e Ingeniería, Campus León, Universidad de Guanajuato

**Abstract.** Unanswered and interesting questions arise in the leptonic sector of the standard model of particle physics: Are there new sources of CP violation? Are the neutrinos Dirac or Majorana particles? Are there more than one Higgs doublet that modify the leptonic or semileptonic decays of mesons? All these questions may be answered by a detailed study of low energy experiments. In this talk we present a short review of some examples where using present or future data on D meson decays, neutrino scattering on electron or coherent scattering on nucleus and neutrinos produced in accelerators may help us to answer those questions.

## 1. Introduction

High energy physics experiments, such as LEP, HERA, TEVATRON, and more recently the LHC, have portrayed a very accurate picture of the Standard Model (SM) description of particle interactions up to energies above the electroweak scale. Future runs of the LHC will search for new physics (NP) and probably probe energy scales up to 13 TeV in center of mass energy. On the other hand, low energy scale experiments (below the electroweak scale) or the so-called *intensity frontier* (figure 1) may shed some light in the search for physics beyond the Standard Model due to their possibility of getting high statistics and hence indirect observables of NP (see for example [1]). A remarkable example of how low energy experiments have changed dramatically the SM is the discovery of neutrino oscillations thanks to the study of solar, atmospheric and reactor neutrinos whose energies range from a few GeV down to MeV [2]. Additional low energy tests include rare processes, forbidden or highly suppressed in the SM, such as electric dipole moments, lepton number violation, flavor changing neutral processes, etc. Other studies include particle decays, CP asymmetries and flavor transitions, allowed within the SM but with sufficient precision to test NP.

Many of the low energy processes mentioned above can be studied within a model-independent framework following an effective Lagrangian approach to parametrize the NP effects. Below the electroweak scale, after integrating out the W and Z bosons (high momentum modes), the dynamics are described by a set of effective operators, that include the SM high momentum modes plus the NP high momentum modes. The dominant effects will be generated by the lowest dimensional effective operators, an effective four Fermi interaction usually called Non Standard Interaction (NSI). The addition of those NSI may enhance the decay rate or the cross

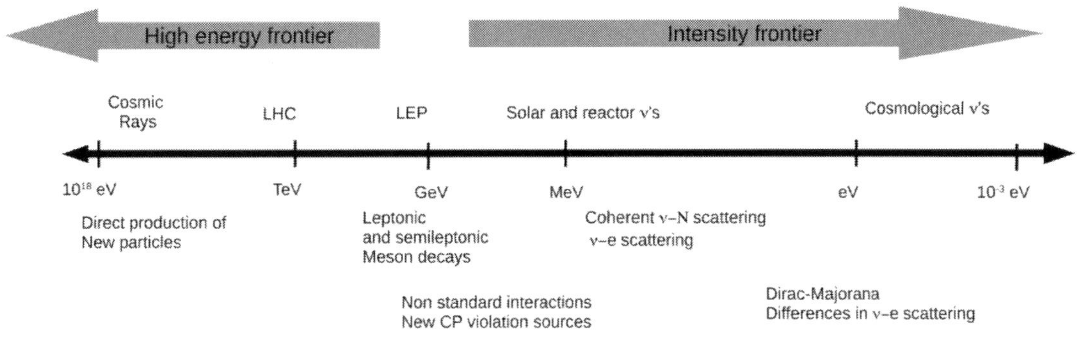


**Figure 1.** Low energy processes such as leptonic and semileptonic decay of mesons, neutrino scattering both in electrons and nucleus may reveal complementary or new constraints on models beyond the SM

section of several processes that can be later compared with the measured values which impose constraints on the Wilson coefficients that parametrize such NSI.

It this conference paper we review several cases that show the usefulness of this approach in the low energy frontier, spanning energies ranges from GeV to eV. We also address other interesting phenomena beyond the standard model, such as novel sources of CP violation or the feasibility of distinguishing the Dirac or Majorana nature of the neutrino by studying low energy $\nu{-}$e scattering. The article is organized as follows: In Sec. 2 we review a set of model independent constraints imposed on NSI by studying different processes. Firstly, we recall constraints on NSI for a $c \to s$ transition process using both leptonic and semileptonic D meson decays [3]. It is found that those particular processes impose restrictive constraints over leptoquark models for the second generation of quarks. Secondly, at even lower energies, of order $\sim 10$ MeV $- 10^{-1}$ MeV, the $\nu - e$ elastic scattering in nuclear reactor, accelerator and solar $\nu$-detectors, are a useful probe. We recall a global analysis done considering non-universal and flavor changing NSI including all relevant experiments and future low energy solar neutrino detector prospects [9, 10]. As a final example, for extremely low threshold energies, the coherent neutrino scatterings may be very sensitive to NSI of neutrinos with quarks, that is neutral quark and lepton operators products [12], due to the enhancement of the cross section and the fact that neutrinos can scatter coherently not only on the nucleons inside the nucleus but also on the atom itself. Final part of Sec. 2 revise its potential to make precision tests of NP.

The possibilities of finding new CP violating sources are explored in section 3. It is a study of the feasibility of generating new sources of CP and flavor violation (coming from NP) on neutrino oscillation experiments at low energies, of order $\sim 1$ GeV, using a standard S-matrix formalism with the relevant set of effective operators [8]. We summarize the relevant results using a generic framework based on quantum field theory to get a simple expression for the CP asymmetry in neutrino oscillations without imposing any assumptions on the effective operators generated by NP.

Finally, once the neutrino polarization evolution is considered, the $\nu{-}e$ scattering process may shed some light into the neutrino nature: Dirac or Majorana. The change of polarization may be achieved in astrophysical environments with strong magnetic fields. Section 4 summarizes this interesting feature, examined in [11].

Journal of Physics: Conference Series **651** (2015) 012022 doi:10.1088/1742-6596/651/1/012022

## 2. Constraining NSI with low energy experiments

There have been significant improvements in the experimental measurements of the meson decays branching ratios, and we are now entering in the precision measurement of the neutrino properties (mixings and masses). This experimental data in combination with accurate theoretical calculations may be used to determine the accuracy of the SM and to search for NP. This allows us to constrain systematically a set of Wilson coefficients using some low energy data that parametrizes NP. Some examples are:

(i) **D meson decays (GeV range):** D meson physics gained special interest when it was first pointed out that the leptonic decay branching ratio measurement was in minor disagreement with the SM [20]. Recent measurements are now within agreement but there is however a consistent small deviation in each decay channel (semileptonic and leptonic) (CLEO [13] , [14, 15, 16], Belle [17]) . For the $\Delta C = \Delta S$ leptonic and semileptonic D meson decays, the new particle state should couple to the leptons and the second generation of quarks, leaving an effective four Fermi interaction parametrized by the Wilson coefficients $C^{I,L,R}_{q_1q_2\ell\nu}$ as

$$-\frac{\mathcal{L}_{NP}}{G_F} = \sum_{\substack{c,s,\ell,\nu \\ I=S,V,T \\ P_{1,2}=L,R}} C^{I,P_1P_2}_{q_1q_2\ell\nu}(\bar{q}_1\Gamma^I P_1 q_2)\cdot(\bar{\nu}_L\Gamma_I P_2\ell)\,. \tag{1}$$

Any kind of intermediate state, such as scalars, vectors or even tensors, are allowed. A model independent analysis is shown in [3] , combining the total leptonic and semileptonic branching ratios of D meson decays (previous analysis were done for leptonic or semileptonic decays independently [20, 21, 22]). The latest results on the calculations for the relevant form factors were implemented, which have been improved significantly reaching a remarkable precision [18, 19]. The $q^2$ distributions for the $D^+ \to \bar{K}^0 e^+\nu_e$ and $D^0 \to K^- e^+\nu$ decays are also considered. This analysis is restricted to three fermion family models of physics BSM, as the CKM matrix element is deduced from unitarity contraints, $W \to cs$ decay and neutrino-nucleon scattering [23].

(ii) **Constraining $\nu$-e NSI** Neutrino-electron scatterings are described by pure neutral leptonic operator products that induce non standard flavor changing interactions.
Now we will consider NSI in the leptonic sector for neutral currents. The general class of non-standard interactions is described via the effective four fermion Lagrangian,

$$-\mathcal{L}^{eff}_{\rm NSI} = \varepsilon^{fP}_{\alpha\beta}2\sqrt{2}G_F(\bar{\nu}_\alpha\gamma_\rho L\nu_\beta)(\bar{f}\gamma^\rho Pf)\,, \tag{2}$$

where $\varepsilon^{fP}_{\alpha\beta}$ parametrize the strength of the NSI. In this particular case we consider only electrons, hence $f = e$). The current sensitivity on NSI as inferred from a global analysis of processes involving (anti)-neutrinos and electrons combines the relevant experimental "neutrino counting" data from $e^+ + e^- \to \nu + \bar{\nu} + \gamma$ obtained by the four LEP collaborations and with all the $\nu_e + e \to \nu_e + e$ data obtained by LSND, and the $\bar{\nu}_e + e \to \bar{\nu}_e + e$ scattering measured in reactor experiments (Irvine, MUNU and Rovno) together with $\nu_\mu + e \to \nu_\mu + e$ and $\bar{\nu}_\mu + e \to \bar{\nu}_\mu + e$ measured at CHARM II.

(iii) **Constraining NSI with coherent neutrino-Nucleus ($\nu$−N) scatterings (eV range):** It is one of the few processes predicted by the SM that has not been experimentally detected . This process takes place when the momentum transfer, $q$, is small compared with the inverse nucleus or atom size. The typical inverse sizes of most nuclei are in the range from 25 to 150 MeV. Therefore the condition for full coherence in the neutrino-nuclei scattering is well satisfied for reactor neutrinos and also for solar, supernovae neutrinos and artificial neutrino sources. The effective Lagrangian at low energies is:

$$\mathcal{L}^{NC}_{\nu Hadron} = -\frac{G_F}{\sqrt{2}} \sum_{\substack{q=u,d \\ \alpha,\beta=e,\mu,\tau}} \left[\bar{\nu}_\alpha \gamma^\mu (1-\gamma^5)\nu_\beta\right] \left( f^{qL}_{\alpha\beta} \left[\bar{q}\gamma_\mu(1-\gamma^5)q\right] + f^{qR}_{\alpha\beta} \left[\bar{q}\gamma_\mu(1+\gamma^5)q\right] \right), \tag{3}$$

where the NSI parameters have been added to the SM effective coupling constants: $f^{uL}_{\alpha\alpha} = \rho^{NC}_{\nu N}\left(\frac{1}{2} - \frac{2}{3}\hat{\kappa}_{\nu N}\hat{s}^2_Z\right) + \lambda^{uL} + \varepsilon^{uL}_{\alpha\alpha}$, $f^{dL}_{\alpha\alpha} = \rho^{NC}_{\nu N}\left(-\frac{1}{2} + \frac{1}{3}\hat{\kappa}_{\nu N}\hat{s}^2_Z\right) + \lambda^{dL} + \varepsilon^{dL}_{\alpha\alpha}$ and $f^{qP}_{\alpha\beta} = \varepsilon^{qP}_{\alpha\beta}$. $(\alpha \neq \beta;\ q = u, d;\ P = L, R)$
Here $\hat{s}^2_Z = \sin^2\theta_W = 0.23120$, $\rho^{NC}_{\nu N} = 1.0086$, $\hat{\kappa}_{\nu N} = 0.9978$, $\lambda^{uL} = -0.0031$, $\lambda^{dL} = -0.0025$ and $\lambda^{dR} = 2\lambda^{uR} = 7.5 \times 10^{-5}$ are the radiative corrections [23].

With the help of Lagrangians (1,2,3) the relevant decay rates and cross sections are computed and the relevant Wilson parameters are fitted through a standard $\chi^2$ analysis. In table 1 we summarize the constraints on the relevant Wilson coefficients for the previous observables.

**Table 1.** Constrains on the Wilson coefficients that parameterize non standard interactions obtained with different low energy experiments

| D meson bounds | Neutrino NSI | Coherent bounds |
|---|---|---|
| $0.072 < C^{VLL}_{scl\nu} < 0.14$ | $-0.03 < \varepsilon^L_{ee} < 0.08$ | $\lvert\varepsilon^{dV}_{ee}\rvert < 0.003$ |
| $0.057 < C^{VRL}_{scl\nu} < 0.13$ | $0.004 < \varepsilon^R_{ee} < 0.151$ | $\lvert\varepsilon^{uV}_{ee}\rvert < 0.032$ |
| $0.00 < C^{SRR}_{scl\nu} < 0.13$ | $\lvert\varepsilon^L_{\mu\mu}\rvert < 0.03$ | $\lvert\varepsilon^{dV}_{\tau e}\rvert < 0.032$ |
| $-0.12 < C^{SLR}_{scl\nu} < 0.00$ | $\lvert\varepsilon^R_{\mu\mu}\rvert < 0.03$ | $\lvert\varepsilon^{uV}_{\tau e}\rvert < 0.036$ |
| | $-0.46 < \varepsilon^L_{\tau\tau} < 0.24$ | |
| | $-0.25 < \varepsilon^R_{\tau\tau} < 0.43$ | |

Under some simplifying assumptions, the model independent constraints can be mapped to some particular models, exemplifying the usefulness of this kind of analysis.

The bounds on the Wilson coefficients coming from current neutrino experiments are loose. Nevertheless, the bounds obtained from the coherent neutrino-nuclei scattering are promising. In particular, we have estimated the sensitivity for a detector's mass of 1Kg and a time of 1 year of data taking, i.e. a facility such as the Texono detector proposal. The sensitivity for the mass of an extra gauge boson $Z'$ in the particular *chi*-model arising from our constraint on $\epsilon$ corresponds to $M_{Z'} < 792$GeV at 90%C.L. Another interesting where we can apply our bounds is the leptoquark model. By fixing the coupling of the leptoquark $\lambda_q = 4\pi/137$ we got an estimation on the leptoquark mass of $m_q > 894$ GeV at 90% C.L. which is slightly above the current Tevatron limit. Similarly, the case of SUSY with R-parity breaking terms, with s-quark masses normalized to 100 GeV we can constrain the parameter $\frac{|\lambda'_{1j1}|}{m_{\tilde{d}^2_{jL}}} - \frac{|\lambda'_{11k}|}{m_{\tilde{d}^2_{kR}}} < 0.002$ at 90% C.L.

Now, if we see the model independent bounds from the D meson decays, they can be mapped to constraints on the parameters of Two Higgs Doublet Model. It is found that the allowed region at 68% C.L. is: $-1.8\times10^{-3}\ \mathrm{GeV}^{-1} < (m_c - m_s\tan^2\beta)/M^2_H < 0.023\ \mathrm{GeV}^{-1}$. Similarly, the D meson decays constrain Leptoquark couplings. If one considers leptoquark states that only couple to the second generation of left handed quarks (chiral generation leptoquarks) and the first generation of left handed leptons stringent constraints are found. The allowed region at 95% C.L. from the semileptonic decays of the D mesons is given by: $0.05\left(\frac{m_{S_0}}{300\mathrm{GeV}}\right)^2 < |\kappa'_{12}|^2 < 0.11\left(\frac{m_{S_0}}{300\mathrm{GeV}}\right)^2$, while previous bounds are reported to be $\kappa'^2 < 5 \times (M_{LQ}/300\mathrm{GeV})^2$ [23], where $S_0$ is a scalar leptoquark with charge 1/3 and (3, 1, 2/3) gauge numbers.

## 3. Constraining NSI in neutrino oscillations

Neutrino interactions on the other hand, either at the source or at the detector site, in oscillation experiments are described by effective Lagrangians that include both quark and lepton operators products, that in general induce new sources of flavor and CP violation. Hence, NSI may modify the oscillation pattern and enhancing the CP asymmetry (see for example [4, 5, 6, 7]), even if neutrino masses and lepton mixing matrices solve the solar and atmospheric neutrinos anomalies. There is an extraordinary effort and a large number of experiments planned to increase the precision measurement of the oscillation parameters and hence allowing us to test NP effects in different oscillation channels (or flavor transitions). Going beyond the usual plane-wave propagation approximation, a generic framework based on quantum field theory is proposed in [8], to get a simple expression for the CP asymmetry without imposing any assumptions on the operators generated by NP. Indeed: Consider a *virtual neutrino* that is produced at the space-time location $(x, t)$, travels to $(x^{'}, t^{'})$ and is detected there because it interacts with a target producing a charged lepton $l$. For definiteness, let us illustrate this process with the production of the neutrino in a $\pi^+$ decay and its later detection via its weak interaction with a target nucleon $N$ in an accelerator experiment. If we assume that light neutrinos are left-handed, lepton flavor (LF)-violating semileptonic interactions can be described by the Lagrangian in eq. (1) for the quarks $u$ and $d$ and addding an index for the neutrino flavor $k$.

The CP asymmetry is defined as $a_{CP}(\tau) = \frac{|T_{\nu^s_\mu - \nu_l}(\tau)|^2 - |T_{\bar{\nu}^s_\mu - \bar{\nu}_l}(\tau)|^2}{|T_{\nu^s_\mu - \nu_l}(\tau)|^2 + |T_{\bar{\nu}^s_\mu - \bar{\nu}_l}(\tau)|^2}$. Where $T_{\overset{(-)}{\nu}_\mu \to \overset{(-)}{\nu}_l}(\tau)$ is the time dependent S-matrix amplitude for the evolution of the system from initial state (at the source) to the final state (at the detector). In the case where we consider only the $(V-A)(V-A)$ operator, we reproduce the usual results [4], once assuming that $1/2E_{\nu_i} = (1/2E_\nu)(1+\mathcal{O}(m^2_{\nu_i}))$ and keeping the first term of the expansion in $m^2_{\nu_i}$. However neutrino experiments enter in the field of high precision experiments where the effects of NSI coming from $(S-P)(S+P)$ and $(S-P)(S-P)$ operators could have an impact on the determination of lepton mixing parameters including the CP violating phases. In the case where the only sources of CP violating phases come from the scalar operators $C_k^{S,LR(RR)}$, we find that we can equivalently define a flavour neutrino eigenstate $|\nu^s_\mu >= \sum_i \left( U^*_{ei}\epsilon_{\mu e} + U^*_{\mu i}(1+\epsilon_{\mu\mu}) + U^*_{\tau i}\epsilon_{\mu\tau} \right) |\nu_i >$ and follow the usual plane-wave description of the propagation of neutrinos. The NSI parameters in this case are defined as $\epsilon_{k\mu} = \frac{1}{C_{SM}} \frac{m_\pi}{m_u + m_d} (C_k^{S,LR*} - C_k^{S,RR*})$, and hence enhanced by a factor of $\sim 15$.
Hence, it is shown that without imposing any assumptions on the operators generated by NP, new contributions to the CP asymmetry appear and it could be important to take them into account once we want to constrain NP using experimental data.

## 4. Probing Majorana or Dirac neutrinos with neutrino scatterings (KeV range)

If the neutrino polarization evolution is considered, the $\nu - e$ scattering process may shed some light into the Dirac or Majorana nature of the neutrino. Indeed, for the Majorana and the Dirac cases the neutrino scattering amplitudes off electrons or nucleons are not the same [25] because if the neutrino is a Majorana particle, then the neutrino is identical to its own antiparticle. The equivalence in the cross sections arise because the neutrino mass is extremely small ($m_\nu < 2$ eV [23]), thus these are almost completely chiral states, that is, almost fully polarized particles due to the left-handed nature of the charged weak interaction. For this reason, an extra state preparation factor $(1-\gamma_5)/2$ is usually added such that the amplitudes become identical. Even if the initial neutrino polarization is considered, since all neutrinos from terrestrial experiments are produced via charged currents, it is extremely difficult to observe significant differences between Dirac and Majorana neutrinos [11]. Indeed, for detectable neutrino energies, the difference is negligible. It becomes significant only for unreasonable (of the order of eV) energies of the neutrino. However, a change of the neutrino polarization may be achieved in astrophysical environments with strong magnetic fields. Indeed, any particle possessing a magnetic moment, as the neutrino does, interacts with external electromagnetic fields and consequently, its spin may rotate around the direction imposed by this external field. Actually, neutrinos can have a non-negligible magnetic moment $\mu_\nu$ (current constraints give $\mu_\nu < 5 \times 10^{-12}\mu_B$ Significant difference between Dirac and Majorana scattering cross section appears in the cross section as function of $s_{||}$, i.e. the longitudinal polarization of the neutrino. In particular, a 25% deviation from the original neutrinos helicity is required for a 10% difference between Dirac and Majorana scattering for any value of the neutrino energy, while a more precise experiment able to detect astrophysical neutrinos with a 2% accuracy will need only a 5% deviation in the neutrinos original helicity. The change of polarization can be achieved in very hot ($T \sim 20$ MeV ) and large (such as AGN or SNR) astrophysical environments with strong magnetic fields if the neutrino magnetic moment is bigger than the expected SM prediction but smaller than the current experimental limit. More details can be found in [11].

**Acknowledgements**
This work has been supported by CONACyT and CONACyT-SNI. VGM was supported in part by a postdoctoral fellowship grant from UC Mexus.

[1] V. Cirigliano and M. J. Ramsey-Musolf, Prog. Part. Nucl. Phys. **71**, 2 (2013) [arXiv:1304.0017 [hep-ph]].
[2] M. C. Gonzalez-Garcia and M. Maltoni, Phys. Rept. **460**, 1 (2008) [arXiv:0704.1800 [hep-ph]].
[3] J. Barranco, D. Delepine, V. Gonzalez Macias and L. Lopez-Lozano, Phys. Lett. B **731**, 36 (2014) [arXiv:1303.3896 [hep-ph]].
[4] M. C. Gonzalez-Garcia, Y. Grossman, A. Gusso and Y. Nir, Phys. Rev. D **64**, 096006 (2001)
[5] T. Ota, J. Sato and N. a. Yamashita, Phys. Rev. D **65**, 093015 (2002)
[6] M. Maltoni, J. Phys. Conf. Ser. **136**, 022024 (2008)
[7] S. Davidson, C. Pena-Garay, N. Rius and A. Santamaria, JHEP **0303**, 011 (2003)
[8] D. Delepine, V. G. Macias, S. Khalil and G. Lopez Castro, Phys. Rev. D **79**, 093003 (2009) [arXiv:0901.1460 [hep-ph]].
[9] J. Barranco, O. G. Miranda, C. A. Moura and J. W. F. Valle, Phys. Rev. D **77**, 093014 (2008) [arXiv:0711.0698 [hep-ph]].
[10] J. Barranco, O. G. Miranda, C. A. Moura and J. W. F. Valle, Phys. Rev. D **73**, 113001 (2006) [hep-ph/0512195].
[11] J. Barranco, D. Delepine, V. G. Macias, C. Lujan-Peschard and M. Napsuciale, Phys. Lett. B **739**, 343 (2014) [arXiv:1408.3219 [hep-ph]].
[12] J. Barranco, O. G. Miranda and T. I. Rashba, JHEP **0512**, 021 (2005) [hep-ph/0508299].
[13] D. Besson *et al.* [CLEO Collaboration], Phys. Rev. D **76**, 072008 (2007) [arXiv:0706.2813 [hep-ex]].
[14] J. M. Link *et al.* [FOCUS Collaboration], Phys. Lett. B **607**, 233 (2005) [hep-ex/0410037].
[15] J. Y. Ge *et al.* [CLEO Collaboration], Phys. Rev. D **79**, 052010 (2009) [arXiv:0810.3878 [hep-ex]].
[16] D. Besson *et al.* [CLEO Collaboration], Phys. Rev. D **80**, 032005 (2009) [arXiv:0906.2983 [hep-ex]].
[17] L. Widhalm *et al.* [Belle Collaboration], Phys. Rev. Lett. **100**, 241801 (2008) [arXiv:0709.1340 [hep-ex]].
[18] C. T. H. Davies, C. McNeile, E. Follana, G. P. Lepage, H. Na and J. Shigemitsu, Phys. Rev. D **82**, 114504 (2010) [arXiv:1008.4018 [hep-lat]]. C. Aubin *et al.* [Fermilab Lattice and MILC and HPQCD Collaborations], Phys. Rev. Lett. **94**, 011601 (2005) [hep-ph/0408306].
[19] J. Koponen, C. T. H. Davies, G. C. Donald, E. Follana, G. P. Lepage, H. Na and J. Shigemitsu, arXiv:1305.1462 [hep-lat].
[20] B. A. Dobrescu and A. S. Kronfeld, Phys. Rev. Lett. **100**, 241802 (2008)
[21] M. Carpentier and S. Davidson, Eur. Phys. J. C **70**, 1071 (2010) [arXiv:1008.0280 [hep-ph]].
[22] A. S. Kronfeld, PoS LATTICE **2008**, 282 (2008) [arXiv:0812.2030 [hep-lat]].
[23] J. Beringer *et al.* [Particle Data Group Collaboration], Phys. Rev. D **86**, 010001 (2012).
[24] O. Deschamps, S. Descotes-Genon, S. Monteil, V. Niess, S. T'Jampens and V. Tisserand, Phys. Rev. D **82**, 073012 (2010) [arXiv:0907.5135 [hep-ph]].
[25] B. Kayser and R. E. Shrock, Phys. Lett. B **112** (1982) 137.
[26] O. G. Miranda, T. I. Rashba, A. I. Rez and J. W. F. Valle, Phys. Rev. Lett. **93** (2004) 051304 [hep-ph/0311014].

Journal of Physics: Conference Series **651** (2015) 012023 doi:10.1088/1742-6596/651/1/012023

# Overview of searches for dark matter at the LHC

**Vasiliki A Mitsou**

Instituto de Física Corpuscular (IFIC), CSIC – Universitat de València,
Parc Científic de la U.V., C/ Catedrático José Beltrán 2,
E-46980 Paterna (Valencia), Spain

E-mail: `vasiliki.mitsou@ific.uv.es`

**Abstract.** Dark matter remains one of the most puzzling mysteries in Fundamental Physics of our times. Experiments at high-energy physics colliders are expected to shed light to its nature and determine its properties. This review talk focuses on recent searches for dark-matter signatures at the Large Hadron Collider, either within specific theoretical scenarios, such as supersymmetry, or in a model-independent scheme looking for mono-$X$ events arising in WIMP-pair production.

## 1. Introduction

Both Astroparticle and Particle Physics pursue the exploration of the nature of dark matter (DM) [1]. Among the long list of well-motivated candidates, the most popular particles are *cold* and weakly interacting, typically predicting missing-energy signals at particle colliders. Supersymmetry [2] and models with extra dimensions [3] are theoretical ideas that inherently provide such a dark matter candidate. High-energy colliders, such as the Large Hadron Collider [4] at CERN, are ideal machines for producing and eventually detecting DM [5].

The structure of this paper is as follows. Section 1 provides a brief introduction to the relevance of colliders, and in particular the LHC experiments ATLAS and CMS, for the production of dark matter. In Section 2, the strategy, methods, and results of the LHC experiments as far as model-independent DM-production is concerned are discussed. In Section 3, the latest results in searches for supersymmetry at the LHC are presented. The paper concludes with a summary and an outlook in Section 4.

### *1.1. Dark matter and colliders*

The nature of the dark sector of the Universe constitutes one of the major mysteries in fundamental physics. According to recent observations from anisotropies of the cosmic microwave background made by the Planck mission team [6], most of our Universe energy budget consists of unknown entities: $\sim$26.8% is dark matter and $\sim$68.3% is dark energy, a form of ground-state energy. Dark energy is believed to be responsible for the current-era acceleration of the Universe. Dark matter, on the other hand, is matter inferred to exist from gravitational effects on visible matter, being undetectable by emitted or scattered electromagnetic radiation.

Evidence from the formation of large-scale structure (galaxies and their clusters) strongly favour cosmologies where non-baryonic DM is entirely composed of cold dark matter (CDM), i.e. non-relativistic particles. CDM particles, in turn, may be weakly interacting massive particles (WIMPs), a class of DM candidates that arise naturally in models which attempt to explain the

Journal of Physics: Conference Series **651** (2015) 012023 doi:10.1088/1742-6596/651/1/012023

origin of electroweak symmetry breaking. Furthermore, the typical (weak-scale) cross sections characterising these models are of the same order of magnitude as the WIMP annihilation cross section, thus establishing the so-called *WIMP miracle*; this is precisely where the connection between Cosmology and Particle Physics lies [7].

WIMP dark matter candidates include the lightest neutralino in models with weak-scale supersymmetry [2], Kaluza-Klein photons arise in scenarios with universal extra dimensions (UED) [3], while lightest $T$-odd particles are predicted in Little Higgs models [8] with a conserved $T$-parity. The common denominator in these theories is that they all predict the existence of an electrically neutral, colorless and *stable* particle, whose decay is prevented by a kind of symmetry: $R$-parity, connected to baryon and lepton number conservation in SUSY models; KK-parity, the four-dimensional remnant of momentum conservation in extra dimension scenarios; and a $Z_2$ discrete symmetry called $T$-parity in Little Higgs models.

Weakly interacting massive particles do not interact neither electromagnetically nor strongly with matter and thus, once produced, they traverse the various detectors layers without leaving a trace, just like neutrinos. However by exploiting the hermeticity of the experiments, we can get a hint of the WIMP presence through the balance of the energy/momentum measured in the various detector components, the so-called *missing energy*. In hadron colliders, in particular, since the longitudinal momenta of the colliding partons are unknown, only the *transverse missing energy*, $E_{\mathrm{T}}^{\mathrm{miss}}$, can be reliably used to 'detect' DM particles.

### *1.2. The ATLAS and CMS experiments at the LHC*

The Large Hadron Collider (LHC) [4], situated at CERN, the European Laboratory for Particle Physics, outside Geneva, Switzerland, started its physics program in 2010 colliding two counter-rotating beams of protons or heavy ions. Before the scheduled 2013–2014 long shutdown, the LHC succeeded in delivering $\sim 5$ fb$^{-1}$ of integrated luminosity at centre-of-mass energy of 7 TeV during 2010–2011 and another $\sim 23$ fb$^{-1}$ at $\sqrt{s} = 8$ TeV in 2012. The LHC has already extended considerably the reach of its predecessor hadron machine, the Fermilab Tevatron, both in terms of instantaneous luminosity and energy, despite the fact that it has not arrived yet to its design capabilities.

The two general-purpose experiments, ATLAS (A Toroidal LHC ApparatuS) [9] and CMS (Compact Muon Solenoid) [10], have been constructed and operate with the aim of exploring a wide range of possible signals of New Physics that LHC renders accessible, on one hand, and performing precision measurements of Standard Model (SM) parameters, on the other. It is worth mentioning that the MoEDAL [11] experiment is specifically designed to explore high-ionisation signatures that may also arise in some theoretical scenarios of dark matter [12].

The ATLAS [9] and CMS [10] detectors are designed to overcome difficult experimental challenges: high radiation levels, large interaction rate and extremely small production cross sections of New Physics signals with respect to known SM processes. To this end, both experiments feature separate subsystems to measure charged particle momentum, energy deposited by electromagnetic showers from photons and electrons, energy from hadronic showers of strongly-interacting particles and muon-track momentum.

The most remarkable highlight of ATLAS and CMS operation so far is undoubtedly the discovery of a new particle [13] that so far seems to have all the features pinpointing to a SM(-like) Higgs boson [14]. The observation of this new boson has strong impact not only on our understanding of the fundamental interactions of Nature, as encoded in the SM, but on the proposed theoretical scenarios of Physics beyond the SM (BSM).

## 2. Model-independent DM production at the LHC

Collider searches for dark matter are highly complementary to direct [15,16] and indirect [16,17] DM detection methods. The main advantage of collider searches is that they do not suffer from

Journal of Physics: Conference Series **651** (2015) 012023 doi:10.1088/1742-6596/651/1/012023

astrophysical uncertainties and that there is no lower limit to their sensitivity on DM masses. The leading generic diagrams responsible for DM production [18, 19] at hadron colliders involve the pair-production of WIMPs plus the initial- or final-state radiation (ISR/FSR) of a gluon, photon or a weak gauge boson $Z$, $W$. The ISR/FSR particle is necessary to balance the two WIMPs' momentum, so that they are not produced back-to-back resulting in negligible $E_{\mathrm{T}}^{\mathrm{miss}}$. Therefore the search is based on selecting events high-$E_{\mathrm{T}}^{\mathrm{miss}}$ events, due to the WIMPs, and a single jet, photon or boson candidate.

The search results are interpreted in terms of a largely model-independent effective-field-theory framework, in which the interactions between a DM Dirac fermion $\chi$ and SM fermions $f$ are described by contact operators. Some of the possible operators are listed in Table 1. In this framework, the interaction between SM and DM particles is determined by only two parameters, namely the DM-particle mass, $m_\chi$, and the suppression scale, $M_*$, which is related to the mediator mass and to its coupling to SM and DM particles. The derived limits are independent of the theory behind the WIMP (SUSY, extra dimensions, etc), yet it *has* been assumed that other hypothetical particles are too heavy to be produced directly in $pp$ collisions. Henceforth, Dirac DM fermions are considered, however conclusions for Majorana fermions can also be drawn, since their production cross section only differs by a factor of two.

**Table 1.** Effective interaction operators of WIMP pair production considered in the mono-$X$ analyses, following the formalism of Ref. [18].

| Name | Initial state | Type | Operator |
|---|---|---|---|
| D1 | $qq$ | scalar | $\frac{m_q}{M_*^3}\bar{\chi}\chi\bar{q}q$ |
| D5 | $qq$ | vector | $\frac{1}{M_*^2}\bar{\chi}\gamma^\mu\chi\bar{q}\gamma_\mu q$ |
| D8 | $qq$ | axial-vector | $\frac{1}{M_*^2}\bar{\chi}\gamma^\mu\gamma^5\chi\bar{q}\gamma_\mu\gamma^\mu q$ |
| D9 | $qq$ | tensor | $\frac{1}{M_*^2}\bar{\chi}\sigma^{\mu\nu}\chi\bar{q}\sigma_{\mu\nu}q$ |
| D11 | $gg$ | scalar | $\frac{1}{4M_*^3}\bar{\chi}\chi\alpha_s(G_{\mu\nu}^s)^2$ |

### *2.1. Monojet searches*

Event topologies with a single high-$E_{\mathrm{T}}$ jet and large $E_{\mathrm{T}}^{\mathrm{miss}}$, hereafter referred to as *monojets*, constitute valuable probes of physics beyond the SM at the LHC. Both ATLAS [20] and CMS [21] experiments have performed searches for an excess of monojet events over SM expectations in a wide range of signatures. The analyses outlined here use the full 2011 $pp$ LHC dataset at a centre-of-mass energy of $\sqrt{s} = 7$ TeV. The primary SM process yielding a true monojet final state is $Z$-boson production in association with a jet, where the $Z$ decays to two neutrinos. Other known processes acting as background in this search are $Z(\to \ell\ell)$+jets —with $\ell = e$, $\mu$—, $W$+jets, $t\bar{t}$ as well as single-top events. All electroweak backgrounds and multijet events passing the selections criteria, as well as non-collision backgrounds, are in most cases determined by data-driven methods. Top and diboson backgrounds are estimated solely from Monte Carlo (MC) simulation.

The monojet analyses for ATLAS and CMS are based on some general requirements: large $E_{\mathrm{T}}^{\mathrm{miss}}$, with thresholds typically ranging from 120 GeV to 500 GeV and a energetic jet with a variable $p_{\mathrm{T}}$ threshold higher than 110 GeV that fulfils high jet-reconstruction quality criteria. Moreover, events with at least one electron or muon or a third jet are vetoed. Back-to-back dijet events are suppressed by requiring the subleading jet not to be collinear with $\mathbf{p}_{\mathrm{T}}^{\mathrm{miss}}$. The selected data are required to pass a trigger based on high $E_{\mathrm{T}}^{\mathrm{miss}}$ (ATLAS) or large $E_{\mathrm{T}}^{\mathrm{miss}}$ plus one high-$E_{\mathrm{T}}$ jet (CMS).

The data, amounting to $\sim 5$ fb$^{-1}$, are found to be in agreement with the SM expectations. The results are interpreted in a framework of WIMP production with the simulated WIMP-signal MC samples corresponding to various assumptions of the effective field theory, as discussed previously. In this framework, the interaction between SM and DM particles are defined by only two parameters, namely the DM-particle mass, $m_\chi$, and the suppression scale, $M_*$, which is related to the mediator mass and to its coupling to SM and DM particles.

Experimental and theoretical systematic uncertainties are considered when setting limits on the model parameters $M_*$ and $m_\chi$. The experimental uncertainties on jet energy scale and resolution and on $E_{\mathrm{T}}^{\mathrm{miss}}$ range from $1-20\%$ of the WIMP event yield, depending on the $E_{\mathrm{T}}^{\mathrm{miss}}$ and $p_{\mathrm{T}}$ thresholds and the considered interaction operator. Other experimental uncertainties include the ones associated with the trigger efficiency and the luminosity measurement. On the other hand, the parton-distribution-function set, the amount of ISR/FSR, and the factorisation and renormalisation scales assumed lead to theoretical uncertainties on the simulated WIMP signal.

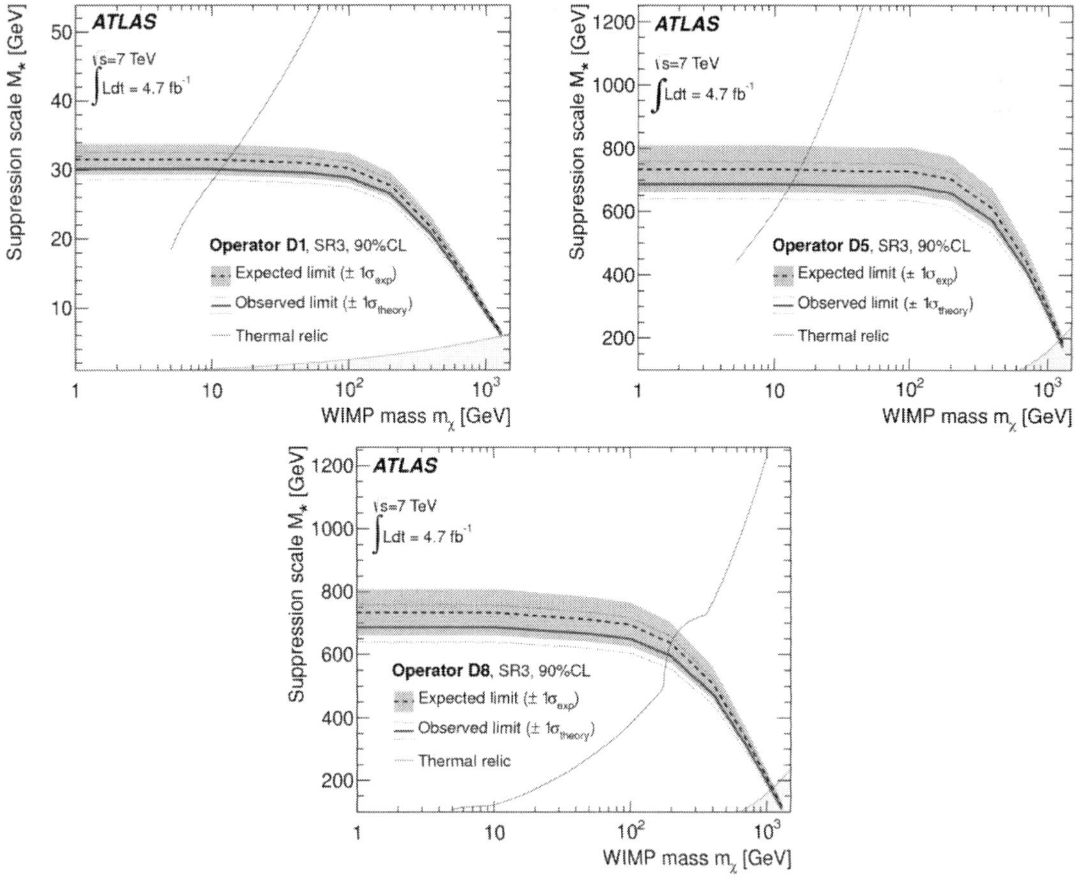


**Figure 1.** ATLAS lower limits at 90% CL on the suppression scale, $M_*$, for different masses of $\chi$ obtained with the monojet analysis for the operators D1 (top left), D5 (top right) and D8 (bottom). The region below the limit lines is excluded. All shown curves and areas are explained in the text. From Ref. [20].

From the limit on the visible cross section of new BSM physics processes, lower limits on the suppression scale as a function of the WIMP mass have been derived by the ATLAS Collaboration [20]. The 90% confidence level (CL) lower limits for the D1, D5 and D8 operators are shown in Fig. 1. The observed limit on $M_*$ includes experimental uncertainties; the effect of theoretical uncertainties is indicated by dotted $\pm 1\sigma$ lines above and below it. Around the expected limit, $\pm 1\sigma$ variations due to statistical and systematic uncertainties are shown as a grey band. The lower limits are flat up to $m_\chi \lesssim 100$ GeV and become weaker at higher mass due to the collision energy. In the bottom-right corner of the $(m_\chi, M_*)$ plane (light-grey shaded area), the effective field theory approach is no longer valid. The rising lines correspond to couplings consistent with the measured thermal relic density [18], assuming annihilation in the early universe proceeded exclusively via the given operator. Similar exclusion limits for all operators listed in Table 1 are given in Ref. [20]. For the operator D1, the limits are much weaker ($\sim 30$ GeV) than for other operators. Nevertheless, if heavy-quark loops are included in the analysis, much stronger bounds on $M_*$ can be obtained [22].

The observed limit on the dark matter-nucleon scattering cross section depends on the mass of the dark matter particle and the nature of its interaction with the SM particles. The limits on the suppression scale as a function of $m_\chi$ can be translated into a limit on the cross section using the reduced mass of $\chi$-nucleon system [19], which can be compared with the constraints from direct and indirect detection experiments, as we shall see at the end of this Section in conjunction with bounds acquired with other mono-$X$ analyses.

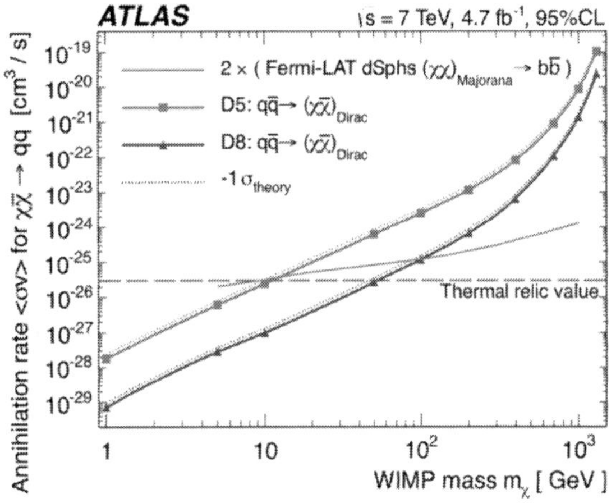


**Figure 2.** ATLAS 95% CL limits on WIMP annihilation rates $\langle\sigma v\rangle$ versus mass $m_\chi$, inferred from the monojet analysis. Explanation of the shown curves is given in the text. From Ref. [20].

The ATLAS collider limits on vector (D5) and axial-vector (D8) interactions are also interpreted in terms of the relic abundance of WIMPs, using the same effective theory approach [18]. The upper limits on the annihilation rate of WIMPs into light quarks, defined as the product of the annihilation cross section $\sigma$ and the relative WIMP velocity $v$ averaged over the WIMP velocity distribution $\langle\sigma v\rangle$, are shown in Fig. 2. The results are compared to limits on WIMP annihilation to $b\bar{b}$, obtained from galactic high-energy gamma-ray observations, measured by the Fermi-LAT telescope [23]. Gamma-ray spectra and yields from WIMPs annihilating to $b\bar{b}$, where photons are produced in the hadronisation of the quarks, are expected to be very similar to those from WIMPs annihilating to light quarks [24]. Under this assumption, the ATLAS and Fermi-LAT limits can be compared, after scaling up the Fermi-LAT values by a factor of two to account for the Majorana-to-Dirac fermion adaptation. Again, the ATLAS bounds are especially important for small WIMP masses: below 10 GeV for vector couplings and below about 100 GeV for axial-vector ones. In this region, the ATLAS limits are below the annihilation cross section needed to be consistent with the thermic relic value, keeping the assumption that

Journal of Physics: Conference Series **651** (2015) 012023 doi:10.1088/1742-6596/651/1/012023

WIMPs have annihilated to SM quarks only via the particular operator in question. For masses of $m_\chi \gtrsim 200$ GeV the ATLAS sensitivity becomes worse than the Fermi-LAT one. In this region, improvements can be expected when going to larger centre-of-mass energies at the LHC.

The case in which the mediator is light enough to be accessible to the LHC has been considered too by the CMS experiment in a monojet search performed with $\sim 20$ fb$^{-1}$ at 8 TeV [25]. Figure 3 shows the observed limits on the contact interaction scale $\Lambda$ as a function of the mass of the mediator $M$, assuming vector interactions and a dark matter mass of 50 GeV and 500 GeV. The width $\Gamma$ of the mediator is varied between $M/3$ and $M/8\pi$ [26]. It shows the resonant enhancement in the production cross section once the mass of the mediator is within the kinematic range and can be produced on-shell. For $m_\chi \gtrsim 100$ GeV, this approach is adequate and quite conservative in the bounds on $\Lambda$. For $m_\chi \lesssim 100$ GeV, the collider bounds are considerably weaker. At large mediator masses, i.e. $M \gtrsim 5$ TeV, the limits on $\Lambda$ approximate to those obtained in the effective theory framework.

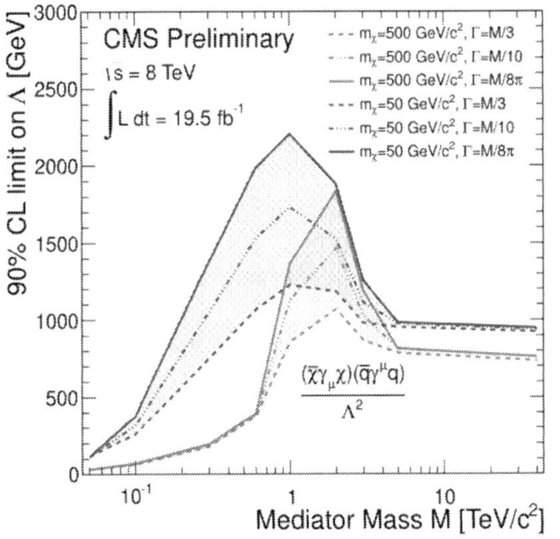


**Figure 3.** Observed limits on the scale $\Lambda$ as a function of the mass $M$, assuming vector interactions and a dark matter mass of 50 GeV (blue) and 500 GeV (red) in a CMS monojet analysis. The width of the mediator was varied between $M/3$, $M/10$ and $M/8\pi$. From Ref. [25].

### *2.2. Monophoton-based probes*

In the same fashion as in the monojet searches, the *monophoton* analyses aim at probing dark matter requiring large $E_T^{miss}$ —from the $\chi$-pair production— and at least one ISR/FSR photon. Searches in monophoton events by ATLAS [27] and CMS [28] also show an agreement with the SM expectations. The limits are derived in a similar fashion as for the monojet search, however the monophoton search is found to be somewhat less sensitive with respect to the monojet topology.

The primary (irreducible) background for a $\gamma+E_T^{miss}$ signal comes from $Z\gamma \to \nu\bar{\nu}\gamma$ production. This together with other SM backgrounds, including $W\gamma$, $W \to e\nu$, $\gamma+$jet multijet, diphoton and diboson events, as well as backgrounds from beam halo and cosmic-ray muons, are taken into account in the analyses. The CMS analysis is based on singe-photon triggers, whilst ATLAS relies on high-$E_T^{miss}$ triggered events. The photon candidate is required to pass tight quality and isolation criteria, in particular in order to reject events with electrons faking photons. The missing transverse momentum of the selected events should be as high as 150 GeV (130 GeV) in the ATLAS (CMS) search. In CMS, events with a reconstructed jet are vetoed, while the ATLAS analysis rejects events with an electron, a muon or a second jet.

Both analyses, observe no significant excess of events over the expected background when applied on $\sim 5$ fb$^{-1}$ of $pp$ collision data at $\sqrt{s} = 7$ TeV. Hence they set lower limits on the suppression scale, $M_*$ versus the DM fermion mass, $m_\chi$, which in turn they are translated into upper limits on the nucleon-WIMP interaction cross section applying the prescription in Ref. [18]. Figure 4 shows the 90% CL upper limits on the nucleon-WIMP cross section as a function of $m_\chi$ derived from the ATLAS search [27]. The results are compared with previous CDF [29], CMS [21,28] and direct WIMP detection experiments [30–34] results. The CMS limit curve generally overlaps the ATLAS curve.

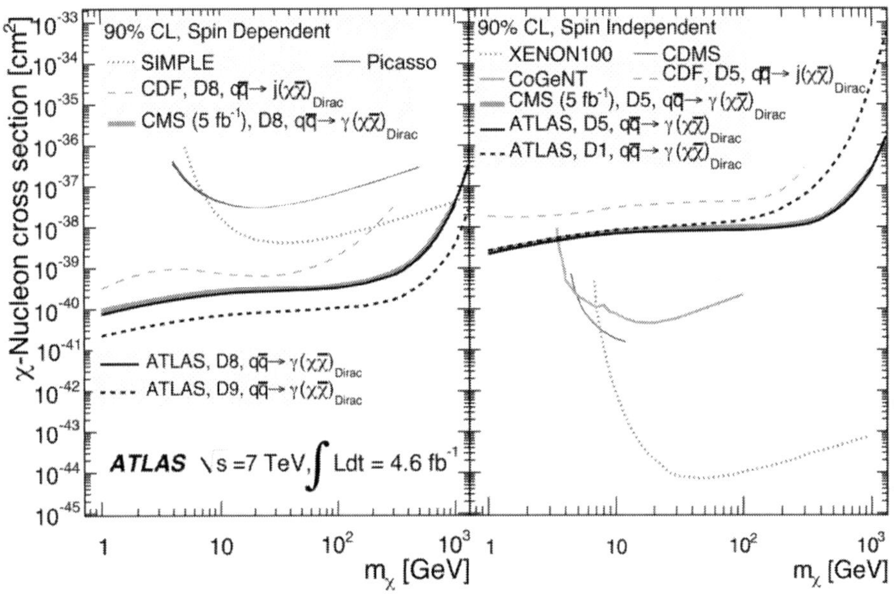


**Figure 4.** ATLAS 90% CL upper limits on the nucleon-WIMP cross section as a function of $m_\chi$ for spin-dependent (left) and spin-independent (right) interactions, corresponding to D8, D9, D1, and D5 operators, derived from the monophoton analysis. Explanation of the shown curves is given in the text. From Ref. [27].

The observed limits on $M_*$ typically decrease by 2% to 10% if the $-1\sigma$ theoretical uncertainty, resulting from the same sources as the one cited in the monojet analysis, is considered. This translates into a 10% to 50% increase of the quoted nucleon-WIMP cross section limits. To recapitulate, the exclusion in the region 1 GeV $< m_\chi <$ 1 TeV (1 GeV $< m_\chi <$ 3.5 GeV) for spin-dependent (spin-independent) nucleon-WIMP interactions is driven by the results from collider experiments, always under the assumption of the validity of the effective theory, and is still dominated by the monojet results.

### *2.3. Mono-W and mono-Z final states*

As demonstrated in the previous sections, searches for monojet or monophoton signatures have yielded powerful constraints on dark matter interactions with SM particles. Other studies propose probing DM at LHC through a $pp \to \chi\bar{\chi} + W/Z$, with a leptonically decaying $W$ [35] or $Z$ [36]. The final state in this case would be large $E_T^{\rm miss}$ and a single charged lepton (electron or muon) for the *mono-W* signature (*monolepton*) or a pair of charged leptons that reconstruct to the $Z$ mass for the *mono-Z* signature. In either case, the gauge boson radiations off a $q\bar{q}$ initial state and an effective field theory is deployed to describe the contact interactions that couple the SM particle with the WIMP.

In Ref. [37], the existing $W'$ searches from CMS [38] —which share a similar final state with mono-$W$ searches— are used to place a bound on mono-$W$ production at LHC, which for some choices of couplings are better than monojet bounds. This is illustrated in the left (right) panel of Fig. 5, where the spin-dependent (spin-independent) WIMP-proton cross section limits are drawn. The parameter $\xi$ parametrises the relative strength of the coupling to down-quarks with respect to up-quarks: $\xi = +1$ for equal couplings; $\xi = -1$ for opposite-sign ones; and $\xi = 0$ when there is no coupling to down-quarks. Even in cases where the monoleptons do not provide the most stringent constraints, they provide an interesting mechanism to disentangle WIMP couplings to up-type versus down-type quarks. Such an interpretation has also been performed in Ref. [35] yielding similar limits.

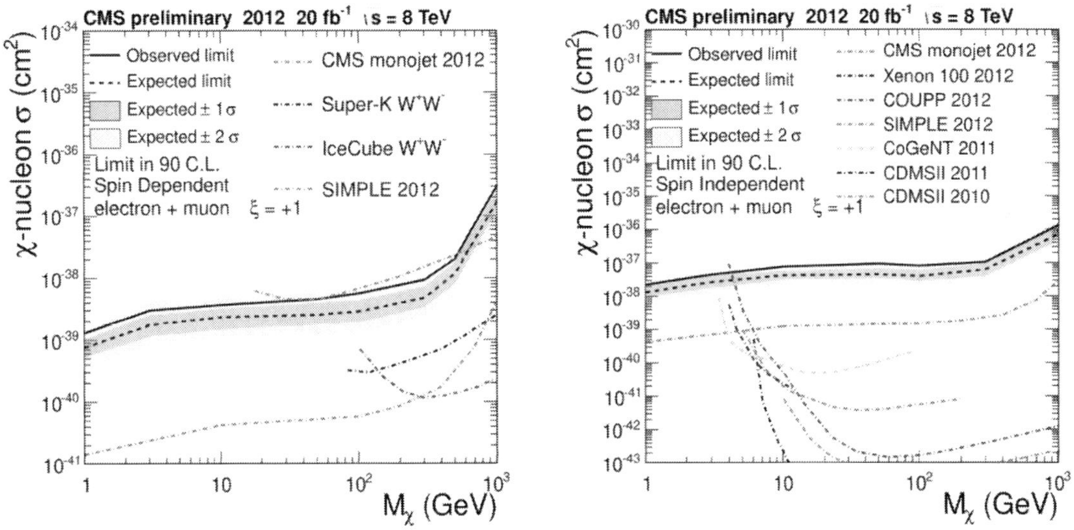


**Figure 5.** CMS monolepton search with 20 fb$^{-1}$ at 8 TeV. Excluded proton-dark matter cross section for axial-vector-like, i.e. spin dependent (left), and vector-like, i.e. spin independent (right), for the combination of electron and muon channels. The CMS monojet result is for 20 fb$^{-1}$ of 2012 data [25]. From Ref. [37].

The ATLAS Collaboration has extended the range of possible mono-$X$ probes by looking for $pp \to \chi\bar{\chi} + W/Z$, when the gauge boson decays to two quarks [39], as opposed to the leptonic signatures discussed so far. The analysis searches for the production of $W$ or $Z$ bosons decaying hadronically and reconstructed as a single massive jet in association with large $E_{\mathrm{T}}^{\mathrm{miss}}$ from the undetected $\chi\bar{\chi}$ particles. For this analysis, the jet candidates are reconstructed using a filtering procedure referred to as *large-radius jets* [40]. This search, the first of its kind, is sensitive to WIMP pair production, as well as to other DM-related models, such as invisible Higgs boson decays ($WH$ or $ZH$ production with $H \to \chi\bar{\chi}$).

Figure 6 shows the 90% CL upper limits on the dark matter-nucleon scattering cross section as a function of the mass of DM particle for the spin-independent (left) and spin-dependent (right) models obtained by the ATLAS mono-$W/Z$ analysis [39]. The new limits are also compared to the limits set by ATLAS in the 7 TeV monojet analysis [20]. Limits from XENON100 [41], CoGent [31], CDMS II [32], SIMPLE [33], COUPP [42], Picasso [34], IceCube [43] are superimposed for comparison. For the spin-independent case with the opposite-sign up-type and

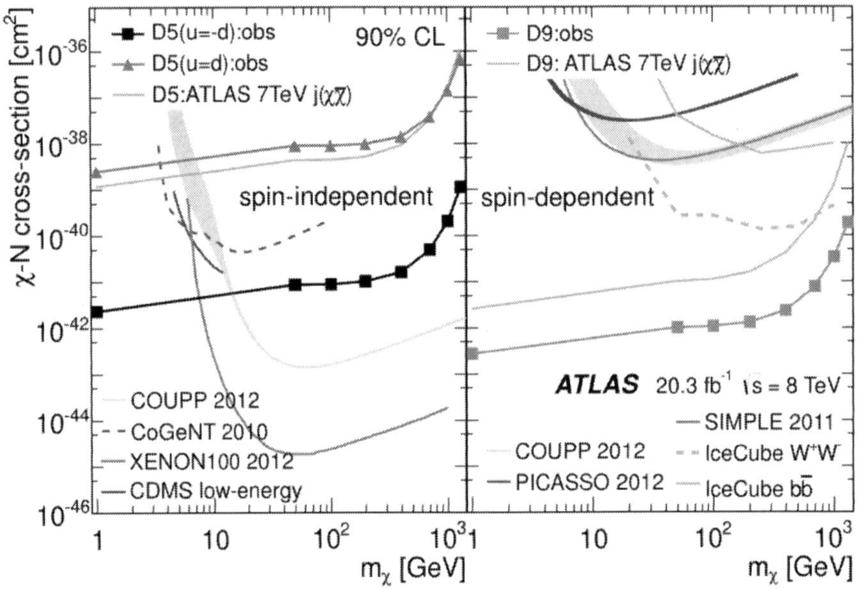


**Figure 6.** ATLAS-derived limits on $\chi$-nucleon cross sections as a function of $m_\chi$ at 90% CL for spin-independent (left) and spin-dependent (right) cases, obtained with the mono-$W/Z$ analysis and compared to previous limits. From Ref. [39].

down-type couplings, the limits are improved by about three orders of magnitude. For other cases, the bounds are similar. Comparable limits have been obtained by the CMS experiment.

It is worth noting that the spin-dependent limits derived from the operator D9, give a smaller, hence better, bound on the WIMP-nucleon cross section throughout the range of $m_\chi$, compared to direct DM experiments. In the spin-independent case the bounds from direct detection experiments are stronger for $m_\chi \gtrsim 10$ GeV, whereas the collider bounds, acquired with the operator D5, get important for the region of low DM masses.

## 3. Searches for supersymmetry

Supersymmetry (SUSY) [44] is an extension of the Standard Model which assigns to each SM field a superpartner field with a spin differing by a half unit. SUSY provides elegant solutions to several open issues in the SM, such as the hierarchy problem and the grand unification. In particular, SUSY predicts the existence of a stable weakly interacting particle —the lightest supersymmetric particle (LSP)— that has the pertinent properties to be a dark matter particle, thus providing a very compelling argument in favour of SUSY [45].

SUSY searches in the ATLAS [9] and CMS [10] experiments typically focus on events with high transverse missing energy, which can arise from (weakly interacting) LSPs, in the case of $R$-parity conserving SUSY, or from neutrinos produced in LSP decays, if $R$-parity is broken (c.f. Section 3.4). Hence, the event selection criteria of inclusive channels are based on large $E_{\mathrm{T}}^{\mathrm{miss}}$, no or few leptons ($e$, $\mu$), many jets and/or $b$-jets, $\tau$-leptons and photons. In addition, kinematical variables such as the transverse mass, $M_{\mathrm{T}}$, and the effective mass, $M_{\mathrm{eff}}$, assist in distinguishing further SUSY from SM events, whilst the *effective transverse energy* [46] can be useful to cross-check results, allowing a better and more robust identification of the SUSY mass scale, should a positive signal is found. Although the majority of the analysis simply look for an excess of events over the SM background, there is an increasing application of distribution shape fitting techniques [47].

Typical SM backgrounds are top-quark production —including single-top—, $W/Z$ in association with jets, dibosons and QCD multijet events. These are estimated using semi- or fully data-driven techniques. Although the various analyses are optimised for a specific SUSY scenario, the interpretation of the results are extended to various SUSY models or topologies.

Analyses exploring $R$-parity conserving SUSY models at LHC are roughly divided into inclusive searches for squarks and gluinos, for third-generation fermions, and for electroweak production (pairs of $\tilde{\chi}^0$, $\tilde{\chi}^{\pm}$, $\tilde{\ell}$). Although these searches are designed and optimised to look for $R$-parity conserving SUSY, interpretation in terms of $R$-parity violating (RPV) models is also possible. Other analyses are purely motivated by oriented by RPV scenarios and/or by the expectation of long-lived sparticles. Recent summary results from each category of ATLAS and CMS searches are presented in this section.

### *3.1. Gluinos and first two generations of quarks*

At the LHC, supersymmetric particles are expected to be predominantly produced hadronically, i.e. through gluino-pair, squark-pair and squark-gluino production. Each of these (heavy) sparticles is going to decay into lighter ones in a cascade decay that finally leads to an LSP, which in most of the scenarios considered is the lightest neutralino $\tilde{\chi}_1^0$. The two LSPs would escape detection giving rise to high transverse missing energy, hence the search strategy followed is based on the detection of high $E_{\mathrm{T}}^{\mathrm{miss}}$, many jets and possibly energetic leptons. The analyses make extensive use of data-driven Standard Model background measurements.

The most powerful of the existing searches are based on all-hadronic final states with large missing transverse momentum [48, 49]. In the 0-lepton search, events are selected based on a jet+$E_{\mathrm{T}}^{\mathrm{miss}}$ trigger, applying a lepton veto, requiring a minimum number of jets, high $E_{\mathrm{T}}^{\mathrm{miss}}$, and large azimuthal separation between the $E_{\mathrm{T}}^{\mathrm{miss}}$ and reconstructed jets, in order to reject multijet background. In addition, searches for squark and gluino production in a final state with one or two leptons have been performed [50,51]. The events are categorised by whether the leptons have higher or lower momentum and are referred to as the *hard* and *soft* lepton channels respectively. The soft-lepton analysis which enhances the sensitivity of the search in the difficult kinematic region where the neutralino and gluino masses are close to each other forming the so-called *compressed spectrum.* [52] Leptons in the soft category are characterised by low lepton-$p_{\mathrm{T}}$ thresholds ($6-10$ GeV) and such events are triggered by sufficient $E_{\mathrm{T}}^{\mathrm{miss}}$. Hard leptons pass a threshold of $\sim 25$ GeV and are seeded with both lepton and $E_{\mathrm{T}}^{\mathrm{miss}}$ triggers. Analyses based on the *razor* [53] variable have also been carried out by both experiments [54,55].

The major backgrounds ($t\bar{t}$, $W$+jets, $Z$+jets) are estimated by isolating each of them in a dedicated control region, normalising the simulation to data in that control region, and then using the simulation to extrapolate the background expectations into the signal region. The multijet background is determined from the data by a matrix method. All other (smaller) backgrounds are estimated entirely from the simulation, using the most accurate theoretical cross sections available. To account for the cross-contamination of physics processes across control regions, the final estimate of the background is obtained with a simultaneous, combined fit to all control regions.

In the absence of deviations from SM predictions, limits for squark and gluino production are set. Figure 7 illustrates the 95% CL limits set by ATLAS under the minimal Supergravity (mSUGRA) model in the ($m_0$, $m_{1/2}$) plane [48, 56]. The remaining parameters are set to $\tan\beta = 30$, $A_0 = -2\,m_0$, $\mu > 0$, so as to acquire parameter-space points where the predicted mass of the lightest Higgs boson, $h^0$, is near 125 GeV, i.e. compatible with the recently observed Higgs-like boson [13, 14]. Exclusion limits are obtained by using the signal region with the best expected sensitivity at each point. By assumption, the mSUGRA model avoids both flavour-changing neutral currents and extra sources of $CP$ violation. For masses in the TeV range, it

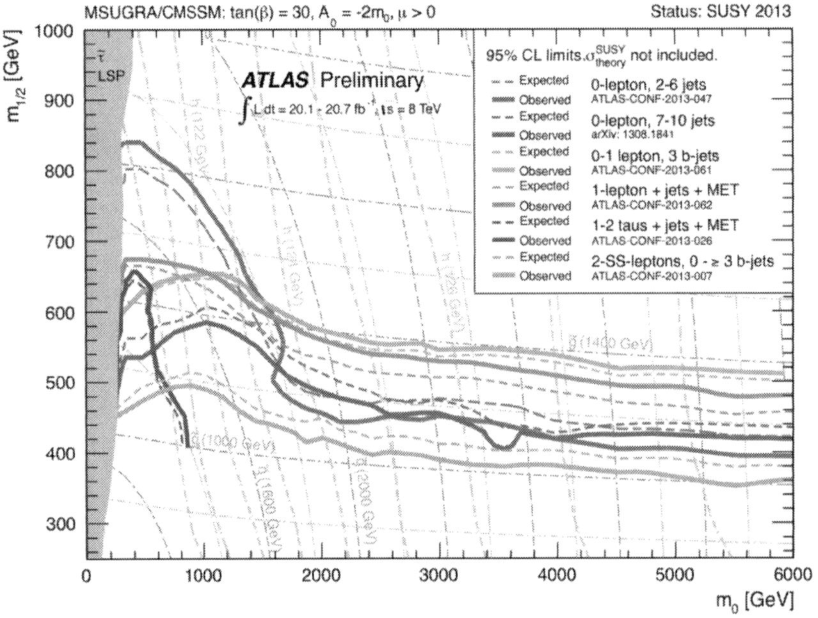


**Figure 7.** Exclusion limits at 95% CL for 8 TeV ATLAS analyses [48, 56] in the $(m_0, m_{1/2})$ plane for the mSUGRA model. From Ref. [57].

typically predicts too much cold dark matter, however these predictions depend of the presence of stringy effects that may dilute [58] or enhance [59] the predicted relic dark matter density. In the mSUGRA case, the limit on squark mass reaches 1750 GeV and on gluino mass is 1400 GeV.

### *3.2. Third-generation squarks*

The previously presented limits from inclusive channels indicate that the masses of gluinos and first/second generation squarks are expected to be above 1 TeV. Nevertheless, in order to solve the hierarchy problem in a *natural* way, the masses of the stops, sbottoms, higgsinos and gluinos have to be below the TeV-scale to properly cancel the divergences in the Higgs mass radiative corrections. Despite their production cross sections being smaller than for the first and second generation squarks, stop and sbottom may well be directly produced at the LHC and could provide the only direct observation of SUSY at the LHC in case the other sparticles are outside of the LHC energy reach. The lightest mass eigenstates of the sbottom and stop particles, $\tilde{b}_1$ and $\tilde{t}_1$, could hence be produced either directly in pairs or through $\tilde{g}$ pair production followed by $\tilde{g} \to \tilde{b}_1 b$ or $\tilde{g} \to \tilde{t}_1 t$ decays. Both cases will be discussed in the following.

For the aforementioned reasons, direct searches for third generation squarks have become a priority in both ATLAS and CMS. Such events are characterised by several energetic jets (some of them $b$-jets), possibly accompanied by light leptons, as well as high $E_{\mathrm{T}}^{\mathrm{miss}}$. A suite of channels have been considered, depending on the topologies allowed and the exclusions generally come with some assumptions driven by the shortcomings of the techniques and variables used, such as the requirement of 100% branching ratios into particular decay modes.

In the case of the gluino-mediated production of stops, a simplified scenario ("Gtt model"), where $\tilde{t}_1$ is the lightest squark but $m_{\tilde{g}} < m_{\tilde{t}_1}$, has been considered. Pair production of gluinos is the only process taken into account since the mass of all other sparticles apart from the $\tilde{\chi}_1^0$ are above the TeV scale. A three-body decay via off-shell stop is assumed for the gluino, yielding a

Journal of Physics: Conference Series **651** (2015) 012023 doi:10.1088/1742-6596/651/1/012023

100% branching ratio for the decay $\tilde{g} \to t\bar{t}\tilde{\chi}_1^0$. The stop mass has no impact on the kinematics of the decay and the exclusion limits [54, 60, 61] set by the CMS experiment are presented in the $(m_{\tilde{g}}, m_{\tilde{\chi}_1^0})$ plane in Fig. 8. For a massless LSP, gluinos with masses from 560 GeV to 1320 GeV are excluded. Similar results are obtained if the decay $\tilde{g} \to b\bar{b}\tilde{\chi}_1^0$ is considered instead, as shown in Fig. 9.

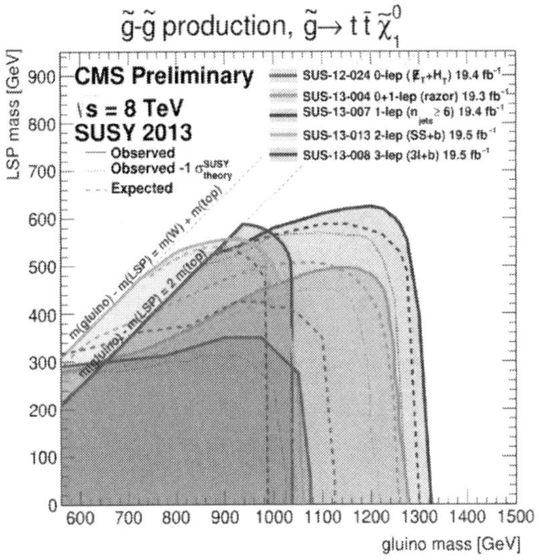


**Figure 8.** Summary of observed and expected limits [54, 60, 61] for gluino pair production with gluino decaying via a 3-body decay to a top, an anti-top and a neutralino. From Ref. [62].

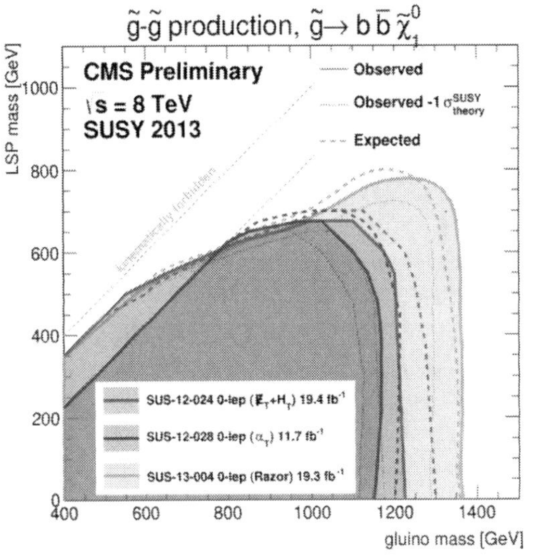


**Figure 9.** Summary of observed and expected limits [54, 60, 61] for gluino pair production with gluino decaying via a 3-body decay to a bottom, an anti-bopttom and a neutralino. From Ref. [62].

If the gluino is too heavy to be produced at the LHC, the only remaining possibility is the direct $\tilde{t}_1\tilde{t}_1$ and $\tilde{b}_1\tilde{b}_1$ production. If stop pairs are considered, two decay channels can be distinguished depending on the mass of the stop: $\tilde{t}_1 \to b\tilde{\chi}_1^{\pm}$ and $\tilde{t}_1 \to t\tilde{\chi}_1^0$. CMS and ATLAS carried out a wide range of different analyses in each of these modes at both 7 TeV and 8 TeV centre-of-mass energy. In all these searches, the number of observed events has been found to be consistent with the SM expectation. Limits have been set by ATLAS on the mass of the scalar

Journal of Physics: Conference Series **651** (2015) 012023 doi:10.1088/1742-6596/651/1/012023

top for different assumptions on the mass hierarchy scalar top-chargino-lightest neutralino [63], as shown in the left panel of Fig. 10. A scalar top quark of mass of up to 480 GeV is excluded at 95% CL for a massless neutralino and a 150 GeV chargino. For a 300 GeV scalar top quark and a 290 GeV chargino, models with a neutralino with mass lower than 175 GeV are excluded at 95% CL.

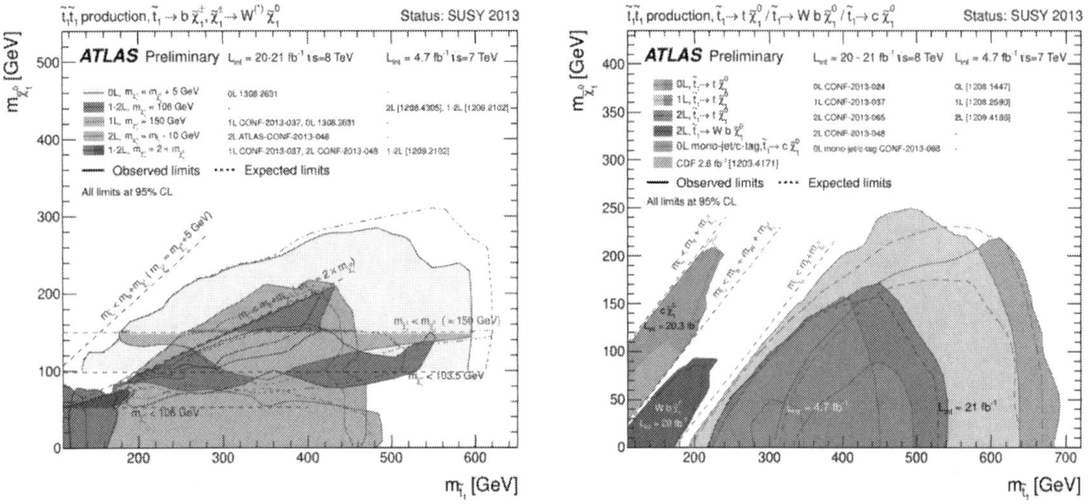


**Figure 10.** Summary of the dedicated ATLAS searches [63] for stop pair production based on $20-21$ fb$^{-1}$ of $pp$ collision data taken at $\sqrt{s}=8$ TeV, and 4.7 fb$^{-1}$ of $pp$ collision data taken at $\sqrt{s}=7$ TeV. Exclusion limits at 95% CL are shown in the $(\tilde{t}_1,\ \tilde{\chi}_1^0)$ mass plane for channels targeting $\tilde{t}_1 \to b\tilde{\chi}_1^{\pm}$, $\tilde{\chi}_1^{\pm} \to W^{\pm}\tilde{\chi}_1^0$ decays (left) and $\tilde{t}_1$ decays to $t\tilde{\chi}_1^0$ or $Wb\tilde{\chi}_1^0$ or $c\tilde{\chi}_1^0$ (right). The dashed and solid lines show the expected and observed limits, respectively, including all uncertainties except the theoretical signal cross section uncertainty. From Ref. [57].

For the case of a high-mass stop decaying to a top and neutralino ($\tilde{t}_1 \to t\tilde{\chi}_1^0$), analyses requiring one, two or three isolated leptons, jets and large $E_{\mathrm{T}}^{\mathrm{miss}}$ have been carried out. No significant excess of events above the rate predicted by the SM is observed and 95% CL upper limits are set on the stop mass in the stop-neutralino mass plane. The region of excluded stop and neutralino masses is shown on the right panel of Fig. 10 for the ATLAS analyses. Stop masses are excluded between 200 GeV and 680 GeV for massless neutralinos, and stop masses around 500 GeV are excluded along a line which approximately corresponds to neutralino masses up to 250 GeV. It is worth noting that a monojet analysis with $c$-tagging is deployed to cover part of the low-$m_{\tilde{t}_1}$, low-$m_{\tilde{\chi}_1^0}$ region through the $\tilde{t}_1 \to c\tilde{\chi}_1^0$ channel.

### *3.3. Electroweak gaugino production*

If all squarks and gluinos are above the TeV scale, weak gauginos with masses of few hundred gigaelectronvolts may be the only sparticles accessible at the LHC. As an example, at $\sqrt{s}=7$ TeV, the cross-section of the associated production $\tilde{\chi}_1^{\pm}\tilde{\chi}_2^0$ with degenerate masses of 200 GeV is above the 1-TeV gluino-gluino production cross section by one order of magnitude. Chargino pair production is searched for in events with two opposite-sign leptons and $E_{\mathrm{T}}^{\mathrm{miss}}$ using a jet veto, through the decay $\tilde{\chi}_1^{\pm} \to \ell^{\pm}\nu\tilde{\chi}_1^0$. A summary of related analyses [64] performed by CMS is shown in Fig. 11. Charginos with masses between 140 and 560 GeV are excluded for a massless LSP in the chargino-pair production with an intermediate slepton/sneutrino between the $\tilde{\chi}_1^{\pm}$

and the $\tilde{\chi}_1^0$. If $\tilde{\chi}_1^{\pm}\tilde{\chi}_2^0$ production is assumed instead, the limits range from 11 to 760 GeV. The corresponding limits involving intermediate $W$, $Z$ and/or $H$ are significantly weaker.

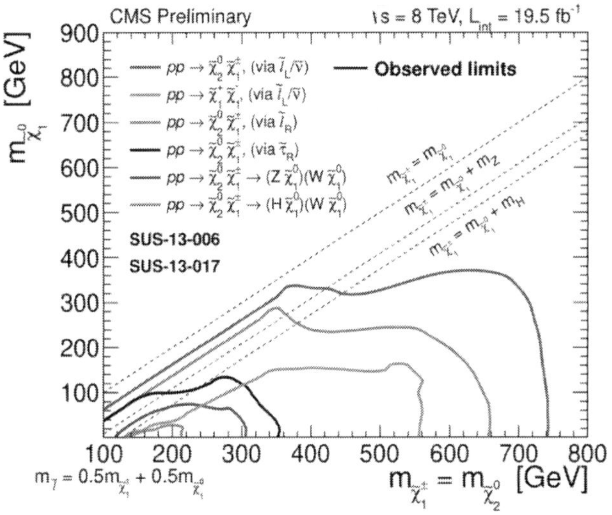


**Figure 11.** Summary of observed limits for electroweak-gaugino production from CMS [64]. From Ref. [62].

### *3.4. R-parity violating SUSY and meta-stable sparticles*

$R$-parity is defined as: $R = (-1)^{3(B-L)+2S}$, where $B$, $L$ and $S$ are the baryon number, lepton number and spin, respectively. Hence $R = +1$ for all Standard Model particles and $R = -1$ for all SUSY particles. It is stressed that the conservation of $R$-parity is an *ad-hoc* assumption. The only firm restriction comes from the proton lifetime: non-conservation of both $B$ and $L$ leads to rapid proton decay. $R$-parity conservation has serious consequences in SUSY phenomenology in colliders: the SUSY particles are produced in pairs and the lightest SUSY particle is absolutely stable, thus providing a WIMP candidate. Here we highlight the status of RPV supersymmetry [65] searches at the LHC.

Both ATLAS and CMS experiments have probed RPV SUSY through various channels, either by exclusively searching for specific decay chains, or by inclusively searching for multilepton events. ATLAS has looked for resonant production of $e\mu$, $e\tau$ and $\mu\tau$ [66], for multijets [67], for events with at least four leptons [68] and for excesses in the $e\mu$ continuum [69]. Null inclusive searches in the one-lepton channel [70] have also been interpreted in the context of a model where RPV is induced through bilinear terms [71].

Recent CMS analyses are focused on studying the lepton number violating terms $\lambda_{ijk}L_iL_j\bar{e}_k$ and $\lambda'_{ijk}L_iQ_j\bar{d}_k$, which result in specific signatures involving leptons in events produced in $pp$ collisions at LHC. A search for resonant production and the following decay of $\tilde{\mu}$ which is caused by $\lambda'_{211} \neq 0$ has been conducted [72]. Multilepton signatures caused by LSP decays due to various $\lambda$ and $\lambda'$ terms in stop production have been probed [73]. Reference [74] discusses the possibility of the generic model independent search for RPV SUSY in 4-lepton events. A summary of the limits set by several CMS analyses [61, 73, 75] are listed in Fig. 12.

In view of the null results in other SUSY searches, it became mandatory to fully explore the SUSY scenario predicting meta-stable or long-lived particles. These particles, not present in the Standard Model, would provide striking signatures in the detector and rely heavily on a detailed understanding of its performance. In SUSY, non-prompt particle decay can be caused by (i) very weak RPV [76], (ii) low mass difference between a SUSY particle and the LSP [77], or (iii) very weak coupling to the gravitino in GMSB models [78, 79]. A small part of these

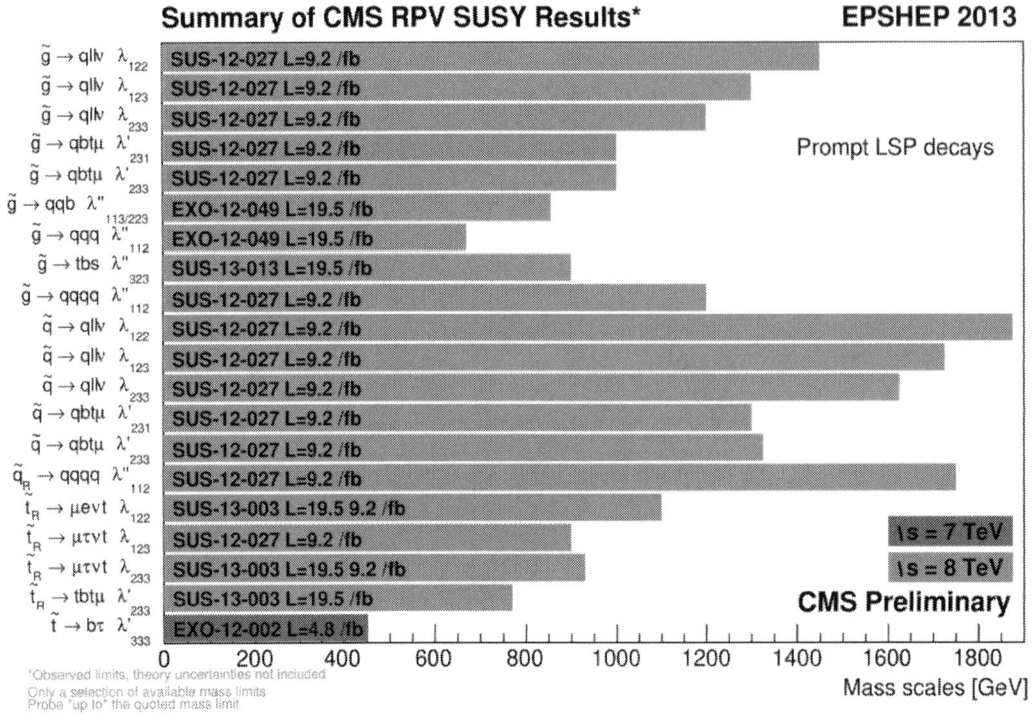


**Figure 12.** Best exclusion limits for the masses of the mother particles, for RPV scenarios, for each topology, for all CMS results [61, 73, 75]. In this plot, the lowest mass range is $m_{\rm mother} = 0$, but results are available starting from a certain mass depending on the analyses and topologies. Branching ratios of 100% are assumed, values shown in plot are to be interpreted as upper bounds on the mass limits. From Ref. [62].

possibilities have been explored by the ATLAS [57] and CMS [62] experiments covering specific cases, difficult to summarise here. There is still a wide panorama of signatures to be explored, in view of various proposed SUSY scenarios pointing towards this direction.

As a last remark, we address the issue of (not necessarily cold) dark matter in RPV SUSY models. These seemingly incompatible concepts *can* be reconciled in models with a gravitino [80, 81] or an axino [82, 83] LSP with a lifetime exceeding the age of the Universe. In both cases, RPV is induced by bilinear terms in the superpotential that can also explain current data on neutrino masses and mixings without invoking any GUT-scale physics [71]. Decays of the next-to-lightest superparticle occur rapidly via RPV interaction, and thus they do not upset the Big-Bang nucleosynthesis, unlike the $R$-parity conserving case. Such gravitino DM is proposed in the context of $\mu\nu$SSM [84] with profound prospects for detecting $\gamma$ rays from their decay [85].

Recent evidence on the four-year Fermi data that have found excess of a 130 GeV gamma-ray line from the Galactic Center [86] have been studied in the framework of $R$-parity breaking SUSY. A decaying axino DM scenario based on the SUSY KSVZ axion model with the bilinear $R$-parity violation explains the Fermi 130 GeV gamma-ray line excess from the GC while satisfying other cosmological constraints [83]. On the other hand, gravitino dark matter with trilinear RPV —in particular models with the $LLE$ RPV coupling— can account for the gamma-ray line, since there is no overproduction of anti-proton flux, while being consistent with big-bang nucleosynthesis and thermal leptogenesis [81].

## 4. Summary and outlook

The nature of dark matter remains one of the mysteries of Particle Physics and Cosmology. Mono-$X$ searches at the LHC provide strong constraints on dark matter properties in an effective field theory formalism. Colliders are superior to direct searches if dark matter is very light ($< 10$ GeV) or if interactions are spin-dependent. Extensive efforts are currently in progress on the validity of the effective field-theory approach, the proper comparison with the results from direct detection experiments and the use of simplified models with light mediators. Analyses looking for specific models providing DM candidates, such as Supersymmetry, are ongoing. Searches continue with the full 2012 dataset but a new discovery might eventually require more energy and more data coming up in 2015.

## Acknowledgments

The author is grateful to the XIV MWPF organisers for the kind invitation and support that gave her the opportunity to present this plenary talk. She acknowledges support by the Spanish Ministry of Economy and Competitiveness (MINECO) under the projects FPA2009-13234-C04-01 and FPA2012-39055-C02-01, by the Generalitat Valenciana through the project PROMETEO II/2013-017 and by the Spanish National Research Council (CSIC) under the JAE-Doc program co-funded by the European Social Fund (ESF).

## References

[1] For a pedagogical introduction, see e.g.: Profumo S 2013 TASI 2012 Lectures on Astrophysical Probes of Dark Matter *Preprint* arXiv:1301.0952 [hep-ph]

[2] Ellis J R, Falk T, Ganis G and Olive K A 2000 Supersymmetric dark matter in the light of LEP and the Tevatron collider *Phys. Rev.* D **62** 075010 (*Preprint* hep-ph/0004169)
Drees M, Kim Y G, Nojiri M M, Toya D, Hasuko K and Kobayashi T 2001 Scrutinizing LSP dark matter at the CERN LHC *Phys. Rev.* D **63** 035008 (*Preprint* hep-ph/0007202)
Chattopadhyay U and Nath P 2001 Upper limits on sparticle masses from $g-2$ and the possibility for discovery of SUSY at colliders and in dark matter searches *Phys. Rev. Lett.* **86** 5854 (*Preprint* hep-ph/0102157)
Lahanas A B, Mavromatos N E and Nanopoulos D V 2003 WMAPing the universe: Supersymmetry, dark matter, dark energy, proton decay and collider physics *Int. J. Mod. Phys.* D **12** 1529 (*Preprint* hep-ph/0308251)
Baer H, Belyaev A, Krupovnickas T and O'Farrill J 2004 Indirect, direct and collider detection of neutralino dark matter *JCAP* **0408** 005 (*Preprint* hep-ph/0405210)
Allanach B C, Belanger G, Boudjema F and Pukhov A 2004 Requirements on collider data to match the precision of WMAP on supersymmetric dark matter *JHEP* **0412** 020 (*Preprint* hep-ph/0410091)
Roszkowski L, Ruiz de Austri R and Choi K-Y 2005 Gravitino dark matter in the CMSSM and implications for leptogenesis and the LHC *JHEP* **0508** 080 (*Preprint* hep-ph/0408227])
Baer H, Mustafayev A, Profumo S, Belyaev A and Tata X 2005 Direct, indirect and collider detection of neutralino dark matter in SUSY models with non-universal Higgs masses *JHEP* **0507** 065 (*Preprint* hep-ph/0504001)
Arbey A, Battaglia M and Mahmoudi F 2012 Light Neutralino Dark Matter in the pMSSM: Implications of LEP, LHC and Dark Matter Searches on SUSY Particle Spectra *Eur. Phys. J.* C **72** 2169 (*Preprint* arXiv:1205.2557 [hep-ph])

[3] Hooper D and Profumo S 2007 Dark matter and collider phenomenology of universal extra dimensions *Phys. Rept.* **453** 29 (*Preprint* hep-ph/0701197)

[4] Evans L and Bryant P 2008 LHC Machine *JINST* **3** S08001

[5] Mitsou V A 2013 Shedding Light on Dark Matter at Colliders *Int. J. Mod. Phys.* A **28** 1330052 (*Preprint* arXiv:1310.1072 [hep-ex])

[6] Ade P A R *et al.* [Planck Collab.] 2013 Planck 2013 results. XVI. Cosmological parameters *Preprint* arXiv:1303.5076 [astro-ph.CO]

[7] Mavromatos N E 2012 Some Aspects of String Cosmology and the LHC *Preprint* arXiv:1210.0211 [hep-ph] in *Proc. 1st Int. Conf. on New Frontiers in Physics*, to appear in *Eur. Phys. J. Web Conf.*

[8] Birkedal A, Noble A, Perelstein M and Spray A 2006 Little Higgs dark matter *Phys. Rev.* D **74** 035002 (*Preprint* arXiv:hep-ph/0603077)

[9] Aad G *et al.* [ATLAS Collab.] 2008 The ATLAS Experiment at the CERN Large Hadron Collider *JINST* **3** S08003
[10] Adolphi R *et al.* [CMS Collab.] 2008 The CMS experiment at the CERN LHC *JINST* **3** S08004
[11] MoEDAL Collaboration 2009 Technical Design Report of the MoEDAL Experiment *CERN Preprint* CERN-LHC-2009-006, MoEDAL-TDR-1.1
[12] Acharya B *et al.* 2014 The Physics Program of the MoEDAL Experiment at the LHC, to appear in *Int. J. Mod. Phys.* A
[13] Aad G *et al.* [ATLAS Collab.] 2012 Observation of a new particle in the search for the Standard Model Higgs boson with the ATLAS detector at the LHC *Phys. Lett.* B **716** 1 (*Preprint* arXiv:1207.7214 [hep-ex])
Chatrchyan S *et al.* [CMS Collab.] 2012 Observation of a new boson at a mass of 125 GeV with the CMS experiment at the LHC *Phys. Lett.* B **716** 30 (*Preprint* arXiv:1207.7235 [hep-ex])
[14] Chatrchyan S *et al.* [CMS Collab.] 2013 Study of the Mass and Spin-Parity of the Higgs Boson Candidate Via Its Decays to Z Boson Pairs *Phys. Rev. Lett.* **110** 081803 (*Preprint* arXiv:1212.6639 [hep-ex])
Aad G *et al.* [ATLAS Collab.] 2013 Measurements of Higgs boson production and couplings in diboson final states with the ATLAS detector at the LHC *Phys. Lett.* B **726** 88 (*Preprint* arXiv:1307.1427 [hep-ex])
Aad G *et al.* [ATLAS Collab.] 2013 Evidence for the spin-0 nature of the Higgs boson using ATLAS data *Phys. Lett.* B **726** 120 (*Preprint* arXiv:1307.1432 [hep-ex])
[15] Baer H and Brhlik M 1998 Neutralino dark matter in minimal supergravity: Direct detection versus collider searches *Phys. Rev.* D **57** 567 (*Preprint* hep-ph/9706509)
Pato M, Baudis L, Bertone G, Ruiz de Austri R, Strigari L E and Trotta R 2011 Complementarity of Dark Matter Direct Detection Targets *Phys. Rev.* D **83** 083505 (*Preprint* arXiv:1012.3458 [astro-ph.CO])
[16] Rajaraman A, Shepherd W, Tait T M P and Wijangco A M 2011 LHC Bounds on Interactions of Dark Matter *Phys. Rev.* D **84** 095013 (*Preprint* arXiv:1108.1196 [hep-ph])
Cheung K, Tseng P-Y, Tsai Y-L S and Yuan T-C 2012 Global Constraints on Effective Dark Matter Interactions: Relic Density, Direct Detection, Indirect Detection, and Collider *JCAP* **1205** 001 (*Preprint* arXiv:1201.3402 [hep-ph])
Bauer D *et al.* 2013 Dark Matter in the Coming Decade: Complementary Paths to Discovery and Beyond *Preprint* arXiv:1305.1605 [hep-ph]
[17] Bertone G, Cerdeno D G, Fornasa M, Pieri L, Ruiz de Austri R and Trotta R 2012 Complementarity of Indirect and Accelerator Dark Matter Searches *Phys. Rev.* D **85** 055014 (*Preprint* arXiv:1111.2607 [astro-ph.HE])
[18] Goodman J, Ibe M, Rajaraman A, Shepherd W, Tait T M P and Yu H-B 2010 Constraints on Dark Matter from Colliders *Phys. Rev.* D **82** 116010 (*Preprint* arXiv:1008.1783 [hep-ph])
[19] Bai Y, Fox P J and Harnik R 2010 The Tevatron at the Frontier of Dark Matter Direct Detection *JHEP* **1012** 048 (*Preprint* arXiv:1005.3797 [hep-ph])
Beltran M, Hooper D, Kolb E W, Krusberg Z A C and Tait T M P 2010 Maverick dark matter at colliders *JHEP* **1009** 037 (*Preprint* arXiv:1002.4137 [hep-ph])
[20] Aad G *et al.* [ATLAS Collab.] 2013 Search for dark matter candidates and large extra dimensions in events with a jet and missing transverse momentum with the ATLAS detector *JHEP* **1304** 075 (*Preprint* arXiv:1210.4491 [hep-ex])
[21] Chatrchyan S *et al.* [CMS Collab.] 2012 Search for dark matter and large extra dimensions in monojet events in $pp$ collisions at $\sqrt{s} = 7$ TeV *JHEP* **1209** 094 (*Preprint* arXiv:1206.5663 [hep-ex])
[22] Haisch U, Kahlhoefer F and Unwin J 2013 The impact of heavy-quark loops on LHC dark matter searches *JHEP* **1307** 125 (*Preprint* arXiv:1208.4605 [hep-ph])
[23] Ackermann M *et al.* [Fermi-LAT Collab.] 2011 Constraining Dark Matter Models from a Combined Analysis of Milky Way Satellites with the Fermi Large Area Telescope *Phys. Rev. Lett.* **107** 241302 (*Preprint* arXiv:1108.3546 [astro-ph.HE])
[24] Bergstrom L, Ullio P and Buckley J H 1998 Observability of gamma-rays from dark matter neutralino annihilations in the Milky Way halo *Astropart. Phys.* **9** 137 (*Preprint* astro-ph/9712318)
Cirelli M *et al.* 2011 PPPC 4 DM ID: A Poor Particle Physicist Cookbook for Dark Matter Indirect Detection *JCAP* **1103** 051 [Erratum-ibid. 2012 **1210** E01] (*Preprint* arXiv:1012.4515 [hep-ph])
[25] CMS Collaboration 2012 Search for new physics in monojet events in $pp$ collisions at $\sqrt{s} = 8$ TeV *CMS Note* CMS-PAS-EXO-12-048
[26] Fox P J, Harnik R, Kopp J and Tsai Y 2012 Missing Energy Signatures of Dark Matter at the LHC *Phys. Rev.* D **85** 056011 (*Preprint* arXiv:1109.4398 [hep-ph])
[27] Aad G *et al.* [ATLAS Collab.] 2013 Search for dark matter candidates and large extra dimensions in events with a photon and missing transverse momentum in $pp$ collision data at $\sqrt{s} = 7$ TeV with the ATLAS detector *Phys. Rev. Lett.* **110** 011802 (*Preprint* arXiv:1209.4625 [hep-ex])
[28] Chatrchyan S *et al.* [CMS Collab.] 2012 Search for Dark Matter and Large Extra Dimensions in pp Collisions

Yielding a Photon and Missing Transverse Energy *Phys. Rev. Lett.* **108** 261803 (*Preprint* arXiv:1204.0821 [hep-ex])

[29] Aaltonen T *et al.* [CDF Collab.] 2012 A Search for dark matter in events with one jet and missing transverse energy in $p\bar{p}$ collisions at $\sqrt{s} = 1.96$ TeV *Phys. Rev. Lett.* **108** 211804 (*Preprint* arXiv:1203.0742 [hep-ex])

[30] Aprile E *et al.* [XENON100 Collab.] 2011 Dark Matter Results from 100 Live Days of XENON100 Data *Phys. Rev. Lett.* **107** 131302 (*Preprint* arXiv:1104.2549 [astro-ph.CO])

[31] Aalseth C E *et al.* [CoGeNT Collab.] 2011 Results from a Search for Light-Mass Dark Matter with a P-type Point Contact Germanium Detector *Phys. Rev. Lett.* **106** 131301 (*Preprint* arXiv:1002.4703 [astro-ph.CO])

[32] Ahmed Z *et al.* [CDMS-II Collab.] 2010 Dark Matter Search Results from the CDMS II Experiment *Science* **327** 1619 (*Preprint* arXiv:0912.3592 [astro-ph.CO])
Ahmed Z *et al.* [CDMS-II Collab.] 2011 Results from a Low-Energy Analysis of the CDMS II Germanium Data *Phys. Rev. Lett.* **106** 131302 (*Preprint* arXiv:1011.2482 [astro-ph.CO])

[33] Felizardo M *et al.* 2012 Final Analysis and Results of the Phase II SIMPLE Dark Matter Search *Phys. Rev. Lett.* **108** 201302 (*Preprint* arXiv:1106.3014 [astro-ph.CO])

[34] Archambault S *et al.* [PICASSO Collab.] 2012 Constraints on Low-Mass WIMP Interactions on $^{19}F$ from PICASSO *Phys. Lett.* B **711** 153 (*Preprint* arXiv:1202.1240 [hep-ex])

[35] Bai Y and Tait T M P 2013 Searches with Mono-Leptons *Phys. Lett.* B **723** 384 (*Preprint* arXiv:1208.4361 [hep-ph])

[36] Bell N F, Dent J B, Galea A J, Jacques T D, Krauss L M and Weiler T J 2012 Searching for Dark Matter at the LHC with a Mono-Z *Phys. Rev.* D **86** 096011 (*Preprint* arXiv:1209.0231 [hep-ph])
Carpenter L M, Nelson A, Shimmin C, Tait T M P and Whiteson D 2013 Collider searches for dark matter in events with a $Z$ boson and missing energy *Phys. Rev.* D **87** 074005 (*Preprint* arXiv:1212.3352 [hep-ex])

[37] CMS Collaboration 2013 Search for dark matter in the mono-lepton channel with $pp$ collision events at $\sqrt{s} = 8$ TeV *CMS Note* CMS-PAS-EXO-13-004

[38] CMS Collaboration 2012 Search for new physics in the final states with a lepton and missing transverse energy at $\sqrt{s} = 8$ TeV *CMS Note* CMS-PAS-EXO-12-060

[39] Aad G *et al.* [ATLAS Collab.] 2014 Search for dark matter in events with a hadronically decaying $W$ or $Z$ boson and missing transverse momentum in $pp$ collisions at $\sqrt{s} = 8$ TeV with the ATLAS detector *Phys. Rev. Lett.* **112** 041802 (*Preprint* arXiv:1309.4017 [hep-ex])

[40] Aad G *et al.* [ATLAS Collab.] 2013 Performance of jet substructure techniques for large-$R$ jets in proton-proton collisions at $\sqrt{s} = 7$ TeV using the ATLAS detector *JHEP* **1309** 076 (*Preprint* arXiv:1306.4945 [hep-ex])

[41] Aprile E *et al.* [XENON100 Collab.] 2012 Dark Matter Results from 225 Live Days of XENON100 Data *Phys. Rev. Lett.* **109** 181301 (*Preprint* arXiv:1207.5988 [astro-ph.CO])

[42] Behnke E *et al.* [COUPP Collab.] 2012 First Dark Matter Search Results from a 4-kg $CF_3I$ Bubble Chamber Operated in a Deep Underground Site *Phys. Rev.* D **86** 052001 (*Preprint* arXiv:1204.3094 [astro-ph.CO])

[43] Aartsen M G *et al.* [IceCube Collab.] 2013 Search for dark matter annihilations in the Sun with the 79-string IceCube detector *Phys. Rev. Lett.* **110** 131302 (*Preprint* arXiv:1212.4097 [astro-ph.HE])

[44] Nilles H P 1984 Supersymmetry, Supergravity and Particle Physics *Phys. Rept.* **110** 1

[45] Mitsou V A 2013 Experimental status of particle and astroparticle searches for supersymmetry *J. Phys. Conf. Series* **447** 012019 (*Preprint* arXiv:1304.1414 [hep-ph])

[46] Cabrera M E and Casas J A 2012 Understanding and improving the Effective Mass for LHC searches *Preprint* arXiv:1207.0435 [hep-ph]

[47] Cabrera M E, Casas J A, Mitsou V A, Ruiz de Austri R and Terron J 2012 Histogram comparison as a powerful tool for the search of new physics at LHC. Application to CMSSM *JHEP* **1204** 133 (*Preprint* arXiv:1109.3759 [hep-ph])

[48] Aad G *et al.* [ATLAS Collab.] 2013 Search for new phenomena in final states with large jet multiplicities and missing transverse momentum at $\sqrt{s} = 8$ TeV proton-proton collisions using the ATLAS experiment *JHEP* **1310** 130 (*Preprint* arXiv:1308.1841 [hep-ex])

[49] Chatrchyan S *et al.* [CMS Collab.] 2012 Search for supersymmetry in hadronic final states using MT2 in $pp$ collisions at $\sqrt{s} = 7$ TeV *JHEP* **1210** 018 (*Preprint* arXiv:1207.1798 [hep-ex])
Chatrchyan S *et al.* [CMS Collab.] 2012 Search for new physics in the multijet and missing transverse momentum final state in proton-proton collisions at $\sqrt{s} = 7$ TeV *Phys. Rev. Lett.* **109** 171803 (*Preprint* arXiv:1207.1898 [hep-ex])

[50] Aad G *et al.* [ATLAS Collab.] 2012 Further search for supersymmetry at $\sqrt{s} = 7$ TeV in final states with jets, missing transverse momentum and isolated leptons with the ATLAS detector *Phys. Rev.* D **86** 092002 (*Preprint* arXiv:1208.4688 [hep-ex])

[51] Chatrchyan S *et al.* [CMS Collab.] 2013 Search for supersymmetry in $pp$ collisions at $\sqrt{s} = 7$ TeV in events with a single lepton, jets, and missing transverse momentum *Eur. Phys. J.* C **73** 2404 (*Preprint*

arXiv:1212.6428 [hep-ex])
Chatrchyan S *et al.* [CMS Collab.] 2013 Search for supersymmetry in events with opposite-sign dileptons and missing transverse energy using an artificial neural network *Phys. Rev.* D **87** 072001 (*Preprint* arXiv:1301.0916 [hep-ex])

[52] Baer H, Box A, Park E-K and Tata X 2007 Implications of compressed supersymmetry for collider and dark matter searches *JHEP* **0708** 060 (*Preprint* arXiv:0707.0618 [hep-ph])
LeCompte T J and Martin S P 2012 Compressed supersymmetry after 1 $\mathrm{fb}^{-1}$ at the Large Hadron Collider *Phys. Rev.* D **85** 035023 (*Preprint* arXiv:1111.6897 [hep-ph])

[53] Rogan C 2010 Kinematical variables towards new dynamics at the LHC *Preprint* arXiv:1006.2727 [hep-ph]

[54] Chatrchyan S *et al.* [CMS Collab.] 2013 Inclusive search for supersymmetry using the razor variables in *pp* collisions at $\sqrt{s} = 7$ TeV *Phys. Rev. Lett.* **111** 081802 (*Preprint* arXiv:1212.6961 [hep-ex])

[55] Aad G *et al.* [ATLAS Collab.] 2013 Multi-channel search for squarks and gluinos in $\sqrt{s} = 7$ TeV *pp* collisions with the ATLAS detector *Eur. Phys. J.* C **73** 2362 (*Preprint* arXiv:1212.6149 [hep-ex])

[56] ATLAS Collaboration 2013 Search for strongly produced superpartners in final states with two same sign leptons with the ATLAS detector using 21 $\mathrm{fb}^{-1}$ of proton-proton collisions at $\sqrt{s} = 8$ TeV *ATLAS Note* ATLAS-CONF-2013-007
ATLAS Collaboration 2013 Search for Supersymmetry in Events with Large Missing Transverse Momentum, Jets, and at Least One Tau Lepton in 21 $\mathrm{fb}^{-1}$ of $\sqrt{s} = 8$ TeV Proton-Proton Collision Data with the ATLAS Detector *ATLAS Note* ATLAS-CONF-2013-026
ATLAS Collaboration 2013 Measurement of multi-jet cross-section ratios and determination of the strong coupling constant in proton-proton collisions at $\sqrt{s} = 7$ TeV with the ATLAS detector *ATLAS Note* ATLAS-CONF-2013-041
ATLAS Collaboration 2013 Search for strong production of supersymmetric particles in final states with missing transverse momentum and at least three *b*-jets using 20.1 $\mathrm{fb}^{-1}$ of *pp* collisions at $\sqrt{s} = 8$ TeV with the ATLAS Detector *ATLAS Note* ATLAS-CONF-2013-061
ATLAS Collaboration 2013 Search for squarks and gluinos in events with isolated leptons, jets and missing transverse momentum at $\sqrt{s} = 8$ TeV with the ATLAS detector *ATLAS Note* ATLAS-CONF-2013-062

[57] ATLAS Supersymmetry Searches Public Results:
`https://twiki.cern.ch/twiki/bin/view/AtlasPublic/SupersymmetryPublicResults`

[58] Lahanas A B, Mavromatos N E and Nanopoulos D V 2007 Smoothly evolving supercritical-string dark energy relaxes supersymmetric-dark-matter constraints *Phys. Lett.* B **649** 83 (*Preprint* hep-ph/0612152)

[59] Mavromatos N E, Sarkar S and Vergou A 2011 Stringy Space-Time Foam, Finsler-like Metrics and Dark Matter Relics *Phys. Lett.* B **696** 300 (*Preprint* arXiv:1009.2880 [hep-th])
Mavromatos N E, Mitsou V A, Sarkar S and Vergou A 2012 Implications of a Stochastic Microscopic Finsler Cosmology *Eur. Phys. J.* C **72** 1956 (*Preprint* arXiv:1012.4094 [hep-ph])

[60] Chatrchyan S *et al.* [CMS Collab.] 2013 Search for gluino mediated bottom- and top-squark production in multijet final states in *pp* collisions at 8 TeV *Phys. Lett.* B **725** 243 (*Preprint* arXiv:1305.2390 [hep-ex])
CMS Collaboration 2013 Search for supersymmetry in *pp* collisions at a center-of-mass energy of 8 TeV in events with a single lepton, multiple jets and *b*-tags *CMS Note* CMS-PAS-SUS-13-007
CMS Collaboration 2013 Search for supersymmetry in *pp* collisions at $\sqrt{s} = 8$ TeV in events with three leptons and at least one *b*-tagged jet *CMS Note* CMS-PAS-SUS-13-008

[61] CMS Collaboration 2013 Search for new physics in events with same-sign dileptons and jets in *pp* collisions at $\sqrt{s} = 8$ TeV *CMS Note* CMS-PAS-SUS-13-013

[62] CMS Supersymmetry Physics Results:
`https://twiki.cern.ch/twiki/bin/view/CMSPublic/PhysicsResultsSUS`

[63] Aad G *et al.* [ATLAS Collab.] 2012 Search for a supersymmetric partner to the top quark in final states with jets and missing transverse momentum at $\sqrt{s} = 7$ TeV with the ATLAS detector *Phys. Rev. Lett.* **109** 211802 (*Preprint* arXiv:1208.1447 [hep-ex])
Aad G *et al.* [ATLAS Collab.] 2012 Search for direct top squark pair production in final states with one isolated lepton, jets, and missing transverse momentum in $\sqrt{s} = 7$ TeV *pp* collisions using 4.7 $\mathrm{fb}^{-1}$ of ATLAS data *Phys. Rev. Lett.* **109** 211803 (*Preprint* arXiv:1208.2590 [hep-ex])
Aad G *et al.* [ATLAS Collab.] 2012 Search for light scalar top quark pair production in final states with two leptons with the ATLAS detector in $\sqrt{s} = 7$ TeV proton-proton collisions *Eur. Phys. J.* C **72** 2237 (*Preprint* arXiv:1208.4305 [hep-ex])
Aad G *et al.* [ATLAS Collab.] 2012 Search for a heavy top-quark partner in final states with two leptons with the ATLAS detector at the LHC *JHEP* **1211** 094 (*Preprint* arXiv:1209.4186 [hep-ex])
Aad G *et al.* [ATLAS Collab.] 2013 Search for light top squark pair production in final states with leptons and *b*-jets with the ATLAS detector in $\sqrt{s} = 7$ TeV proton-proton collisions *Phys. Lett.* B **720** 13 (*Preprint* arXiv:1209.2102 [hep-ex])

Aad G *et al.* [ATLAS Collab.] 2013 Search for direct third-generation squark pair production in final states with missing transverse momentum and two $b$-jets in $\sqrt{s} = 8$ TeV $pp$ collisions with the ATLAS detector JHEP **1310** 189 (*Preprint* arXiv:1308.2631 [hep-ex])
ATLAS Collaboration 2013 Search for direct production of the top squark in the all-hadronic $t\bar{t} + E_T^{miss}$ final state in 21 fb$^{-1}$ of $pp$ collisions at $\sqrt{s} = 8$ TeV with the ATLAS detector *ATLAS Note* ATLAS-CONF-2013-024
ATLAS Collaboration 2013 Search for direct top squark pair production in final states with one isolated lepton, jets, and missing transverse momentum in $\sqrt{s} = 8$ TeV $pp$ collisions using 21 fb$^{-1}$ of ATLAS data *ATLAS Note* ATLAS-CONF-2013-037
ATLAS Collaboration 2013 Search for direct top squark pair production in final states with two leptons in $\sqrt{s} = 8$ TeV $pp$ collisions using 20 fb$^{-1}$ of ATLAS data *ATLAS Note* ATLAS-CONF-2013-048
ATLAS Collaboration 2013 Searches for direct scalar top pair production in final states with two leptons using the stransverse mass variable and a multivariate analysis technique in $\sqrt{s} = 8$ TeV $pp$ collisions using 20.3 fb$^{-1}$ of ATLAS data *ATLAS Note* ATLAS-CONF-2013-065
ATLAS Collaboration 2013 Search for pair-produced top squarks decaying into charm quarks and the lightest neutralinos using 20.3 fb$^{-1}$ of $pp$ collisions at $\sqrt{s} = 8$ TeV with the ATLAS detector at the LHC *ATLAS Note* ATLAS-CONF-2013-068

[64] CMS Collaboration 2013 Search for electroweak production of charginos, neutralinos, and sleptons using leptonic final states in $pp$ collisions at $\sqrt{s} = 8$ TeV *CMS Note* CMS-PAS-SUS-13-006
CMS Collaboration 2013 Search for electroweak production of charginos and neutralinos in final states with a Higgs boson in $pp$ collisions at $\sqrt{s} = 8$ TeV *CMS Note* CMS-PAS-SUS-13-017

[65] Barbier R *et al.* 2005 $R$-parity violating supersymmetry *Phys. Rept.* **420** 1 (*Preprint* hep-ph/0406039)

[66] Aad G *et al.* [ATLAS Collab.] 2011 Search for a heavy particle decaying into an electron and a muon with the ATLAS detector in $\sqrt{s} = 7$ TeV $pp$ collisions at the LHC *Phys. Rev. Lett.* **106** 251801 (*Preprint* arXiv:1103.5559 [hep-ex])
Aad G *et al.* [ATLAS Collab.] 2011 Search for a heavy neutral particle decaying into an electron and a muon using 1 fb$^{-1}$ of ATLAS data *Eur. Phys. J.* C **71** 1809 (*Preprint* arXiv:1109.3089 [hep-ex])
Aad G *et al.* [ATLAS Collab.] 2013 Search for a heavy narrow resonance decaying to $e\mu$, $e\tau$, or $\mu\tau$ with the ATLAS detector in $\sqrt{s} = 7$ TeV $pp$ collisions at the LHC *Phys. Lett.* B **723** 15 (*Preprint* arXiv:1212.1272 [hep-ex])

[67] Aad G *et al.* [ATLAS Collab.] 2012 Search for pair production of massive particles decaying into three quarks with the ATLAS detector in $\sqrt{s} = 7$ TeV $pp$ collisions at the LHC *JHEP* **1212** 086 (*Preprint* arXiv:1210.4813 [hep-ex])

[68] Aad G *et al.* [ATLAS Collab.] 2012 Search for $R$-parity-violating supersymmetry in events with four or more leptons in $\sqrt{s} = 7$ TeV $pp$ collisions with the ATLAS detector *JHEP* **1212** 124 (*Preprint* arXiv:1210.4457 [hep-ex])

[69] Aad G *et al.* [ATLAS Collab.] 2012 Search for lepton flavour violation in the $e\mu$ continuum with the ATLAS detector in $\sqrt{s} = 7$ TeV $pp$ collisions at the LHC *Eur. Phys. J.* C **72** 2040 (*Preprint* arXiv:1205.0725 [hep-ex])

[70] Aad G *et al.* [ATLAS Collab.] 2012 Search for supersymmetry in final states with jets, missing transverse momentum and one isolated lepton in $\sqrt{s} = 7$ TeV $pp$ collisions using 1 fb$^{-1}$ of ATLAS data *Phys. Rev.* D **85** 012006 (*Preprint* arXiv:1109.6606 [hep-ex])
ATLAS Collaboration 2012 Search for supersymmetry at $\sqrt{s} = 7$ TeV in final states with large jet multiplicity, missing transverse momentum and one isolated lepton with the ATLAS detector *ATLAS Note* ATLAS-CONF-2012-140

[71] Romao J C, Diaz M A, Hirsch M, Porod W and Valle J W F 2000 A Supersymmetric solution to the solar and atmospheric neutrino problems *Phys. Rev.* D **61** 071703 (*Preprint* hep-ph/9907499)
Hirsch M, Diaz M A, Porod W, Romao J C and Valle J W F 2000 Neutrino masses and mixings from supersymmetry with bilinear $R$ parity violation: A Theory for solar and atmospheric neutrino oscillations *Phys. Rev.* D **62** 113008 [Erratum-ibid. 2002 D **65** 119901] (*Preprint* hep-ph/0004115)
Porod W, Hirsch M, Romao J and Valle J W F 2001 Testing neutrino mixing at future collider experiments *Phys. Rev.* D **63** 115004 (*Preprint* hep-ph/0011248)
Diaz M A, Hirsch M, Porod W, Romao J C and Valle J W F 2003 Solar neutrino masses and mixing from bilinear $R$ parity broken supersymmetry: Analytical versus numerical results *Phys. Rev.* D **68** 013009 [Erratum-ibid. 2005 D **71** 059904] (*Preprint* hep-ph/0302021)
Hirsch M and Valle J W F 2004 Supersymmetric origin of neutrino mass *New J. Phys.* **6** 76 (*Preprint* hep-ph/0405015)

[72] CMS Collaboration 2013 Search for RPV SUSY resonant second generation slepton production in same-sign dimuon events at $\sqrt{s} = 7$ TeV *CMS Note* CMS-PAS-SUS-13-005

[73] Chatrchyan S *et al.* [CMS Collab.] 2013 Search for top squarks in $R$-parity-violating supersymmetry using three or more leptons and $b$-tagged jets *Phys. Rev. Lett.* **111** 221801 (*Preprint* arXiv:1306.6643 [hep-ex])

[74] CMS Collaboration 2013 Search for RPV SUSY in the four-lepton final state *CMS Note* CMS-PAS-SUS-13-010

[75] Chatrchyan S *et al.* [CMS Collab.] 2013 Search for pair production of third-generation leptoquarks and top squarks in $pp$ collisions at $\sqrt{s} = 7$ TeV *Phys. Rev. Lett.* **110** 081801 (*Preprint* arXiv:1210.5629 [hep-ex])
CMS Collaboration 2012 Search for RPV supersymmetry with three or more leptons and $b$-tags *CMS Note* CMS-PAS-SUS-12-027
CMS Collaboration 2012 Search for Light- and Heavy-flavour Three-Jet Resonances In Multijet Final States at 8 TeV *CMS Note* CMS-PAS-EXO-12-049

[76] Aad G *et al.* [ATLAS Collab.] 2013 Search for long-lived, heavy particles in final states with a muon and multi-track displaced vertex in proton-proton collisions at $\sqrt{s} = 7$ TeV with the ATLAS detector *Phys. Lett.* B **719** 280 (*Preprint* arXiv:1210.7451 [hep-ex])

[77] Aad G *et al.* [ATLAS Collab.] 2013 Search for direct chargino production in anomaly-mediated supersymmetry breaking models based on a disappearing-track signature in $pp$ collisions at $\sqrt{s} = 7$ TeV with the ATLAS detector *JHEP* **1301** 131 (*Preprint* arXiv:1210.2852 [hep-ex])

[78] Aad G *et al.* [ATLAS Collab.] 2013 Searches for heavy long-lived sleptons and R-Hadrons with the ATLAS detector in $pp$ collisions at $\sqrt{s} = 7$ TeV *Phys. Lett.* B **720** 277 (*Preprint* arXiv:1211.1597 [hep-ex])
Chatrchyan S *et al.* [CMS Collab.] 2013 Searches for long-lived charged particles in $pp$ collisions at $\sqrt{s} = 7$ and 8 TeV *JHEP* **1307** 122 (*Preprint* arXiv:1305.0491 [hep-ex])

[79] Aad G *et al.* [ATLAS Collab.] 2013 Search for non-pointing photons in the diphoton and $E_{\mathrm{T}}^{\mathrm{miss}}$ final state in $\sqrt{s} = 7$ TeV proton-proton collisions using the ATLAS detector *Phys. Rev.* D **88** 012001 (*Preprint* arXiv:1304.6310 [hep-ex])
Chatrchyan S *et al.* [CMS Collab.] 2013 Search for long-lived particles decaying to photons and missing energy in proton-proton collisions at $\sqrt{s} = 7$ TeV *Phys. Lett.* B **722** 273 (*Preprint* arXiv:1212.1838 [hep-ex])

[80] Takayama F and Yamaguchi M 2000 Gravitino dark matter without $R$-parity *Phys. Lett.* B **485** 388 (*Preprint* hep-ph/0005214)
Hirsch M, Porod W and Restrepo D 2005 Collider signals of gravitino dark matter in bilinearly broken $R$-parity *JHEP* **0503** 062 (*Preprint* hep-ph/0503059)
Buchmuller W, Covi L, Hamaguchi K, Ibarra A and Yanagida T 2007 Gravitino Dark Matter in $R$-Parity Breaking Vacua *JHEP* **0703** 037 (*Preprint* hep-ph/0702184)

[81] Liew S P 2013 Gamma-ray line from radiative decay of gravitino dark matter *Phys. Lett.* B **724** 88 (*Preprint* arXiv:1304.1992 [hep-ph])

[82] Chun E J and Kim H B 2006 Axino Light Dark Matter and Neutrino Masses with $R$-parity Violation *JHEP* **0610** 082 (*Preprint* hep-ph/0607076)

[83] Endo M, Hamaguchi K, Liew S P, Mukaida K and Nakayama K 2013 Axino dark matter with $R$-parity violation and 130 GeV gamma-ray line *Phys. Lett.* B **721** 111 (*Preprint* arXiv:1301.7536 [hep-ph])

[84] Lopez-Fogliani D E and Munoz C 2006 Proposal for a new minimal supersymmetric standard model *Phys. Rev. Lett.* **97** 041801 (*Preprint* hep-ph/0508297)
Escudero N, Lopez-Fogliani D E, Munoz C and Ruiz de Austri R 2008 Analysis of the parameter space and spectrum of the $\mu\nu$SSM *JHEP* **0812** 099 (*Preprint* arXiv:0810.1507 [hep-ph])
Ghosh P, Lopez-Fogliani D E, Mitsou V A, Munoz C and Ruiz de Austri R 2013 Probing the $\mu$ from $\nu$ supersymmetric standard model with displaced multileptons from the decay of a Higgs boson at the LHC *Phys. Rev.* D **88** 015009 (*Preprint* arXiv:1211.3177 [hep-ph])

[85] Choi K-Y, Lopez-Fogliani D E, Munoz C and Ruiz de Austri R 2010 Gamma-ray detection from gravitino dark matter decay in the $\mu\nu$SSM *JCAP* **1003** 028 (*Preprint* arXiv:0906.3681 [hep-ph])

[86] Bringmann T, Huang X, Ibarra A, Vogl S and Weniger C 2012 Fermi LAT Search for Internal Bremsstrahlung Signatures from Dark Matter Annihilation *JCAP* **1207** 054 (*Preprint* arXiv:1203.1312 [hep-ph])
Weniger C 2012 A Tentative Gamma-Ray Line from Dark Matter Annihilation at the Fermi Large Area Telescope *JCAP* **1208** 007 (*Preprint* arXiv:1204.2797 [hep-ph])
Tempel E, Hektor A and Raidal M 2012 Fermi 130 GeV gamma-ray excess and dark matter annihilation in sub-haloes and in the Galactic centre *JCAP* **1209** 032 [Addendum-ibid. 2012 **1211** A01] (*Preprint* arXiv:1205.1045 [hep-ph])

**Institute of Physics**
Dirac House, Temple Back
Bristol BS1 6BE UK

ISSN: 1742-6588
ISBN 978-1-5108-1477-6